AF356182

Effect of Climate Change on Salmonid Fishes in Rivers

Effect of Climate Change on Salmonid Fishes in Rivers

Editor

Bror Jonsson

Basel • Beijing • Wuhan • Barcelona • Belgrade • Novi Sad • Cluj • Manchester

Editor
Bror Jonsson
Norwegian Institute for Nature Research
Oslo
Norway

Editorial Office
MDPI
St. Alban-Anlage 66
4052 Basel, Switzerland

This is a reprint of articles from the Special Issue published online in the open access journal *Fishes* (ISSN 2410-3888) (available at: https://www.mdpi.com/journal/fishes/special_issues/3JHJ00DXS4).

For citation purposes, cite each article independently as indicated on the article page online and as indicated below:

Lastname, A.A.; Lastname, B.B. Article Title. *Journal Name* **Year**, *Volume Number*, Page Range.

ISBN 978-3-7258-0309-5 (Hbk)
ISBN 978-3-7258-0310-1 (PDF)
doi.org/10.3390/books978-3-7258-0310-1

Cover image courtesy of Nina Jonsson

Contents

About the Editor

Bror Jonsson

Dr. Bror Jonsson is a Professor at the Oslo Department of the Norwegian Institute for Nature Research. His research interests include life history evolution, bio-energetic allocations, resource polymorphism, fish migration, the effects of escaped farmed salmon on conspecific wild populations and the effects of climate change on salmonid fishes. His most recent studies pertain to how temperature during the embryo stage of Atlantic salmon and brown trout affects the life history and behavioural traits of these species. Jonsson has published more than 230 peer-reviewed scientific papers, as well as a book, *"Ecology of Atlantic salmon and brown trout: Habitat as template for life histories"*. He is an elected member of the Royal Norwegian Society of Science, The Royal Danish Academy of Science and a Knight of the Icelandic Falcon Order

Preface

Salmon and trout both originated in the Northern Hemisphere and evolved in Europe and North America. In addition, they have been deliberately released around the world since the 1850s; this is not just because they are loved by anglers and fishermen, they are also high in nutritional value and a popular choice for many dishes around the world. However, because of climate change, their existence is threatened not only at the southern edge of their distribution, but over large parts of the entire area in which they are present.

For this Special Issue, leading scientists from Europe and North America working with salmonid ecologies were asked to share their knowledge and scientific findings about how climate change, with its increasing temperatures and resultant changes in the flow regimes of rivers, affects the ecology of natural salmonid populations. This is important both because of their economic value and iconic status as indicators of clean, healthy habitats, and their importance in aquatic ecosystems over large parts of the planet. Most salmon and trout species spawn in rivers, but many migrate long distances in the ocean to find food, before returning to their river of origin for breeding. Therefore, the effects of climate change on their ecology, both in fresh water and at sea, have become pressing issues.

The aim of collating these papers is, partly, to show how temperature changes, through their effects on metabolic rates and behaviors, affect important biological characteristics, such as muscle growth, body size, age of maturity and reproductive traits, and thereby their ability to adapt to varying environments and to produce offspring under natural conditions. Temperature also influences behaviors related to feeding, competition and predation, as well as migration and movement between breeding, feeding and wintering habitats. Salmonids have very good spatial cognition, and, during warm periods, they can locate cold-water refuges where they are able to escape from hostile conditions. These relationships have been illustrated in case studies, modelling work and literature reviews. This volume includes papers spanning a large geographic area from the Svalbard Islands in the north ($\sim$80 °N) to Spain ($\sim$40 °N) in the south.

With climate warming, global precipitation patterns have changed. For instance, there is now more precipitation at higher latitudes and less at lower latitudes. In the north, more precipitation falls as rain and less as snow, and the durations in which ice sheets cover water have become shorter. Another aim of this volume was to exhibit how these alterations may influence the ecology of salmonids through behavioral studies and modelling exercises.

Some populations have declined and become extinct. Yet, habitat managers have enhanced salmonid populations through supportive breeding practices and the release of hatchery-produced fish. With such releases, however, genetic diversity has decreased, with negative effects on the fishes' adaptability and reproduction. In this Special Issue, the authors provide advice on improving the management of these populations in an altered climate. In addition, they also describe gaps in the research related to species interactions, behavioral responses and biological characteristics, which, as the authors reveal, could be important to students, researchers and research administrators.

Knowledge exchanges are a fundamental part of scientific investigations. This Special Issue will hopefully contribute to this practice. Its target readers include students of fish ecology, fishery managers, fellow researchers and interested naturalists, but also anglers who have gained a serious interest in the natural history of their species of interest. This collection of papers will hopefully be of help to salmonid research as well as their management and conservation in a changing climate.

I wish to thank the Managing Editor, Ms. Gemee Cheng, and the staff of *Fishes* for their excellent co-operation and support as well as all of the authors that have willingly shared their knowledge and results and responded positively and quickly to the call for papers. It was a pleasure to work with them all.

Bror Jonsson
Editor

Editorial

The Effect of Climate Change on Salmonid Fishes in Rivers

Bror Jonsson

Norwegian Institute for Nature Research, Sognsveien 68, 0855 Oslo, Norway; bror.jonsson@nina.no

Citation: Jonsson, B. The Effect of Climate Change on Salmonid Fishes in Rivers. *Fishes* **2024**, *9*, 29. https://doi.org/10.3390/fishes9010029

Received: 3 January 2024
Accepted: 5 January 2024
Published: 8 January 2024

Climate warming is a serious threat to many organisms, such as cold-adapted ectotherms. Among these, salmonid fishes are highly threatened not only at their southern edge of distribution, but also in large parts of the temperate climate zone. With climate warming, populations go extinct, population sizes decline, and there are sub-lethal effects with genetic, epigenetic, and phenotypic changes as a result. These phenotypic changes include changes in physiology, behaviour, life history, and distributions. For instance, metabolic rates, growth and body size changes, and the tendencies and timings of migrations are altered. In addition, exotic species spread and affect interspecific competition and change threats from parasites, contagious diseases, and predators. Habitats also change in an altered climate. In rivers, this means that flow regimes, feeding opportunities, vegetation, substrate stability, oxygen concentration, particle transport, and the turbidity of the water will change. It is important to document how these alterations influence salmonid populations to reveal causes and effects and make predictions to foresee how salmonid communities will be modified in the future to improve our ability to perform proper management of habitats and populations.

The climate, i.e., the long-term pattern of the weather in a particular area, is constantly changing. We all know how the weather can vary over a day, week, month, and year. In a longer perspective, the weather is also changing, with cold and dry periods followed by mild and wet ones. At least five major ice ages have occurred through the history of the earth, and 10,000 years ago, in the early Holocene at the end of the last Ice Age, surface temperatures in the Arctic were ca. 7 °C warmer than they are today. This was because of the high energy flux from the sun in the atmosphere (radiative forcing) and the intensified inflow of warm Atlantic waters towards the north [1]. The ocean north of Siberia was open, and fish, such as walleye pollock *Gadus chalcogrammus* Pallas and Pacific herring *Clupea pallasi* Valenciennes, spread from the Pacific to the Atlantic Ocean. Then, between 9500 and 8000 years ago, temperatures began to drop in response to freshwater fluxes from melting ice. After that, the climate slowly became colder until a thermal low ca. 200 years before the present, when the climate again changed and has since become gradually warmer. After 1880, the surface temperature of the Earth increased by ca. 1.2 °C; a major part of this increase was after 1975. Thus, the climate has always changed, and will do so in the future, and fish have mechanisms that make them able to sustain climatic variations if they are not extreme. In this volume of *Fishes*, ecologists present knowledge about how salmonids spawning in fresh water react to climate warming, and what may be done to mitigate negative population effects.

Temperature affects the metabolic rates and aerobic scope of fishes; the latter is the difference between the maximum metabolic rate and the standard metabolic rate. Jonsson ([2], this volume) reviewed how metabolic rates are associated with growth, body size, behaviour, and reproduction in salmonid species. In a warmer climate, adult body size is expected to decrease, and the fish attain maturity at a younger age [3]. A younger age at maturity is associated with the faster growth of juvenile fish. In addition to this direct thermal effect, water temperature has an indirect effect, induced during early development, and expressed in a later life stage as a knock-on effect. Among these indirect thermal effects are changed egg sizes and effects on swimming and migratory activities. These phenotypically plastic responses may pre-adapt offspring to perform better in the expected future thermal environment. The mechanism allowing this flexibility in ecology is not well known, but epigenetic mechanisms have been a suggested cause [4].

The vulnerability of salmonids to climate warming is associated with their high oxygen demand. The oxygen content in water decreases with increasing temperatures. Thus, the production of salmonids in rivers at low latitudes and elevations has decreased because of warm and stressful thermal regimes. To mitigate these production losses, fishery managers release artificially bred offspring in rivers (supportive breeding). The fitness in nature of the released fish is often low because of their inadequate behaviour [5]. Also, releases may reduce the genetic diversity of natural populations. This may be because a few captive parents produce large numbers of offspring that are released (the Ryman–Laikre effect [6]). Almodóvar et al. ([7], this volume) show how the genetic diversity of Atlantic salmon populations in Spain, at the southern edge of distribution in Europe, has decreased dramatically in the last 70 years because of population reductions associated with climate change and supportive stocking of artificially bred offspring from various, often foreign sources.

Elevated temperature levels also influence the transcriptome (protein coding part of the genome) of the skeletal muscles, thereby influencing the growth and activity of fish. Molina et al. ([8], this volume) show how high temperatures upregulate many of the genes associated with autophagy (breakdown and reuse of cell parts), amino acid transport, and the glutamine metabolic process of rainbow trout *Oncorhynchus mykiss*, but they downregulate several other genes associated with digestion and muscle contraction, which are important for the adaptive processes of fish.

Many salmonids are anadromous, i.e., they migrate between rivers and the open ocean and are particularly sensitive to climate warming at the time when they migrate to sea at a life-stage called smolts. Vehanen et al. ([9], this volume) summarise how higher temperatures lead to earlier smolting, which influences their survival at sea. In warmer rivers, the salmonids smolt younger and smaller, and they leave earlier in spring, which may result in a mismatch between migration timing and optimal conditions for survival and growth at sea.

River flow changes with climate. The amount of precipitation increases in the northern and decreases in the southern part of the Northern Hemisphere, and in the north, more precipitation comes as rain instead of snow, with effects on the water flow and thermal climate. Watz et al. ([10], this volume) investigated, with the use of replicated simulations, the effects of increased temperature and discharge on the overwinter growth and mortality of one- and two-summers-old freshwater resident Atlantic salmon and brown trout in a regulated Swedish river. While increased water temperature had a positive effect on growth, the effect of a changed flow depended on the winter temperature and differed between the two species. Thus, climate change may affect the competition between salmon and trout. Ecological models that predict the effects of different environmental conditions may offer insight into such effects when in situ experiments are not feasible.

In temperate parts of Europe, the growth of Atlantic salmon during the first year at sea has decreased [11,12]. Typically, this decrease is accompanied by decreased production caused by younger age at maturity and a decrease in fish size [13]. Less is known about the climatic effects of growth in sub-Arctic parts of Europe. However, Alioravainen et al. ([14], this volume) found that freshwater growth has decreased in the River Teno (Tana), a border river between northern Finland and Norway, which probably holds the largest natural salmon population in the world. However, growth has increased during their first year at sea, contrasting observations farther south in Europe. Post-smolts of northern and southern salmon populations feed in different oceanic areas and may be differently affected by recent environmental changes.

In rivers, salmonids avoid stressful temperatures in thermal refuges. Linnansaari et al. ([15], this volume) summarise how young salmon seek cold tributary plumes, groundwater springs, alcoves and hyporheic upwellings, deep lakes, and artificial impoundments.

Salmonids have very good spatial cognition to locate and re-locate cold-water refuges. There, they may stay close together, although their distribution within the refuge can be hierarchic. With continued climate warming, managers may create new thermal refuges in rivers to protect fish populations.

Climate change may have a dramatic effect on the growth of Arctic charr in the high Arctic. This species is the northernmost freshwater fish in the world and the only fish species present in Arctic lakes on the Svalbard Islands (74–81 °N). Svenning et al. ([16], this volume) investigated the growth of Arctic charr in two Svalbard lakes between 1960 and 2008. The growth rate of young Arctic charr correlated positively with the air temperature, but negatively with the annual snow accumulation. This suggests that Arctic charr may grow better in a future with warmer Arctic lakes. On the other hand, the authors speculate that a loss of glacier rivers in a warmer future may affect the opportunity of anadromous Arctic charr to migrate to and from the sea, where the sub-adults and adults in these populations grow the most. New fish species may also enter northern rivers in a warmer future, influencing the success of Arctic charr because of interspecific interactions [17].

In a warmer northern climate, the ice-covered period will decrease in length, and salmonids will be more exposed to predators in winter. Filipsson et al. ([18], this volume) used experimental flumes to examine how surface ice and light affected the anti-predator behaviour of juvenile brown trout (*Salmo trutta*) in relation to piscivorous burbot *Lota lota* and northern pike *Esox lucius* at 4 °C. Trout had lower foraging and swimming activity and spent more time sheltering when predators were present than when absent. In daylight, the trout's swimming activity was not affected by predators, but in darkness trout were less active when predators were present. Trout consumed more drifting prey during the day when ice was present and positioned themselves further upstream when ice was not present. In the presence of pike, the trout stayed closer to conspecifics under ice. Thus, one may consider not only the potential for adaptation to changes in abiotic factors during climate change, but also how changes in environmental factors can affect behavioural species interactions which influence their survival.

In the last paper of this volume, Crozier and Siegel [19] performed a systematic literature review of climate impacts on Pacific and Atlantic salmon as a resource for stakeholders, managers, and researchers. They review published studies that address climate impacts on salmon from 2010 to 2021. They summarise expected phenotypic and genetic responses and management actions by life stage. They also show the largest research gaps related to species interactions, behavioural responses, and effects across life stages. With this literature collection, one may better perform salmonid management in a changing climate.

Conflicts of Interest: The author declares no conflicts of interest.

References

1. Van der Bilt, W.G.M.; D'Andrea, W.J.; Werner, J.P.; Bakke, J. Early Holocene temperature oscillations exceed amplitude of observed and projected warming in Svalbard lakes. *Geophys. Res. Lett.* **2019**, *46*, 14732–14741. [CrossRef]
2. Jonsson, B. Thermal effects on ecological traits of salmonids. *Fishes* **2023**, *8*, 337. [CrossRef]
3. Jonsson, B.; Jonsson, N.; Finstad, A.G. Effects of temperature and food quality on age at maturity of ectotherms: An experimental test of Atlantic salmon. *J. Anim. Ecol.* **2013**, *82*, 201–210. [CrossRef] [PubMed]
4. Jonsson, B.; Jonsson, N.; Hansen, M.M. Knock-on effects of environmental factors on ectothermic vertebrates with special reference to embryo temperature. *Q. Rev. Biol.* **2022**, *97*, 95–139. [CrossRef]
5. Jonsson, B.; Jonsson, N.; Jonsson, M. Supportive breeders of Atlantic salmon *Salmo salar* have reduced fitness in nature. *Cons. Sci. Pract.* **2019**, *1*, c85. [CrossRef]
6. Hagen, I.J.; Ugedal, O.; Jensen, A.J.; Lo, H.; Hothe, E.; Bjøru, B.; Florø-Larsen, B.; Sægrov, H.; Skoglund, H.; Karlsson, S. Evaluation of genetic effects on wild salmon populations from stock enhancement. *ICES J. Mar. Sci.* **2021**, *78*, 900–909. [CrossRef]
7. Almodóvar, A.; Nicola, G.G.; Allyón, D.; Leal, S.; Marchán, D.F.; Elvira, B. A benchmark for Atlantic salmon conservation: Genetic diversity and structure in a southern European glacial refuge before the climate changed. *Fishes* **2023**, *8*, 321. [CrossRef]
8. Molina, A.; Dettleff, P.; Valenzuela-Muños, V.; Gallardo-Escarate, C.; Valdés, J.A. High-temperature stress induced autophagy in rainbow trout skeletal muscle. *Fishes* **2023**, *8*, 303. [CrossRef]
9. Vehanen, T.; Sutela, T.; Huusko, A. Potential impact of climate change on salmonid smolt ecology. *Fishes* **2023**, *8*, 382. [CrossRef]

10. Watz, J.; Schill, J.; Addo, L.; Piccolo, J.J.; Hajiesmaeili, M. Increased temperature and discharge influence overwinter growth and survival of juvenile salmonids in a hydropeaking river: Simulating effects of climate change using individual-based modelling. *Fishes* **2023**, *8*, 323. [CrossRef]
11. Friedland, K.D.; Shank, B.V.; Todd, C.D.; McGinnity, P.; Nye, J.A. Differential response of continental stock complexes of Atlantic salmon to the Atlantic Multidecadal Oscillation. *J. Mar. Syst.* **2005**, *133*, 77–87. [CrossRef]
12. Vollset, K.W.; Urdal, K.; Utne, K.; Thorstad, E.B.; Sægrov, H.; Raunsgård, A.; Skagseth, Ø.; Lennox, R.J.; Østborg, G.M.; Ugedal, O.; et al. Ecological regime shift in the northeast Atlantic Ocean revealed from unprecedented reduction in marine growth. *Sci. Adv.* **2022**, *8*, eabk2542. [CrossRef] [PubMed]
13. Jonsson, B.; Jonsson, N.; Albretsen, J. Environmental change influences the life history of salmon in the North Atlantic. *J. Fish Biol.* **2016**, *88*, 618–637. [CrossRef] [PubMed]
14. Alioravainen, N.; Orell, P.; Erkinaro, J. Long-term trends in Freshwater and marine growth patterns in three Sub-Arctic Atlantic salmon populations. *Fishes* **2023**, *8*, 441. [CrossRef]
15. Linnansaari, T.; O'Sullivan, A.M.; Breau, C.; Corey, E.M.; Collet, E.N.; Curry, R.A.; Cunjak, R.A. The role of cold-water thermal refuges for stream salmonids in a changing climate—Experiences from Atlantic Canada. *Fishes* **2023**, *8*, 471. [CrossRef]
16. Svenning, M.-A.; Bjørvik, E.T.; Godiksen, J.A.; Hammar, J.; Kohle, J.; Borgstrøm, R.; Yoccoz, N.G. Expected climate change in the high Arctic—Good or bad for Arctic charr? *Fishes* **2024**, *9*, 8. [CrossRef]
17. Finstad, A.G.; Forseth, T.; Jonsson, B.; Bellier, E.; Hesthagen, T.; Jensen, A.J.; Hessen, D.O.; Foldvik, A. Competition exclusion along climate gradients: Energy efficiency influences the distribution of two salmonid fishes. *Glob. Chang. Biol.* **2011**, *17*, 1703–1711. [CrossRef]
18. Filipsson, K.; Åsman, V.; Greenberg, L.; Östling, M.; Watz, J.; Bergman, E. Winter behavior of juvenile brown trout: How do light and ice affect encounters with instream predators? *Fishes* **2023**, *8*, 323. [CrossRef]
19. Crozier, L.G.; Siegel, J.E. A comprehensive review of the impacts of climate change on salmon: Strengths and weaknesses of the literature by life stage. *Fishes* **2023**, *8*, 319. [CrossRef]

Review

Thermal Effects on Ecological Traits of Salmonids

Bror Jonsson

Norwegian Institute for Nature Research, Sognsveien 68, N-0855 Oslo, Norway; bror.jonsson@nina.no

Abstract: Here, I review thermal influences on metabolic rates and aerobic scope; growth; adult body size; and reproductive and behavioural traits, such as tendency and timing of the migration of salmonid fishes. A thermal window bounded by the upper and lower incipient lethal temperatures (UILT and LILT) determines where salmonids can survive. For most salmonids, LILT is close to 0 and UILT is between 20 and 30 °C. UILT and LILT are influenced by the acclimation temperature. Thermal tolerance is affected by fish size and ambient oxygen content, which decreases with increasing temperature. Standard metabolic rate (SMR), the energy required to maintain essential functions, increases with temperature, whereas maximum metabolic rate (MMR) increases with temperature until reaching a peak (pejus). Then, it decreases gradually to zero, i.e., the upper critical limit (T_{Crit}). Aerobic scope (AS = MMR-SMR) reaches its maximum at the pejus temperature. Metabolic rates and aerobic scope can be modified by temperatures that the fish experiences during embryogenesis and possibly also as larvae and young fry. At maximum feeding, maximum growth increases to a point at or below the pejus temperature. The optimum temperature for growth decreases with reduced food intake and increased body size. As for metabolic rate, the growth rate is influenced by the temperature during embryonic development. In a warmer climate, adult body size is expected to decrease chiefly because of a younger age at maturity. Parental fish retained at a higher temperature during maturation produce larger eggs, and this change in egg size may also be transferred to next-generation offspring. Furthermore, embryogenesis in warmer water leads to larger gonad and egg sizes at maturity. Water temperature influences locomotion, foraging and migratory activity. In a warmer climate, juveniles migrate to the sea earlier in spring. In addition, higher embryo temperature leads to delayed return of adult salmon from the ocean. Thus, temperature affects life history traits of salmonid fishes, partly as a direct effect on metabolic rates and food consumption and partly induced as a phenotypically plastic effect. The phenotypically plastic response may preadapt offspring to perform better in the expected future thermal environment.

Keywords: adaptive developmental programming; behaviour; climate; life history traits; metabolism; Salmonidae; temperature

Key Contribution: This paper summarizes the literature on how water temperature—via influences on metabolic rates and growth—affects life history traits of salmonid fishes. In addition, temperature during embryogenesis and early life of the organisms may have phenotypically plastic knock-on effects on these traits through epigenetic mechanisms such as DNA methylation.

Citation: Jonsson, B. Thermal Effects on Ecological Traits of Salmonids. *Fishes* **2023**, *8*, 337. https://doi.org/10.3390/fishes8070337

Academic Editor: Claus Wedekind

Received: 20 May 2023
Revised: 22 June 2023
Accepted: 23 June 2023
Published: 25 June 2023

1. Introduction

The global climate is gradually becoming more variable and warmer. Mean surface temperature has increased by ca. 1 °C during the last 100 years and is expected to increase even faster towards 2100 [1,2]. Climate change is one of the reasons why animal populations decline, and local extinctions occur at the warmest part of their distribution areas [3]. Fish are vulnerable to temperature increases because their body temperature varies with that of the surrounding water, and the oxygen content in water is low and decreases when the water becomes warmer [4]. Higher temperatures also have sublethal effects, such as changes in biochemical reactions in the body and ecological traits such as growth, body size, age at maturity and behaviour [5]. Many of the ecological changes in species are linked to changes in life history and behavioural traits. Herein, I review how water temperature affects these traits of salmonid fishes.

Salmonidae, with subfamilies Coregoninae (freshwater whitefishes), Thymallinae (graylings), and Salmoninae (trout, salmon, charr, lenoks, and taimens), are globally distributed in fresh water and as anadromous fish migrating in sea water [6]. This is an important fish family not only economically but also culturally and ecologically [7]. These species provide food for millions of people, as well as recreation and sport, and they play key roles in ecosystem functioning and health [8,9]. However, many salmonid species are in decline because of a multitude of human-induced pressures including climate warming [10–13]. For instance, brown trout (*Salmo trutta* L., 1759) has been in strong decline in southern Europe because of climate warming and an increased prevalence of extremely warm events [14–16]. Many other salmonids also show strong climate-associated declines, such as chinook salmon (*Oncorhynchus tshawytscha* Walbaum, 1792) [17], bull trout (*Salvelinus confluentus* Suckley, 1859) [18], Arctic charr (*Salvelinus alpinus* L., 1759) [19] and Atlantic salmon (*Salmo salar* L., 1759) [20,21].

Climate warming concerns Salmonidae because they are cold water species with high oxygen demands. Oxygen has low solubility in water, and dissolved O_2 content in fully saturated water decreases by approximately 2% per 1°C increase in temperature within the thermal niche of salmonids [22]. The species become stressed when the temperature increases above their normal thermal niche and go extinct if temperatures increase too quickly or too much [23]. The optimal temperatures for growth of the juveniles of most species are around 15 °C, and thermal stress and death occur between 20 and 30 °C (Table 1).

Table 1. Thermal sensitivity of selected salmonids. Optimal temperature for growth, upper critical maximum temperature at which death is almost instantaneous and upper incipient critical temperature over which the juveniles do not feed and eventually die with sources of reference.

Species	Optimal Temperature for Growth (°C)	Upper Critical Maximum Temperature (°C)	Upper Incipient Critical Temperature (°C)	References
Atlantic salmon *Salmo salar*	16–20	28–33	27	[24,25]
Brown trout *Salmo trutta*	13–17	26–30	22–25	[25–28]
Rainbow trout *Oncorhynchus mykiss* (Walbaum, 1792)	15–19	30	24–27	[29–32]
Chinook salmon *Oncorhynchus tshawytscha*	15–19	29	25	[33]
Sockeye salmon *Oncorhynchus nerka* (Walbaum, 1759)	15		23–26	[34,35]
Cutthroat trout *Oncorhynchus clarkia* (Richardson, 1836)	13–14	28	19–25	[29,36,37]
Coho salmon *Oncorhynchus kisutch* (Walbaum, 1792)	12–15	29	25–26	[33,38]
Chum salmon *Oncorhynchus keta* (Walbaum, 1792)	12–14	32–34	22–24	[33,38,39]
Brook trout *Salvelinus fontinalis* (Mitchill, 1814)	12–16	28–31	25	[36,40–42]
Arctic charr *Salvelinus alpinus* (L.)	14–17	26–27	22–23	[25,43,44]
Whitespotted charr *Salvelinus leucomaenis* (Pallas, 1814)			26–28	[45]
Lake trout *Salvelinus namaycush* (Walbaum, 1792)	12	28–29	24	[46–49]

Table 1. *Cont.*

Species	Optimal Temperature for Growth (°C)	Upper Critical Maximum Temperature (°C)	Upper Incipient Critical Temperature (°C)	References
Bull trout *Salvelinus confluentus*	12–16	25–29	21	[41,42,44]
Dolly varden *Salvelinus malma* (Walbaum, 1792)			22–23	[45]
European grayling *Thymallus thymallus* (L., 1759)	17		21	[50]
Arctic grayling *Thymallus arcticus* (Pallas, 1776)		29	23–25	[51,52]

Individuals have a limited ability to face thermal stress and adjust to rapidly changing temperature. Consequently, populations may decline, extirpate, or even go extinct when the climate changes quickly [53–56]. Fish are ectotherms, and their metabolic and developmental rates, as well as behaviour, are strongly affected by their ambient temperature. However, owing to heritability and phenotypic plasticity, vulnerability varies among individuals and populations [57]. Variations in genetic structure, composition and function influence the capacity of individuals and populations to endure warmer conditions [55,58], and individual variations in thermal tolerance have important implications for the vulnerability to both short-term extreme heat waves and long-term, gradual warming [53]. In addition, thermal sensitivity varies ontogenetically. Young, highly oxygen-requiring stages are more sensitive to warming than older stages [59]. Furthermore, the rate of change in body temperature is inversely related to the mass of the fish, making small salmonids more susceptible to variations in water temperature than larger conspecifics [60]. Embryos are also more sensitive to thermal stress than advanced stages because of their rapid formation of tissues, structures, and organs [61].

Thermal limits increase with increasing acclimation temperature but only to a certain point. For instance, the upper incipient critical temperature (cf. Table 1) of brown trout increases from 20 to 25 °C as the acclimation temperature increases from 0 to 18 °C, but it does not increase any further at higher acclimation temperatures [60]. The incipient critical temperature defines a tolerance zone that is stressful, but the fish can stay alive for a considerable length of time in even warmer water. However, outside the tolerance zone, the thermal stress is lethal. The upper critical maximum temperature is the temperature at which death is almost instantaneous, i.e., the limit of the critical range.

Phenotypic plasticity can buffer against the immediate impacts of thermal stress and reduce the sensitivity of individuals [3]. Phenotypic plasticity may develop as a knock-on effect, which is the ability of a genotype to later express alternative phenotypes in response to environmental differences [46]. Knock-on effects consist of cues or imprints experienced in a sensitive phase, transferred as a parental effect, or induced early in life, which change developmental rates, activities, or resource use [62]. Salmonids appear to be sensitive to environmental knock-on effects during embryogenesis or at the alevin stage, with later effects on morphology, life history, physiology, and behaviour. This plasticity may be an epigenetic effect [62]. Epigenetic effects are transcriptional regulators of DNA. Methylations of CpG sites in DNA sequences are the most extensively studied mechanism of epigenetic effects in ecology [63]. Methylation of CpG sites in regulatory regions of DNA downregulates genetic expression, and demethylation upregulates genetic expressions [64]. Other epigenetic mechanisms include histone modifications and microRNA, which alter the transcriptional capacities of genomes [63]. However, little is known about how phenotypic plasticity is genetically or epigenetically regulated.

Herein, I review thermal influences on metabolic rates; aerobic scope; growth; adult body size; and reproductive and behavioural traits such as tendency and timing of migration of salmonid fishes. Many examples from studies of brown trout and Atlantic salmon are included. These fishes have been used for scientific studies for more than 100 years, but examples are also available from other genera and species. My view is that temperature acts as a controller of these traits by governing the metabolic rate. There are both maximum and minimum temperatures outside which the metabolic rate is no longer sufficient to maintain life of the fish. I provide examples of direct thermal effects and how the temperature during early life stages causes phenotypic changes later.

2. Metabolic Rates and Aerobic Scope

Metabolic rates, which are the sum of all energy-yielding processes, vary with temperature and reflect the energetic cost of living [65,66]. The standard metabolic rate (SMR) is a measure of the energy required to maintain essential functions, such as breathing and blood circulation. SMR increases approximately proportionally with water temperature [67] and decreases with increasing body mass. Mass-specific SMR declines as a negative power function of body mass as organisms grow to maturity [68]. SMR should be measured in unfed and not growing fish, as both digestion of food and growth use energy, which may influence metabolic measurements [69]. For shorter time intervals, such as weeks or a few months, the mass-specific SMR of a salmonid is stable and repeatable and may hold even under variable thermal conditions [66,70]. For instance, McCarthy [71] demonstrated the stability of SMR by correlating the mass-specific SMR of individual Atlantic salmon measured 5 and 22 weeks after first feeding (June and October, respectively). This stability makes SMR a useful measurement when considering physiological traits underlying organismal performance [66].

The maximum metabolic rate (MMR) is the maximum rate of oxygen consumption that fish can achieve and use to oxidize matter for ATP generation without accumulating oxygen debt. MMR increases with temperature until reaching a peak called the pejus temperature; then, it decreases gradually to zero, which is known as the upper critical limit (T_{Crit}) [60,72]. Pejus temperature corresponds to the point at which individuals start to lose individual performance capacity. At T_{Crit}, the survival of fish is time-limited, and they live in a passive state [3]. The difference between the maximum and standard metabolic rates is called the aerobic scope (AS = MMR-SMR) [72]. AS corresponds to the highest level of energy available for activity. Individuals with higher aerobic scope are better able to take advantages of high food abundance [66] and have improved locomotor ability [73], boldness and competitive dominance, as well as increased levels of territorial aggression [74]. The optimal temperature of a species or population is the temperature resulting in the highest AS and determines their capacity to carry out functions such as foraging, growth, competition, patrolling, immune reactions, and predator defence. As these activities are temperature-dependent and influence spatial distributions and phenology of populations, they are important in contexts of climate change [75,76].

Fish can only survive for long periods of time within temperature ranges where AS is positive. The upper thermal limit is set by the physiological limits of aerobic capacity. Thermal limit diversity among populations with different adaptive histories is likely a result of adaptations in aerobic capacity to different environmental temperature regimes. Thus, thermal tolerance may vary among populations within species as a response to past selection. The ability to cope with global warming is determined by the upper thermal tolerance limit, and populations exposed to high temperatures over their evolutionary history exhibit higher thermal tolerance than conspecific populations developed under colder thermal regimes [77]. For instance, Eliason et al. [78] reported that sockeye salmon (*Oncorhynchus nerka*) in the Fraser River that experienced more challenging migratory environments have greater AS than those with less arduous migrations and that variations in AS are consistent with the historic river temperature ranges for each local population. Thus, thermal adaptation appears to occur at a local scale, with population-specific thermal limits set by physiological limitations in aerobic performance.

Variable environmental conditions influence metabolic rates. Oligney-Hébert et al. [79] compared the metabolic rates of juvenile Atlantic salmon from two rivers with different thermal regimes and acclimated the fish to either 15 or 20 °C and constant ($\pm$0.5 °C) or diel fluctuating ($\pm$2.5 °C) water temperature. Fluctuating temperature at 15 $\pm$ 2.5 °C did not influence SMR relative to stable temperature (15 $\pm$ 0.5 °C). However, diel temperature fluctuation at 20 $\pm$ 2.5 °C increased the SMR of Atlantic salmon from the warmer river by 33.7% and in the colder river by 8 % compared with the same fish acclimated to a constant temperature of 20 $\pm$ 0.5 °C. Thus, the mean temperature to which the juveniles is exposed may affect their responses to diel temperature fluctuation, and this response may vary between populations originating from rivers with different natural thermal regimes.

On the other hand, intraspecific variations in AS need not be caused by genetic differences. Instead, this may be a phenotypically plastic response induced by previously experienced differences in thermal climate [80]. For instance, Cook et al. [81] reported that temperatures experienced by brook trout (*Salvelinus fontinalis*) embryos affected body mass and routine metabolic rates as free-swimming fry. Furthermore, prehatching temperature influenced the metabolic rate of brown trout. Durtsche et al. [82] found that the SMR, MMR and AS of young brown trout (parr) were reduced when incubated as embryos in 3°C warmer water. This result is consistent with the counter-gradient variation hypothesis (CGV), according to which phenotypic variation—in this case, variation in metabolic rates—is inversely related to thermal conditions experienced by the organisms in early life [83]. This hypothesis was originally proposed in relation to altitudinal or latitudinal gradients [84]. Thus, the temperature experienced when the fish develop within the eggshell may preadapt individuals to life in either colder or warmer temperatures. Trout experiencing cold environments as embryos prepare for life in a cold environment and have higher metabolic rates at the same temperature than those that developed in warmer water. Accordingly, those that develop in cold water compensate for negative effects of low temperatures. A warm early environment favours low metabolic rates later, enabling fishes to conserve energy in an otherwise costly environment. Thus, direct environmental influences counteract inherited differences among natural populations growing up under different thermal conditions through a process of thermal plasticity. There may also be sensitive periods later in life during which SMR is programmed. For instance, Álvarez et al. [85] found a negative correlation between the temperature experienced by brown trout fry during the first 2 months after yolk resorption and SMR later. Thus, exposure to low temperatures at an early stage in life increases the temperature-dependent SMR. Such an early influence on metabolic rate has consequences for later growth, feeding and locomotor activity.

3. Growth

The aerobic scope represents the capacity of organisms to concurrently supply oxygen and energy for swimming, food digestion, absorption, assimilation (specific dynamic action SDA) and growth. High energy intake leads to faster growth, although SDA also increases with higher SMR food consumption and assimilation [86,87]. Typically, increased growth is advantageous because it protects against gape-limited predators and increases competitive ability and reproductive capacity [88]. However, a cost of faster growth may be reduced life span. There is still little information about how individual fish share their resources between these functions and restrict meal sizes to maximize growth and minimize the probability of death.

Like AS, growth rate and food consumption increase with temperature to a maximum point (optimal temperature for growth (T_{Opt})) at which oxygen availability starts limiting a further increase and the growth rate starts to decline [24,68]. The optimal temperature depends on the oxygen content in the water. For individual fish, T_{Opt} is reduced if the water is not fully saturated and increased if the water is supersaturated [89]. Temperature-dependent reaction norms for growth and food consumption are maximized at approximately the same temperature [24], and the maximum point decreases with decreasing food consumption [90,91]. Therefore, maximum growth of brown trout is reached at 13 °C for invertebrate and pellet feeding and 16 °C for fish feeding on conspecifics [26,27].

There are small differences in T_{Opt} among salmonid species, and all have relatively low thermal tolerances associated with their high oxygen requirements (Table 1). Typically, the optimal temperature for growth is round 15 °C and is lowest in lake trout (12 °C) and highest in Atlantic salmon (16–20 °C). There are intraspecific variations in thermal performance among studies, which may be partly due to methodological variation across studies, such as variation in size of test fish, acclimation temperature, oxygen content in water and other stressful conditions [89]. In addition, there may be some genetic variation in thermal performance [92,93]; however, when experimental conditions are similar, intraspecific variation in thermal performance is small. Debes et al. [94] investigated population differences and within-population genetic variation and plasticity in thermal performance traits of Atlantic salmon reared under common-garden conditions and found heritability for growth, condition and CT_{Max}. However, with increasing acclimation temperature, differences in the heritability of CT_{Max} diminished. CT_{Max} and body size were negatively correlated at the genetic and phenotypic levels, and there was indirect evidence of a positive correlation between maximum growth and thermal performance breadth for growth. Thus, population differences in thermal performance and plasticity may represent a genetic resource, in addition to the within-population genetic variance, to facilitate thermal adaptation.

Optimal temperature for growth decreases with increasing fish size [73,95]. Therefore, in lakes and at sea, large individuals often tend to live deeper and in colder water than smaller conspecifics [95,96], and small individuals may show increased growth at the same temperature as larger conspecifics experience negative growth because of lower individual optimal temperature [97]. The effect of rearing temperature on the relationship between growth and the metabolic rate of brown trout was studied by Archer et al. [98]. For 15 months, they kept study groups in either cold water ranging between 5.9 °C and 16.4 °C or in 1.8 °C warmer water (7.9–18.2 °C). They found that SMR was positively related to growth in the cool water but negatively related to growth in the warmer water. The opposite patterns were found for MMR and growth associations (positive in warm and negative in the cool regime). Mean SMR but not MMR was lower in warm regimes within both populations. Thus, there appears to be a phenotypic plastic reaction in the relationship between growth and metabolic rate depending on the thermal regime of the fish. Furthermore, a study by Finstad and Jonsson [99] demonstrated that embryo temperature had a knock-on effect on the growth of young Atlantic salmon. Young juveniles (parr) grew faster at the optimal temperature when the eggs were incubated in 7.2 ± 0.6 SD instead of 2.6 °C ± 0.4 SD water. A higher temperature during egg incubation also increases smolt size at 1 year of age and size at maturity at 2 years of age in Atlantic salmon, but it showed no effect on mass specific growth at sea after smolting [100,101] (Appendix A). Higher egg incubation temperature appeared to stimulate the fish to feed more at 1 year of age and therefore grow faster as young juveniles; however, this growth effect in the salmon disappeared after smolting.

Although the optimal temperature for growth declines with increasing body size, embryos, and alevins, which are very small, have narrower thermal limits and are more vulnerable to high temperatures than larger fish. Early life stages are highly oxygen-demanding, and high temperatures may have a negative effect on cellular functions through thermally induced oxygen diffusion limitation [102]. In addition, cell proliferation, migration, differentiation, and apoptosis (programmed cell death) are adversely affected by elevated embryo temperature. In particular, the development of the central nervous system and the notochord is highly susceptible to high temperatures [103]. The development of the notochord is thermally sensitive because of effects on the sheath cells [104]. These cells accumulate misfolded protein at elevated temperatures, leading to structural failure of the notochord and other anatomic defects in the embryo, causing malformations and death. Thus, both oxygen limitations and malformations during foetal development are causes of the high temperature sensitivity of embryos and larvae.

There are inherited differences in reaction norms of temperature-dependent growth among conspecific populations of Atlantic salmon, brown trout and Arctic charr [24,27,43] that may also hold for other salmonid species. Juveniles from large-sized, late-maturing salmonids grow better at the same temperature than those from populations of small-sized, late-maturing conspecifics. Growth differences possibly reflect different personalities of the fish, as offspring of large, late-maturing fish also feed more at the same temperature than those from populations of earlier-maturing conspecifics [24]. The optimal temperature for growth is similar among Norwegian populations of Atlantic salmon, although the thermal regimes of the rivers vary. Thus, differences in maximum growth among conspecific populations appear to reflect habitat differences rather than differences in thermal regimes [43,105,106].

4. Adult Size

According to the temperature–size rule for ectotherms, individuals maintained at a lower temperature grow more slowly but become larger at sexual maturity than those maintained at a higher temperature [107]. This is at least partly because age at maturity is growth-dependent, and slower growth means delayed maturation [108–110]. However, this does not necessarily mean that those that live in warm water are smaller in mean size than those from colder environments. This depends on the difference in annual length increment at the two temperatures and the fraction of the population that mature younger in the warmer water. Experimentally, Jonsson et al. [109,110] showed that the probability of that Atlantic salmon attained maturity for the first time during their second year in sea water increased with increasing growth rate during the last winter before maturation. Increased summer temperature had no additional effect. Atlantic salmon reared at elevated temperature attained maturity at a larger body mass and exhibited higher mass–length ratios than those of similar age reared in colder water. Temperature functions similarly to the accelerator of a motor, and higher temperatures induce faster growth if the oxygen supply is sufficient, i.e., below the pejus temperature.

Faster growth requires increased energetic assimilation, and recent findings indicate poorer feeding opportunities of Atlantic salmon in the North Atlantic Ocean resulting in poorer survival, reduced production, and smaller size for their age. However, size at maturity may, on average, be larger in many rivers because the fish attain maturity at an older age because of poorer growth [111,112]. Pacific salmon along the west coast of North America, on the other hand, mature younger with decreased production because of ocean warming, as found in large-scale investigations in Alaska [113,114]. The same declining trends hold for chinook, chum, coho and sockeye salmon. Because of the smaller fish size and reduced production, the effect is reduced nutrient transport from the ocean to rivers and riparian and terrestrial ecosystems [115,116], reduced fisheries value and fewer meals for rural people [114].

Polymorphism with sympatric morphs of different sizes occurs in several salmonids, such as brown trout [117], Sevan trout (*Salmo ischchan* Kessler, 1877) [118], Arctic charr [119] and freshwater whitefish (*Coregonus* spp.) [120]. Sympatric phenotypes often occur in pairs, exhibiting a large and a small adult morphotype of the same species [121,122]; however, in some systems, there are more than two sympatric forms. Sevan trout in Lake Sevan, Armenia [118] and Arctic charr in Lake Thingvallavatn, Iceland, exhibit four sympatric morphotypes [118,123]. The morph variation is partly inherited [121], and in several cases, clear genetic foundations of morph differentiation have been demonstrated, along with divergent life histories [120,124]. However, differences in egg incubation temperature may also influence phenotypic differentiation.

Two forms of European whitefish (*Coregonus lavaretus*) segregate vertically in Traunsee, Austria. The forms exhibit different metabolic adaptations and behavioural preferences for different temperatures [125]. In the lake, the two forms diverge by incubating embryos at either 2 °C or 6 °C, i.e., the typical temperature during embryogenesis of the two. Offspring of the two forms were reared and subjected to similar thermal conditions after hatching. The offspring differentiated in muscle growth and body size depending on the egg temperature; offspring incubated as eggs in 2 °C water grew larger than those incubated at 6 °C, regardless of whether their parents were large or small whitefish. The experiment also revealed that muscle hypertrophy (increased fibre size) and hyperplasia (increased fibre number) were affected by the thermal histories. Immunolabeling showed that the cellular mechanisms leading to increased growth after cold incubation were increased proliferation and reduced differentiation rates of muscle precursor cells, most probably associated with epigenetic differences. Thermal plasticity possibly arises from changes in physiological and endocrinological pathways, in which epigenetic regulation is likely to play an essential role [126].

Many salmonids are anadromous in addition to having freshwater living forms. This is, for instance, observed in sockeye salmon, Arctic charr, brown trout and masu salmon (*Oncorhynchus masou* Brevoort, 1856). For masu salmon, Morita et al. [127] showed that these alternative tactics were associated with temperature gradients. The occurrence of mature resident males increased, and the proportion of immature migrant males decreased with increasing temperature in Japanese rivers. They suggested that the change in the ratio of anadromous to freshwater resident males resulted from improved growth opportunities in warmer water. According to Morán and Pérez-Figueroa [128], resident and anadromous male Atlantic salmon differ in DNA methylation, although they are genetically similar. Earlier maturation and freshwater residency may be mediated by epigenetic processes rather than by genetic differences between young fish. How these differences develop is still obscure.

5. Reproductive Traits

Reproductive processes of fish are affected by the environmental temperature. Moderate thermal variation affects endocrine functions and either advance or retard gametogenesis and maturation. Above-normal temperatures may have deleterious effects on reproductive functions, and low temperatures can arrest the maturation process. For instance, in Atlantic salmon females, exposure to elevated temperatures during gametogenesis may impair both gonadal steroid synthesis and hepatic vitellogenin production, alter hepatic oestrogen receptor dynamics and ultimately result in reduced maternal investment and gamete viability [129]. High temperatures during maturation also impair gonadal steroidogenesis and delay or inhibit the preovulatory shift from production of androgens to maturation-inducing steroids. Similar effects are observed in rainbow trout and Arctic charr [129]. In Atlantic salmon, higher temperature may increase maturation of male parr [130,131], although in another study, Baum et al. [132] observed no effect of high temperature on parr maturation. In male Arctic charr and rainbow trout, high temperatures can inhibit spermiation (maturation-inducing steroids [129]), and it is reasonable to assume that the same effect also holds for other salmonid species.

Furthermore, both egg size and fecundity tend to increase with female body size. Thus, in a warmer climate with smaller females, egg sizes decrease. On the other hand, egg size is larger for similarly sized conspecifics spawning in warmer streams [133–135]. The transformation from yolk to tissue is less effective under warmer conditions. Large eggs are also favourable under poorer oxygen conditions [136]. Thus, increased egg size may give offspring an adaptive benefit in a warmer climate and should be favoured by natural selection. This is probably the reason why the egg size of salmonids decreases with increasing latitude and altitude [133,135,137,138]. Egg size differences appear to diminish when fish from different populations are reared under common thermal conditions, showing that this trait is phenotypically plastic [133]. Furthermore, egg size is influenced by the temperature that females experience during their own embryogenesis. A higher incubation temperature stimulates females to produce larger eggs as a phenotypically plastic knock-on effect [139,140]. Total ovary mass but not fecundity increases with incubation temperature years earlier. Male gonad mass is also larger in fish incubated in warmer water.

Females retained in warmer water during maturation produce larger eggs. There is also a transgenerational effect of temperature on egg size. Experimentally, Jonsson and Jonsson [140] exhibited that the mass of eggs produced by next-generation females was larger when their mothers experienced warmer water during the last two months of egg maturation relative to similar fish that experienced unheated water. This is possibly caused by an epigenetic modification of the parental fish. In brook trout, using whole-genome bisulphite sequencing, Venney et al. [141] found 188 differentially methylated DNA regions due to parental maturation temperature. Stable intergenerational inheritance of DNA methylation may transfer the epigenetic states to offspring, priming them for a warming environment. This has implications regarding the role of intergenerational epigenetic inheritance in response to climate change.

6. Behavioural Traits

Temperature influences fish behaviour, such as swimming activity and foraging, exploratory behaviour, prey capture and predator avoidance [142,143].

The timing of smolt migration, i.e., when young anadromous salmonids migrate from fresh water to the sea, occurs coincidentally with environmental changes such as increasing temperature and day length in spring. The warmer the water is and faster the temperature increase, the earlier in the season the smolts migrate downstream and out to sea [144–146]. Earlier outmigration from a warmer river appears to decrease smolt size and increase the mortality of the fish, as found for brown trout and Atlantic salmon [147–149]. Thus, the body size at migration is influenced by river temperature, with smaller smolts in warmer rivers, as found for brown trout along the Norwegian coast [150]. However, river temperature may not be the only factor that influences smolt size. Smolts are selected to survive in sea water [151]. As the ionic stress in sea water increases with decreasing temperature, smolts entering colder seas should be relatively large, as observed in anadromous brown trout in Europe [152]. There is evidence that reduced survival is associated with earlier smolt migration, as found for brown trout and Atlantic salmon in Norway [147,148].

During the spawning migration, high temperatures may lead to anaerobic locomotion, energy losses and prespawning mortality [153–155]. River-dwelling salmonids may avoid high temperatures by entering cooler water (thermal refuging) to maintain a body temperature close to optimal levels and minimize energetic costs associated with high temperature [156–159]. Thermal refuges include cold water tributaries, groundwater seeps, deep pools, and cold alcoves [159–162]. This holds for both juveniles and adults of various species and at both low and high latitudes [163,164]. In addition, salmonids experiencing suboptimal temperature during migration may reduce energy use by burst-and-coast swimming [165]. Burst-and-coast swimming, whereby bursts of fast swimming are in cyclic alternation with phases of coasting and the body is kept straight and motionless, can be an energetically advantageous behaviour. It allows the fish to gain fast swimming speeds during short bursts while preventing the effects of fatigue by allowing metabolic recovery of muscle fibres during the coast phases [166].

Higher temperature may also affect salmonid spawning migrations. These fish feed little or not at all during the migration [167,168], as they rely on endogenous energy stores to fuel the migration back to their home river and spawning sites and the development of sexual characters and reproduction [165]. Adipose tissue reserves are the primary source of energy used for upriver migration and gonad production [169], and protein from muscular tissue fuels the development of secondary sexual characters and metabolism during spawning activities [170]. Migratory energy costs increase with temperature and reduce the capacity of the fish to recover from exhaustive exercises [171].

Temperature may also influence where in the ocean the salmon feed. During warmer summers, Atlantic salmon in the Baltic Sea fed closer to their home river in the Gulf of Bothnia, while in colder summers, they fed farther south in the Baltic Main Basin [172]. Furthermore, the temperature influences the timing of the return migration. Elevated water temperature may induce either earlier [173,174] or later arrival in the spawning area [175]. Salmon may migrate outside the peak summer temperatures [175] and respond to suboptimal water temperatures by delaying migration in cool thermal refuges [176–178]. In addition, temperatures during embryogenesis may influence phenological decisions such as when to return from the ocean and spawn in rivers. Jonsson and Jonsson [179] released groups of juvenile Atlantic salmon (smolts) produced from eggs incubated at either ambient (~4 °C) or 3 °C warmer water temperature. After hatching, both experimental groups were reared under similar thermal conditions until smolting and released. The fish migrated concurrently to the sea as juveniles, and after feeding in the ocean for one or two years, they returned to the experimental river for spawning. Atlantic salmon that were developed from eggs incubated in warmer water returned from the sea ca. 2 weeks later in the summer/autumn than adults of the same age developed from colder eggs. The later return was independent of the body size of the juveniles at outmigration and similar for offspring of three different tested populations. Hence, thermal conditions during early development appeared to prepare the offspring, when adult, to return for spawning later in the year. Later return to a warmer (or earlier return to a colder) river may be advantageous both because metabolic rates are higher in warmer water, reducing the reserve energy used during migration and spawning faster, but also because egg incubation time decreases with increasing temperature, so the fish may spawn later under warmer conditions. The mechanism driving this phenotypic plastic response has not been investigated but may be linked to the maturation process.

At high latitudes, climate change leads to higher winter temperatures when the eggs of autumn-spawned salmonids develop within their gravel beds [88]. Recent investigations have exhibited that temperature during embryogenesis affects the behaviour of young brown trout. Exposing the fertilized eggs to 1.5–2–5 °C warmer water reduced activity level of the young juveniles after hatching [180]. It is difficult to predict the overall consequences on fitness of this response to warmer egg incubation temperature, as the activity of their ectothermic predators and prey may also change [181]. Nevertheless, the results of this and previous studies [82,182] suggest that juvenile brown trout in a warmer climate have lower metabolic rates and aerobic scopes and are less active, with similar consequences across migratory and non-migratory phenotypes.

Temperature during egg incubation influences later river emigration. Jonsson and Greenberg [182] showed that the proportion of warm-incubated brown trout released in the River Imsa, Norway, moved downstream towards the sea to a greater extent than those incubated as embryos in colder water. Most of the emigrants moved downstream in the autumn. The cold-incubated offspring possibly had a higher metabolic rate and kept their position at low temperature in the fast-flowing river, while more of the warm-incubated fish moved downstream towards the sea. A similar difference was observed with respect to whether the parents were anadromous or lacustrine adfluvial phenotypes or crosses between the two. Thus, the difference in emigration regarding embryonic temperature was phenotypically plastic and may be associated with an epigenetic effect of the thermal conditions during early development. However, the outmigration ratios varied between offspring of geographically isolated populations, suggesting that there are genetic divergences in this trait among populations.

7. Discussion

Temperature has a pervasive and direct effect on biochemical and physiological functions of ectotherms and a strong influence on life history traits. Temperature regulates metabolism and aerobic scope, growth, body size, gonadal size, behavioural locomotion and phenological patterns. Variations in thermal responses are caused by inherited differences, such as that of metabolic rates. Auer et al. [66] measured SMR, MMR and AS of juvenile brown trout and observed that metabolic rates were repeatable over time periods of months, even under changing thermal conditions. The among-individual differences in metabolism appear to have fitness-related effects and be related to individual differences in growth, body mass, reproduction, survival, behaviour, and phenology. Thus, temperature plays an important role in determining evolutionary trajectories of species [183,184].

In addition to the direct phenotypic effect of temperature, recent research suggests that embryo and larval temperature or the temperature experienced by parents may affect life history traits. Although not extensively investigated, these latter findings indicate that past temperatures may prepare offspring for conditions that they may encounter later in life. Some of these effects, such as that on growth, appear to be strongest early in life [100], whereas others, such as that on migration, may also influence the phenology of adults [141]. These latter knock-on effects have been demonstrated for Atlantic salmon and brown trout, and effects on growth have also been demonstrated for other species, for instance, common carp (*Cyprinus carpio* L., 1759) [185], haddock (*Melanogrammus aeglefinus* L., 1759) [186] and Senegalese sole (*Solea senegalensis* Kaup, 1858) [187]. However, population-specific variations are expected, although scarcely investigated [188].

There are also transgenerational effects of temperature. For instance, the temperature during egg maturation prior to fertilization influences egg size, gonad size and the amount of energy available for the offspring after hatching. These thermal effects are transferred to the next-generation offspring of Atlantic salmon [140] and may positively affect early juvenile growth in a warmer climate [189,190]. Knock-on effects of previous temperature appear to parallel changes observed among conspecific populations living under different thermal regimes. These effects reduce phenotypic differences among individuals experiencing different climatic conditions, as explained in the subchapter "Metabolic rates and aerobic scope". However, there is an urgent need for further research on transgenerational, thermal effects on the ecology of fishes and the mechanism by which they are transferred from parents to offspring.

Phenotypic plasticity in thermal response appears to be initiated by influences experienced during a sensitive period, although the consequences may be long-lasting after the sensitive period has ended [139,191]. Such processes initiated by early priming are referred to as adaptive developmental plasticity [192]. Typically, thermal conditions during early development or at a parental phase may have effects that change developmental trajectories, which may be helpful later in life [62]. The influences stimulate genotypes to express different phenotypes in response to thermal differences during early development [193,194]. For instance, knock-on effects that are initiated by cues or imprints experienced in individuals' early life, transferred as a parental effect, or induced at embryo or larval stages may affect developmental trajectories, activities, and resource allocations within an organism's life span [62]. The effects may be short-term or long-term and prepare individuals for conditions that they may encounter later, thereby buffering otherwise detrimental effects of environmental change [195–197]. Such parental or early thermal effects should be advantageous when parents or early environments provide a reliable forecast about the thermal climate that offspring may encounter later [198–200]. Like local adaptation [201], developmental knock-on effects should evolve through natural selection in responses to the environments to which offspring are repeatedly and consistently exposed over evolutionary time, whereas responses to novel or atypical environments may be maladaptive [202,203]. The latter is likely to occur in response to extreme temperatures and severe temperature variation. However, at present, there is no quantitative study demonstrating how beneficial such early programming of salmonids is.

The mechanisms involved in thermal plasticity have not been extensively investigated. However, DNA methylation is sensitive to thermal climate [204] and may be involved. Polar fishes show higher DNA methylation levels than temperate fishes, and Antarctic icefishes (Channichthyidae) have the highest DNA methylation level on record. There is an inverse relationship between DNA methylation and body temperature when maintained over evolutionary time. DNA methylation links thermal conditions to subsequent changes in genetic expressions [205], but the response differs among iso-genetic lines, as shown for rainbow trout.

There are also examples from other fish families suggesting effects of temperature on DNA methylation. For instance, Atlantic cod (*Gadus morhua* L., 1759) embryos exposed to a 4 °C increase in temperature exhibited changes in the expression of genes involved in one carbon pathway [206]. Furthermore, higher temperature affected the DNA methylome of a coral reef fish (*Acanthochromis polyacanthus*; Bleeker, 1855) and influenced phenotypic plasticity, which enabled some populations to maintain their performance under thermal stress [207]. Anistadiadi et al. [208] exposed European sea bass (*Dicentrarchus labrax* L., 1759) larvae to periods of moderate temperature increases. The authors found that a 2 °C increase in temperature changed global DNA methylation and the expression of ecologically relevant genes related to stress response, muscle development and organ formation. DNA methylation changes were more pronounced in larvae previously acclimated to a different temperature.

Both in early and later life, temperature change may lead to DNA methylation. Beemelmanns et al. [209] challenged Atlantic salmon post smolts with increasing temperatures from 12 to 20 °C. They reported that exposure to high temperature affected the methylation of CpG sites. There were distinct CpG methylation profiles for different treatment groups, indicating that each environmental condition may induce different epigenetic signatures.

8. Future Research

Little is known about how temperature influences distributions of salmonid species. Finstad et al. [210] hypothesized that differences in temperature-dependent growth efficiencies were a main reason for differences in distributional patterns of brown trout and Arctic charr. Their thermal optima are similar, but Arctic charr outcompete brown trout in cold and ultraoligotrophic lakes and rivers because they have twice as high growth efficiency in cold water. Little is known about the degree to which a similar effect may influence distributional patterns of other salmonids. They all have similar thermal niches, but differences in metabolic rates and growth efficiencies may still exist. Such knowledge is important in understanding geographical distributions, immigration, and local extinction of species.

Some studies indicate intraspecific variation in thermal growth performance. However, as many factors influence the results of growth experiments, such as light conditions, fish size, food rations, acclimatization temperature, early temperature, stressful environmental conditions and parental temperature during egg development, experimental results may be difficult to compare [29,90,99,100]. Because experimental results differ [60,78,93], there is a need to reveal how much of the variation in thermal performance is due to phenotypic plasticity and how much is because of additive genetic variation.

Research supports the hypothesis that thermal conditions early in life affect life histories of organisms. It is important to identify to what degree developmental plasticity is adaptive. Influences encountered by organisms long before experiments start can affect the results. If not considered, this may lead to incorrect interpretations of trait differences. One may believe that observed differences are genetically adapted, while they may be an early environmental effect. Such misinterpretation may lead to incorrect decisions when managing populations under climate change.

Epigenetics appear to be central in the understanding of how the early thermal environment affects the development of phenotypes. There are examples suggesting a role of epigenetics in developmental plasticity. However, the field of epigenetics is still young, and I expect that many new studies soon will be performed to better understand how environmental temperature influences the ecology of salmonid fishes. Studies using whole-genome approaches should be performed, as such studies may reveal new relationships between phenotypes and epigenetic determinants.

There is strong evidence supporting the hypothesis that thermal cues experienced at an early stage in life can affect the development of organisms, with consequences for life in environments encountered at a later stage. However, at present, there is little if any knowledge about energetic costs involved in thermal plasticity and whether thermal plasticity is adaptive and affects fitness. Such tests are needed, as populations' responses in changing environments are critical to their persistence. Their capacity to exhibit adaptive plasticity to a warmer climate may determine their future success.

9. Conclusions

Temperature has pervasive effects on growth, life history and behavioural traits of salmonid fishes. This is caused by a direct effect on metabolic rates and food consumption. Intraspecific genetic variability in these thermal effects needs further investigation. In addition, there is an indirect, phenotypically plastic effect induced by temperature during embryogenesis and early life, as well as possibly by temperatures experienced by mothers during maturation. This phenotypically plastic response may preadapt offspring to perform better in anticipated future thermal environments. An epigenetic mechanism such as DNA methylation may be responsible for the phenotypic effect. Future research should investigate this and other possible epigenetic mechanisms and how they may influence fitness and induce alternative phenotypes.

Funding: This research received no external funding.

Institutional Review Board Statement: Not applicable.

Data Availability Statement: No new data were created or analyzed in this study. Data sharing is not applicable to this article.

Conflicts of Interest: The author declares no conflict of interest.

Appendix A

Smolting

Smolting is a preparatory physiological adaptation process that occurs in spring prior to the seaward migration of anadromous salmonids. The young fish become slimmer, with a more streamlined body with darker back, more silvery flanks and a whiter belly. Physiologically, the Na^+,K^+-ATPase activity and salinity tolerance increase, as do the density of visual pigment porphyropsin in the retinae of the eyes. On the other hand, the fat density of the muscle tissue decreases, and the activities of metabolic enzymes change. These changes precede downstream migration and prepare the fish for marine life in pelagic waters.

References

1. USGCRP. *Climate Science Special Report: Fourth National Climate Assessment*; Wuebbles, D.J., Fahey, D.W., Hibbard, K.A., Dokken, D.J., Stewart, B.C., Maycock, T.K., Eds.; U.S. Global Change Research Program: Washington, DC, USA, 2017; Volume I, pp. 1–470.
2. NOAA. State of the Climate: Global Climate Report for 2021. *WMO* **2022**, *1290*, 1–54.
3. McKenzie, D.J.; Zhang, Y.; Eliason, E.J.; Schulte, P.M.; Claireaux, G.; Blasco, F.R.; Nati, J.J.H.; Farrell, A.P. Intraspecific variation in tolerance of warming in fishes. *J. Fish Biol.* **2021**, *96*, 1536–1551. [CrossRef]
4. Volkoff, H.; Rønnestad, I. Effects of temperature on feeding and digestive processes in fish. *Temperature* **2020**, *7*, 307–320. [CrossRef]
5. Schulte, P.M.; Healy, T.M.; Fangue, N.A. Thermal performance curves, phenotypic plasticity, and the time scales of temperature exposure. *Integr. Comp. Biol.* **2011**, *51*, 691–702. [CrossRef]
6. Nelson, J.S. *Fishes of the World*, 4th ed.; Wiley: Hoboken, NJ, USA, 2006.
7. Kershner, J.L.; Williams, J.E.; Gresswell, R.E.; Lobón-Cerviá, J. *Trout and Char of the World*; American Fisheries Society: Bethesda, MD, USA, 2019.
8. Bottom, D.L.; Jones, K.K.; Simenstad, C.A.; Smith, C.L. Reconnecting social and ecological resilience in salmon ecosystems. *Ecol. Soc.* **2009**, *14*, 5. [CrossRef]
9. Watz, J.; Alvén, D.; Andreasson, P.; Aziz, K.; Blixt, M.; Calles, O.; Bjørnås, K.L.; Olsson, I.; Österling, M.; Stålhammar, S.; et al. Atlantic salmon in regulated rivers: Understanding river management through the ecosystem services lens. *Fish Fish.* **2022**, *23*, 478–491. [CrossRef]
10. Jonsson, B.; Waples, R.S.; Friedland, K.D. Extinction considerations for diadromous fishes. *ICES J. Mar. Sci.* **1999**, *56*, 405–409. [CrossRef]
11. Comte, L.; Buisson, L.; Daufresne, M.; Grenouillet, G. Climate-induced changes in freshwater fish distribution: Observed and predicted trends. *Freshw. Biol.* **2013**, *58*, 625–639. [CrossRef]
12. Kovach, R.; Jonsson, B.; Jonsson, N.; Arismendi, I.; Williams, J.E.; Kerchner, J.L.; Al-Chokhachy, R.; Letcher, B.; Muhlfeld, C.C. Climate Change and the Future of Trout and Char. In *Trout and Char of the World*; American Fisheries Society: Bethesda, MD, USA, 2019; pp. 685–716.

13. Muhlfeld, C.C.; Dauwalter, D.C.; D'Angelo, V.S.; Ferguson, A.; Giersch, J.J.; Impson, D.; Koizumi, I.; Kovach, R.; McGinnity, P.; Schöffmann, J.; et al. Global Status of Trout and Char: Conservation Challengers in the Twenty-First Century. In *Trout and Char of the World*; American Fisheries Society: Bethesda, MD, USA, 2019; pp. 717–760.

14. Almodóvar, A.; Nicola, G.G.; Ayllón, D.; Elvira, B. Global warming threatens the persistence of Mediterranean brown trout. *Glob. Chang. Biol.* **2012**, *18*, 1549–1560. [CrossRef]

15. Réalis-Doyelle, E.; Pasquet, A.; De Charleroy, D.; Fontaine, P.; Teletchea, F. Strong effects of temperature on the early life stages of a cold stenothermal fish species, brown trout (*Salmo trutta* L.). *PLoS ONE* **2016**, *11*, e0155487. [CrossRef]

16. Frölicher, T.L.; Laufkötter, C. Emerging risks from marine heat waves. *Nat. Commun.* **2018**, *9*, 650. [CrossRef]

17. Crozier, L.G.; Bruke, B.J.; Chasco, B.E.; Widener, D.L.; Zabel, R.W. Climate change threatens Chinook salmon throughout their life cycle. *Commun. Biol.* **2021**, *4*, 222. [CrossRef]

18. Isaak, D.J.; Luce, C.H.; Rieman, B.E.; Nagel, D.E.; Peterson, E.E.; Horan, D.L.; Parkes, S.; Chandler, G.L. Effects of climate and wildfire on stream temperatures and salmonid thermal habitat in a mountain river network. *Ecol. Appl.* **2010**, *20*, 1350–1371. [CrossRef]

19. Kelly, S.; Moore, T.N.; de Eyto, E.; Dillane, M.; Goulon, C.; Guillard, J.; Lasne, E.; McGinnity, P.; Poole, R.; Winfield, I.J.; et al. Warming winters threaten peripheral Arctic charr populations of Europe. *Clim. Chang.* **2020**, *63*, 599–618. [CrossRef]

20. Nicola, G.G.; Elvira, B.; Jonsson, B.; Ayllón, D.; Almodóvar, A. Local and global climatic drivers of Atlantic salmon decline in southern Europe. *Fish. Res.* **2018**, *198*, 78–85. [CrossRef]

21. Almodóvar, A.; Ayllón, D.; Nicola, G.G.; Jonsson, B.; Elvira, B. Climate-driven bio-physical changes in feeding and breeding environments explain the decline of southernmost European Atlantic salmon populations. *Can. J. Fish Aquat. Sci.* **2019**, *76*, 1581–1595. [CrossRef]

22. Atkins, P.W. *Physical Chemistry*, 6th ed.; Oxford University Press: Oxford, UK, 1998.

23. Gillson, J.P.; Bašić, T.; Davison, P.I.; Riley, W.D.; Talks, L.; Walker, A.M.; Russell, I.C. A review of marine stressors impacting Atlantic salmon *Salmo salar*, with an assessment of the major threats to English stocks. *Rev. Fish Biol. Fish.* **2022**, *32*, 879–919. [CrossRef]

24. Jonsson, B.; Forseth, T.; Jensen, A.J.; Næsje, T.F. Thermal performance in juvenile Atlantic salmon, *Salmo salar* L. *Funct. Ecol.* **2001**, *15*, 701–711. [CrossRef]

25. Elliott, J.M.; Elliott, J.A. Temperature requirements of Atlantic salmon *Salmo salar*, brown trout, *Salmo trutta* and Arctic charr *Salvelinus alpinus*: Predicting the effects of climate change. *J. Fish Biol.* **2010**, *77*, 1793–1817. [CrossRef]

26. Forseth, T.; Jonsson, B. The growth and food ration of piscivorous brown trout (*Salmo trutta*). *Funct. Ecol.* **1994**, *8*, 171–177. [CrossRef]

27. Forseth, T.; Larsson, S.; Jensen, A.J.; Jonsson, B.; Näslund, I.; Berglund, I. Thermal growth performance of juvenile brown trout *Salmo trutta*: No support for thermal adaptation hypothesis. *J. Fish Biol.* **2009**, *74*, 133–149. [CrossRef]

28. Lee, R.M.; Rinne, J.N. Critical thermal maxima of five trout species in the south-western United States. *Trans. Am. Fish. Soc.* **1980**, *109*, 632–635. [CrossRef]

29. Bear, E.A.; McMahon, T.E.; Zale, A.V. Comparative thermal requirements of Westslope cutthroat trout and rainbow trout: Implications for species interactions and development of thermal protection standards. *Trans. Am. Fish. Soc.* **2007**, *136*, 1113–1121. [CrossRef]

30. Hokanson, K.E.; Kleiner, C.F.; Thorslund, T.W. Effects of constant temperatures and diel temperatures of specific growth and mortality rates and yield of juvelile rainbow trout, *Salmo gairdneri*. *J. Fish. Res. Bd. Can.* **1977**, *34*, 639–648. [CrossRef]

31. Railsback, S.F.; Rose, K.A. Bioenergetics modelling of stream trout growth: Temperature and food consumption effects. *Trans. Am. Fish. Soc.* **1999**, *128*, 241–256. [CrossRef]

32. Wurtsbaugh, W.A.; Davis, G.E. Effects of temperature and ration level on the growth and food conservation efficiency of *Salmo gairdneri*, Richardson. *J. Fish Biol.* **1977**, *11*, 87–98. [CrossRef]

33. Brett, J.R. Temperature tolerance in young Pacific salmon, genus *Oncorhynchus*. *J. Fish. Res. Bd. Can.* **1952**, *9*, 265–323. [CrossRef]

34. Brett, J.R. Energic responses of salmon to temperature. A study of some thermal relations in the physiology and freshwater ecology of the sockeye salmon (*Oncorhynchus nerka*). *Am. Zool.* **1971**, *11*, 99–113. [CrossRef]

35. Chen, Z.; Anttila, K.; Wu, J.; Whitney, C.K.; Hinch, S.G.; Farrell, A.P. Optimum and maximum temperatures of sockeye salmon (*Oncorhynchus nerka*) populations hatched at different temperatures. *Can. J. Zool.* **2013**, *91*, 265–274. [CrossRef]

36. De Staso, J., III; Rahel, F.J. Influence of water temperature on interactions between juvenile Colorado River cutthroat trout and brook trout in a laboratory stream. *Trans. Am. Fish. Soc.* **1994**, *123*, 289–297. [CrossRef]

37. Dickerson, B.R.; Vinyard, G.L. Effects of high chronic temperatures and diel temperature cycles on the survival and growth of Lahontan cutthroat trout. *Trans. Am. Fish. Soc.* **1999**, *128*, 516–521. [CrossRef]

38. Beschta, R.L.; Bilby, R.E.; Brown, G.W.; Holtby, L.B.; Hofstra, T.D. Stream Temperature and Aquatic Habitat: Fisheries and Forestry Interactions. In *Streamside Management: Forestry and Fishery Interactions*; Salo, E.O., Cundy, T.W., Eds.; University of Washington, Institute of Forest Resources: Seattle, DC, USA, 1987; Volume 57, pp. 191–232.

39. Snyder, G.R.; Blahm, T.H. Effects of increased temperature on cold-water organisms. *J. Water Pollut. Cont. Fed.* **1971**, *43*, 890–899.

40. McCormick, J.H.; Hokanson, K.E.F.; Jones, B.R. Effects of temperature on growth and survival of young brook trout, *Salvelinus fontinalis*. *J. Fish. Res. Bd. Can.* **1972**, *29*, 1107–1112. [CrossRef]

41. Selong, J.H.; McMahon, T.E.; Zale, A.V.; Barrows, F.T. Effect of temperature on growth and survival of bull trout, with application of an improved method for determining thermal tolerance in fishes. *Trans. Am. Fish. Soc.* **2001**, *130*, 1026–1037. [CrossRef]
42. McMahon, T.E.; Zale, A.V.; Barrows, F.T.; Selong, J.H.; Danehy, R.J. Temperature and competition between bull trout and brook trout: A test of the elevation refuge hypothesis. *Trans. Am. Fish. Soc.* **2007**, *136*, 1313–1326. [CrossRef]
43. Larsson, S.; Forseth, T.; Berglund, I.; Jensen, A.J.; Näslund, I.; Elliott, J.M.; Jonsson, B. Thermal adaptation of Arctic charr: Experimental studies of growth in eleven charr populations from Sweden, Norway and Britain. *Freshw. Biol.* **2005**, *50*, 353–368. [CrossRef]
44. Jobling, M.; Jørgensen, E.H.; Arnesen, A.M.; Ringø, E. Feeding, growth and environmental requirements of Arctic charr: A review of aquaculture potential. *Aquacult. Int.* **1993**, *1*, 20–46. [CrossRef]
45. Takami, T.; Kitano, F.; Nakano, S. High water temperature influences on foraging responses and thermal death of dolly varden *Salvelinus malma* and whitespotted charr *S. leucomaenis*. *Fish. Sci.* **1997**, *63*, 6–8. [CrossRef]
46. Kelly, N.I.; Burness, G.; McDermid, J.L.; Wilson, C.C. Ice age fish in a warming world: Minimal variation in thermal acclimation capacity among lake trout (*Salvelinus namaycush*) populations. *Cons. Physiol.* **2014**, *16*, cou025. [CrossRef]
47. Edsall, T.A.; Cleland, J. Optimum temperature for growth and preferred temperatures of age-0 lake trout. *N. Am. J. Fish. Manag.* **2000**, *20*, 804–809. [CrossRef]
48. Gibson, E.S.; Fry, F.E.J. The performance of the lake trout, *Salvelinus namaycush*, at various levels of temperature and oxygen pressure. *Can. J. Zool.* **1954**, *32*, 252–260. [CrossRef]
49. Hasnain, S.S.; Minns, C.K.; Shuter, B.J. *Key Ecological Temperature Metrics for Canadian Freshwater Fishes*; Ontario Ministry of Natural Resources, Applied Research and Development Branch: Sault Ste. Marie, ON, Canada, 2010; pp. 1–42.
50. Mallet, J.P.; Charles, S.; Persat, H.; Auger, P. Growth modelling in accordance with daily water temperature in European grayling (*Thymallus thymallus* L.). *Can. J. Fish. Aquat. Sci.* **1999**, *56*, 994–1000. [CrossRef]
51. McLeay, D.J.; Knox, A.J.; Malick, J.G.; Birtwell, I.K.; Hartman, G.; Ennis, G.L. Effects on Arctic grayling (*Thymallus arcticus*) of short-term exposure to Yukon mining sediments: Laboratory and field studies. *Can. Tech. Rep. Fish. Aquat. Sci. Ott.* **1983**, *1171*, 1–134.
52. Lohr, S.C.; Byorth, P.A.; Kaya, C.M.; Dwyer, W.P. High temperature tolerances of fluvial Arctic grayling and comparisons with summer river temperatures of the Big Hole River, Montana. *Trans. Am. Fish. Soc.* **1996**, *125*, 933–939. [CrossRef]
53. Bennett, S.; Duarte, C.M.; Marbà, N.; Wernberg, T. Integrating within-species variation in thermal physiology into climate change ecology. *Philos. Trans. R. Soc. B* **2019**, *374*, 20180550. [CrossRef]
54. Burggren, W.W. Inadequacy of typical physiological experimental protocols for investigating consequences of stochastic weather events emerging from global warming. *Am. J. Physiol. Regul. Integr. Comp. Physiol.* **2019**, *316*, 318–322. [CrossRef]
55. Pacifici, M.; Foden, W.B.; Visconti, P.; Watson, J.E.M.; Butchart, S.H.M.; Kovacs, K.M.; Rondinini, C. Assessing species vulnerability to climate change. *Nat. Clim. Chang.* **2015**, *5*, 215–225. [CrossRef]
56. Thorstad, E.B.; Bliss, D.; Breau, C.; Damon-Randall, K.; Sundt-Hansen, L.E.; Hatfield, E.M.C.; Horsburgh, G.; Hansen, H.; Maoiléidigh, N.Ó.; Sheehan, T.; et al. Atlantic salmon in a rapidly changing environment—Facing the challenges of reduced marine survival and climate. *Aquat. Cons. Mar. Freshw. Ecosyst.* **2021**, *31*, 2654–2665. [CrossRef]
57. Sparks, M.M.; Westley, P.A.H.; Falke, J.A.; Quinn, T.P. Thermal adaptation and phenotypic plasticity in a warming world: Insights from common garden experiments on Alaskan sockeye salmon. *Glob. Chang. Biol.* **2017**, *23*, 5203–5217. [CrossRef]
58. Moran, E.V.; Hartig, F.; Bell, D.M. Intraspecific trait variation across scales: Implications for understanding global change responses. *Glob. Chang. Biol.* **2016**, *22*, 137–150. [CrossRef]
59. Dahlke, F.T.; Wohlrab, S.; Butzin, M.; Pörtner, H.-O. Thermal bottlenecks in the life cycle define climate vulnerability of fish. *Science* **2020**, *369*, 65–70. [CrossRef]
60. Elliott, J.M. *Quantitative Ecology and the Brown Trout*; Oxford University Press: Oxford, UK, 1994.
61. Edwards, M.J.; Saunders, R.D.; Shiota, K. Effects of heat on embryos and foetuses. *Int. J. Hypertherm.* **2003**, *19*, 295–324. [CrossRef]
62. Jonsson, B.; Jonsson, N.; Hansen, M.M. Knock-on effects of environmental influences during embryonic development of ectothermic vertebrates. *Q. Rev. Biol.* **2022**, *97*, 95–139. [CrossRef]
63. Verhoeven, K.J.F.; von Holdt, B.M.; Sork, V.L. Epigenetics in ecology and evolution: What we know and what we need to know. *Mol. Ecol.* **2016**, *25*, 1631–1638. [CrossRef]
64. Greenberg, M.V.C.; Bourc'his, D. The diverse roles of DNA methylation in mammalian development and disease. *Nat. Rev. Mol. Cell Biol.* **2019**, *20*, 590–607. [CrossRef]
65. Fry, F.E.J. The Effect of Environmental Factors on the Physiology of Fish. In *Fish Physiology VI*; Hoar, W.S., Randall, D.J., Eds.; Academic Press: London, UK, 1971; pp. 1–98.
66. Auer, S.K.; Anderson, G.J.; McKelvey, S.; Bassar, R.D.; McLennan, D.; Armstrong, J.D.; Nislow, K.H.; Downie, H.K.; McKelvey, L.; Morgan, T.A.J.; et al. Nutrients from salmon parents alter selection pressures on their offspring. *Ecol. Lett.* **2018**, *21*, 287–295. [CrossRef]
67. Norin, T.; Malte, H.; Clark, T.D. Aerobic scope does not predict the performance of a tropical eurythermal fish at elevated temperatures. *J. Exp. Biol.* **2014**, *217*, 244–251. [CrossRef]
68. Schmidt-Nielsen, K. *Scaling: Why Is Animal Size So Important?* Cambridge University Press: New York, NY, USA, 1984.
69. Rosenfeldt, J.; Van Leeuwen, T.; Richards, J.; Allen, D. Relationship between growth and standard metabolic rate: Measurement artefacts and implications for habitat use and life-history adaptation in salmonids. *J. Anim. Ecol.* **2015**, *84*, 4–20. [CrossRef]

70. O'Connor, K.I.; Taylor, A.C.; Metcalfe, N.B. The stability of standard metabolic rate during a period of food deprivation in juvenile Atlantic salmon. *J. Fish Biol.* **2005**, *57*, 41–51. [CrossRef]
71. McCarthy, I.D. Temporal repeatability of relative standard metabolic rate in juvenile Atlantic salmon and its relation to life history variation. *J. Fish Biol.* **2005**, *57*, 224–238. [CrossRef]
72. Pörtner, H.-O. Oxygen- and capacity-limitation of thermal tolerance: A matrix for integrating climate-related stressor effects in marine ecosystems. *J. Exp. Biol.* **2010**, *213*, 881–893. [CrossRef]
73. Reidy, S.P.; Kerr, S.R.; Nelson, J.A. Aerobic and anaerobic swimming performance of individual Atlantic cod. *J. Exp. Biol.* **2000**, *203*, 347–357. [CrossRef]
74. Biro, P.A.; Stamps, J.A. Do consistent individual differences in metabolic rate promote consistent individual differences in behavior? *Trends Ecol. Evol.* **2010**, *25*, 653–659. [CrossRef]
75. O'Connor, M.I.; Selig, E.R.; Pinsky, M.L.; Altermatt, T. Toward a conceptual synthesis for climate change responses. *Glob. Ecol. Biogeogr.* **2012**, *21*, 693–703. [CrossRef]
76. Pörtner, H.-O.; Gutt, J. Impacts of climate variability and change on (marine) animals: Physiological underpinnings and evolutionary consequences. *Integr. Comp. Biol.* **2016**, *56*, 31–44. [CrossRef]
77. Nati, J.J.H.; Svendsen, M.B.S.; Marras, S.; Killen, S.S.; Steffensen, J.F.; McKenzie, D.J.; Domenici, P. Intraspecific variation in thermal tolerance differs between tropical and temperate fishes. *Sci. Rep.* **2021**, *11*, 21272. [CrossRef]
78. Eliason, E.J.; Clark, T.D.; Hague, M.J.; Hanson, L.M.; Gallagher, Z.S.; Jeffries, K.M.; Gale, M.K.; Patterson, D.A.; Hinch, S.G.; Farrell, A.P. Differences in thermal tolerance among sockeye salmon populations. *Science* **2011**, *332*, 109–112. [CrossRef]
79. Oliny-Hébert, H.; Senay, C.; Enders, E.C.; Boisclair, D. Effects of diel temperature fluctuation on the standard metabolic rate of juvenile Atlantic salmon (*Salmo salar*): Influence of acclimation temperature and provenience. *Can. J. Fish. Aquat. Sci.* **2015**, *72*, 1306–1315. [CrossRef]
80. Bossdorf, O.; Richards, C.L.; Pigliucci, M. Epigenetics for ecologists. *Ecol. Lett.* **2008**, *11*, 106–115. [CrossRef]
81. Cook, C.J.; Wilson, C.C.; Burness, G. Impacts of environmental matching on the routine metabolic rate and mass of native and mixed-ancestry brook trout (*Salvelinus fontinalis*) fry. *Cons. Physiol.* **2018**, *6*, coy023. [CrossRef]
82. Durtsche, R.D.; Jonsson, B.; Greenberg, L.A. Thermal conditions during embryogenesis influence metabolic rates of juvenile brown trout *Salmo trutta*. *Ecosphere* **2021**, *12*, e03374. [CrossRef]
83. Conover, D.; Schultz, E.T. Phenotypic similarity and the evolutionary significance of countergradient variation. *Trends Ecol. Evol.* **1995**, *10*, 248–252. [CrossRef]
84. Levins, R. *Evolution in Changing Environments*; Princeton University Press: Princeton, NJ, USA, 1968.
85. Álvarez, D.; Cano, J.M.; Nicieza, A.G. Microgeographic variation in metabolic rate and energy storage of brown trout: Counter-gradient selection or thermal sensitivity? *Evol. Ecol.* **2006**, *20*, 345–363. [CrossRef]
86. Millidine, K.J.; Armstrong, J.D.; Metcalfe, N.B. Juvenile salmon with high standard metabolic rates have higher energy costs but can process meals faster. *Proc. R. Soc. Lond. B* **2009**, *276*, 2103–2108. [CrossRef]
87. Norin, T.; Clark, T.D. Fish face a trade-off between 'eating big' for growth efficiency and 'eating small' to retain aerobic capacity. *Biol. Lett.* **2017**, *13*, 20170298. [CrossRef]
88. Jonsson, B.; Jonsson, N. *Ecology of Atlantic Salmon and Brown Trout: Habitat as a Template for Life Histories*; Fish & Fisheries Series 33; Springer: Dordrecht, The Netherlands, 2011.
89. Brett, J.R. Some principles in the thermal requirements of fishes. *Q. Rev. Biol.* **1956**, *31*, 75–87. [CrossRef]
90. Elliott, J.M. The energetics of feeding, metabolism, and growth of brown trout (*Salmo trutta* L.) in relation to body weight, water temperature and ration size. *J. Anim. Ecol.* **1976**, *45*, 923–948. [CrossRef]
91. Elliott, J.M.; Hurley, M.A. Daily energy intake and growth of piscivorous brown trout, *Salmo trutta*. *Freshw. Biol.* **2000**, *44*, 237–245. [CrossRef]
92. Rodnick, K.J.; Gamperl, A.K.; Lizars, K.R.; Bennett, M.T.; Rausch, R.N.; Keele, E.R. Thermal tolerane and metabolic physiology among redband trout population in south-eastern Oregon. *J. Fish Biol.* **2004**, *64*, 310–335. [CrossRef]
93. Verhille, C.E.; English, K.K.; Cocherell, D.E.; Farrell, A.P.; Fangue, N.A. High thermal tolerance of a rainbow trout population near its southern range limit suggests local thermal adjustment. *Cons. Physiol.* **2016**, *4*, cow057. [CrossRef]
94. Debes, P.V.; Solberg, M.F.; Matre, I.H.; Dyrhovden, L.; Glover, K.A. Genetic variation for upper thermal tolerance diminishes within and between populations with increasing acclimation temperature in Atlantic salmon. *Heredity* **2021**, *127*, 455–466. [CrossRef]
95. Morita, K.; Fukuwaka, M.; Tanimata, N.; Yamamura, O. Size-dependent thermal preferences in a pelagic fish. *Oikos* **2010**, *119*, 1265–1272. [CrossRef]
96. Haraldstad, Ø.; Jonsson, B. Age and sex segregation in habitat utilization by brown trout in a Norwegian lake. *Trans. Am. Fish. Soc.* **1983**, *112*, 27–37. [CrossRef]
97. Lindmark, M.; Ohlberger, J.; Gårdmark, A. Optimum growth temperature declines with body size within fish species. *Glob. Chang. Biol.* **2022**, *28*, 2259–2271. [CrossRef]
98. Archer, L.C.; Hutton, S.A.; Harman, L.; Poole, R.W.; Gargan, P.; McGinnity, P.; Reed, T.E. Associations between metabolic traits and growth rate in brown trout (*Salmo trutta*) depend on thermal regime. *Proc. R. Soc. Lond. B* **2021**, *288*, 20211509. [CrossRef]
99. Finstad, A.G.; Jonsson, B. Effect of incubations temperature on growth performance in Atlantic salmon. *Mar. Ecol. Progr. Ser.* **2012**, *454*, 75–82. [CrossRef]

100. Jonsson, B.; Jonsson, N.; Finstad, A.G. Linking embryonic temperature with adult reproductive investment. *Mar. Ecol. Progr. Ser.* **2014**, *515*, 217–226. [CrossRef]

101. Burgerhout, E.; Mommens, M.; Johansen, H.; Aunsmo, A.; Santi, N.; Andersen, Ø. Genetic background and embryonic temperature affect DNA methylation and expression of *myogenin* and muscle development in Atlantic salmon (*Salmo salar*). *PLoS ONE* **2017**, *12*, e179918. [CrossRef]

102. Flewelling, S.; Parker, S.L. Effects of temperature and oxygen on growth and differentiation of embryos of the ground skink, *Scincella lateralis*. *J. Exp. Zool. A* **2015**, *323*, 445–455. [CrossRef]

103. José-Edwards, D.S.; Oda-Ishii, I.; Nibu, Y.; Di Gregorio, A. Tbx2/3 is an essential mediator within the Brachyury gene network during *Ciona* notochord development. *Development* **2013**, *140*, 2422–2433. [CrossRef]

104. Dorrity, M.W.; Saunders, L.M.; Duran, M.; Srivatsan, S.R.; Ewing, B.; Queitsch, C.; Shendure, J.; Raible, D.W.; Kimelman, D.; Trapnell, C. Proteostasis governs differential temperature sensitivity across embryonic cell types. *bioRxiv* **2022**. [CrossRef]

105. Jonsson, N.; Hansen, L.P.; Jonsson, B. Variation in age, size and repeat spawning of adult Atlantic salmon in relation to river discharge. *J. Anim. Ecol.* **1991**, *60*, 937–947. [CrossRef]

106. Jonsson, B.; Jonsson, N. Sexual size dimorphism in anadromous brown trout *Salmo trutta*. *J. Fish Biol.* **2015**, *87*, 187–193. [CrossRef]

107. Atkinson, D. Temperature and organism size—A biological law for ectotherms? *Adv. Ecol. Res.* **1994**, *25*, 1–58.

108. Alm, G. Connection between maturity, size and age in fishes. *Rep. Inst. Freshw. Res. Drottningholm* **1959**, *40*, 5–145.

109. Jonsson, B.; Finstad, A.G.; Jonsson, N. Winter temperature and food quality affect age and size at maturity in ectotherms: An experimental test with Atlantic salmon. *Can. J Fish. Aquat. Sci.* **2012**, *69*, 1817–1826. [CrossRef]

110. Jonsson, B.; Jonsson, N.; Finstad, A.G. Effects of temperature and food quality on age at maturity of ectotherms: An experimental test of Atlantic salmon. *J. Anim. Ecol.* **2013**, *82*, 201–210. [CrossRef]

111. Jonsson, B.; Jonsson, N.; Albretsen, J. Environmental change influences the life history of salmon in the North Atlantic. *J. Fish Biol.* **2016**, *88*, 618–637. [CrossRef]

112. Vollset, K.W.; Urdal, K.; Utne, K.; Thorstad, E.B.; Sægrov, H.; Raunsgard, A.; Skagseth, Ø.; Lennox, R.J.; Østborg, G.M.; Ugedal, O.; et al. Ecological regime shift in the Northeast Atlantic Ocean revealed from the unprecedented reduction in marine growth of Atlantic salmon. *Sci. Adv.* **2022**, *8*, eabk2542. [CrossRef]

113. Bigler, B.S.; Welch, D.W.; Helle, J.H. A review of size trends among North Pacific salmon (*Oncorhynchus* spp.). *Can. J. Fish. Aquat. Sci.* **2011**, *53*, 455–465. [CrossRef]

114. Oke, K.B.; Cunningham, C.J.; Westley, P.A.H.; Baskett, M.L.; Carlson, S.M.; Clark, J.; Hendry, A.P.; Karatayev, V.A.; Kendall, N.W.; Kibele, J.; et al. Recent declines in salmon body size impact ecosystems and fisheries. *Nat. Commun.* **2020**, *11*, 4155. [CrossRef]

115. Naiman, R.J.; Bilby, R.E.; Schindler, D.E.; Helfield, J.M. Pacific salmon, nutrients, and the dynamics of freshwater and riparian ecosystems. *Ecosystems* **2002**, *5*, 399–417. [CrossRef]

116. Walsh, J.C.; Pendray, J.E.; Godwin, S.C.; Artelle, K.A.; Kindsvater, H.K.; Field, R.D.; Harding, J.N.; Swain, N.R.; Reynolds, J.D. Relationships between Pacific salmon and aquatic and terrestrial ecosystems: Implications for ecosystem-based management. *Ecology* **2020**, *101*, e03060. [CrossRef]

117. Ferguson, A.; Prodöhl, P.A. Identifying and conserving sympatric diversity in trout of the genus *Salmo*, with particular reference to Lough Melvin, Ireland. *Ecol. Freshw. Fish* **2022**, *31*, 177–207. [CrossRef]

118. Levin, B.; Simonov, E.; Gabrielyan, B.K.; Mayden, R.L.; Rastorguev, S.M.; Roubenyan, H.R.; Sharko, F.S.; Nedoluzhko, A.V. Caucasian treasure: Genomics sheds light on the evolution of half-extinct Sevan trout, *Salmo ischchan*, species flock. *Mol. Phylogen. Evol.* **2022**, *167*, 107346. [CrossRef]

119. Jonsson, B.; Hindar, K. Reproductive strategy of dwarf and normal Arctic charr (*Salvelinus alpinus*) from Vangsvatnet Lake, western Norway. *Can. J. Fish. Aquat. Sci.* **1982**, *39*, 1404–1413. [CrossRef]

120. Bernatchez, L.; Renaut, S.; Whiteley, A.R.; Derome, N.; Jeukens, J.; Landry, L.; Lu, G.; Nolte, A.W.; Østbye, K.; Rogers, S.M.; et al. On the origin of species: Insights from the ecological genomics of lake whitefish. *Philos. Trans. R. Soc. B* **2010**, *365*, 1783–1800. [CrossRef]

121. Hindar, K.; Jonsson, B. Habitat and food segregation of dwarf and normal Arctic charr (*Salvelinus alpinus*) from Vangsvatnet Lake, western Norway. *Can. J. Fish. Aquat. Sci.* **1982**, *39*, 1030–1045. [CrossRef]

122. Hindar, K.; Jonsson, B. Ecological polymorphism in Arctic charr. *Biol. J. Linn. Soc.* **1993**, *8*, 63–74. [CrossRef]

123. Snorrason, S.S.; Skulason, S.; Jonsson, B.; Malmqvist, H.; Jonasson, P.M.; Sandlund, O.T.; Lindem, T. Trophic specialization in Arctic charr *Salvelinus alpinus* (Pisces: Salmonidae): Morphological divergence and ontogenetic shifts. *Biol. J. Linn. Soc.* **1994**, *52*, 1–18. [CrossRef]

124. Rougeux, C.; Bernatchez, L.; Gagnaire, P.-A. Modeling the multiple facets of speciation-with-gene-flow toward inferring the divergence history of lake whitefish species pairs (*Coregonus clupeaformis*). *Gen. Biol. Evol.* **2017**, *9*, 2057–2074. [CrossRef]

125. Steinbacher, P.; Wanzenböck, J.; Brandauer, M.; Holper, R.; Landertshammer, J.; Mayr, M.; Platzl, C.; Stoiber, W. Thermal experience during embryogenesis contributes to the induction of dwarfism in whitefish *Coregonus lavaretus*. *PLoS ONE* **2017**, *12*, e0185384. [CrossRef]

126. Scott, G.R.; Johnston, I.A. Temperature during embryonic development has persistent effects on thermal acclimation capacity in zebrafish. *Proc. Natl. Acad. Sci. USA* **2012**, *109*, 14247–14252. [CrossRef]

127. Morita, K.; Tamate, T.; Kuroki, M.; Nagasawa, T. Temperature-dependent variation in alternative migratory tactics and its implications for fitness and population dynamics in a salmonid fish. *J. Anim. Ecol.* **2014**, *83*, 1268–1278. [CrossRef]

128. Morán, P.; Pérez-Figueroa, A. Methylation changes associated with early maturation stages in the Atlantic salmon. *BMC Gen.* **2011**, *12*, 86. [CrossRef]
129. Pankhurst, N.W.; King, H. Temperature and salmonid reproduction: Implications for aquaculture. *J. Fish Biol.* **2010**, *76*, 69–85. [CrossRef]
130. Debes, P.V.; Piavchenko, N.; Ruokolainen, A.; Ovaskainen, O.; Moustakas-Verho, J.E.; Parre, N.; Aykanat, T.; Erkinaro, J.; Primmer, C.R. Large single-locus effects for maturation timing are mediated via body condition in Atlantic salmon. *bioRxiv* **2019**. [CrossRef]
131. Fjelldal, P.G.; Hansen, T.; Huang, T.-S. Continuous light and elevated temperature can trigger maturation both during and immediately after smoltification in male Atlantic salmon (*Salmo salar*). *Aquaculture* **2011**, *321*, 93–100. [CrossRef]
132. Baum, D.; Armstrong, J.D.; Metcalfe, N.B. The effect of temperature on growth and early maturation in a wild population of Atlantic salmon parr. *J. Fish Biol.* **2005**, *67*, 1370–1380. [CrossRef]
133. Jonsson, N.; Jonsson, B. Trade-off between egg size and numbers in brown trout. *J. Fish Biol.* **1999**, *55*, 767–783. [CrossRef]
134. Braun, D.C.; Patterson, D.A.; Reynolds, J.D. Maternal and environmental influences on egg size and juvenile life-history traits in Pacific salmon. *Ecol. Evol.* **2013**, *3*, 1727–1740. [CrossRef]
135. Takatsu, K.; Brodersen, J. Repeated elevational clines of early life-history traits and their proximate mechanisms in brown trout. *Freshw. Biol.* **2023**, *68*, 609–620. [CrossRef]
136. Einum, S.; Hendry, A.P.; Fleming, I.A. Egg-size evolution in aquatic environments: Does oxygen availability constrain size? *Proc. R. Soc. Lond. B* **2002**, *269*, 2325–2330. [CrossRef]
137. Fleming, I.A.; Gross, M.R. Latitudinal clines: A trade-off between egg number and size in Pacific salmon. *Ecology* **1990**, *71*, 1–11. [CrossRef]
138. Beacham, T.D.; Murray, C.B. Fecundity and egg size variation in North American Pacific salmon (*Oncorhynchus*). *J. Fish Biol.* **1993**, *42*, 485–508. [CrossRef]
139. Jonsson, B.; Jonsson, N. Early environments affect later performances in fishes. *J. Fish Biol.* **2014**, *85*, 155–188. [CrossRef]
140. Jonsson, B.; Jonsson, N. Trans-generational maternal effect: Temperature influences egg size of the offspring in Atlantic salmon *Salmo salar*. *J. Fish Biol.* **2016**, *89*, 1482–1489. [CrossRef]
141. Venney, C.J.; Wellband, K.W.; Houle, C.; Garant, D.; Audet, C.; Bernatchez, L. Thermal regime during parental sexual maturation, but not during offspring rearing, modulates DNA methylation in brook charr (*Salvelinus fontinalis*). *Proc. R. Soc. Lond. B* **2022**, *289*, 20220670. [CrossRef]
142. Biro, P.A.; Beckman, C.; Stamps, J.A. Small within-day increases in temperature affects boldness and alters personality in coral reef fish. *Proc. R. Soc. Lond. B* **2010**, *277*, 71–77. [CrossRef]
143. Bartolini, T.; Burtail, S.; Porfiri, M. Temperature influences sociality and activity of freshwater fish. *Environ. Biol. Fish.* **2015**, *98*, 825–832. [CrossRef]
144. Jonsson, B.; Ruud-Hansen, J. Water temperature as the primary influence on timing of seaward migrations of Atlantic salmon smolts. *Can. J. Fish. Aquat. Sci.* **1985**, *42*, 593–595. [CrossRef]
145. Sykes, G.E.; Johnson, C.J.; Shrimpton, J.M. Temperature and flow effects on migration timing of Chinook salmon smolts. *Trans. Am. Fish. Soc.* **2009**, *138*, 1252–1265. [CrossRef]
146. Otero, J.; L'Abée-Lund, J.H.; Castro-Santos, T.; Leonardsson, K.; Storvik, G.O.; Jonsson, B.; Dempson, B.; Russell, I.C.; Jensen, A.J.; Baglinière, J.L.; et al. Basin-scale phenology and effects of climate variability on global timing of initial seaward migration of Atlantic salmon (*Salmo salar*). *Glob. Chang. Biol.* **2014**, *20*, 61–75. [CrossRef]
147. Jonsson, B.; Jonsson, N. Migratory timing, marine survival and growth of anadromous brown trout *Salmo trutta* in the River Imsa, Norway. *J. Fish Biol.* **2009**, *74*, 621–638. [CrossRef]
148. Jonsson, N.; Jonsson, B. Time and size at seaward migration influence the sea survival of Atlantic salmon (*Salmo salar* L.). *J. Fish Biol.* **2014**, *84*, 1457–1473. [CrossRef]
149. Russell, I.C.; Aprahamian, M.W.; Barry, J.; Davidson, I.C.; Fiske, P.; Ibbotson, A.T.; Kennedy, R.J.; Maclean, J.C.; Moore, A.; Otero, J.; et al. The influence of the freshwater environment and the biological characteristics of Atlantic salmon smolts on their subsequent marine survival. *ICES J. Mar. Sci.* **2012**, *69*, 1563–1573. [CrossRef]
150. L'Abée-Lund, J.H.; Jonsson, B.; Jensen, A.J.; Sættem, L.M.; Heggberget, T.G.; Johnsen, B.O.; Næsje, T.F. Latitudinal variation in life history characteristics of sea-run migrant brown trout *Salmo trutta*. *J. Anim. Ecol.* **1989**, *58*, 525–542. [CrossRef]
151. Jonsson, B.; Jonsson, M.; Jonsson, N. Optimal size at seaward migration in an anadromous salmonid. *Mar. Ecol. Progr. Ser.* **2016**, *559*, 193–200. [CrossRef]
152. Jonsson, B.; L'Abée-Lund, J.H. Latitudinal clines in life history variables of anadromous brown trout in Europe. *J. Fish Biol.* **1993**, *43* (Suppl. A), 1–16. [CrossRef]
153. Budy, P.; Thiede, G.P.; Bouwes, N.; Petrosky, C.E.; Schaller, H. Evidence linking delayed mortality of snake river salmon to their earlier hydrosystem experience. *N. Am. J. Fish Manag.* **2002**, *22*, 35–51. [CrossRef]
154. Lee, C.G.; Farrell, A.P.; Lotto, A.; Hinch, S.G.; Healey, M.C. Excess post-exercise oxygen consumption in adult sockeye (*Oncorhynchus nerka*) and coho (*O. kisutch*) salmon following critical speed swimming. *J. Exp. Biol.* **2003**, *206*, 3253–3260. [CrossRef]
155. Lee, C.G.; Farrell, A.P.; Lotto, A.; MacNutt, M.J.; Hinch, S.G.; Healey, M.C. The effect of temperature on swimming performance and oxygen consumption in adult sockeye (*Oncorhynchus nerka*) and coho (*O. kisutch)* salmon stocks. *J. Exp. Biol.* **2003**, *206*, 3239–3251. [CrossRef]

156. Mathes, M.T.; Hinch, S.G.; Cooke, S.J.; Crossin, G.T.; Patterson, D.A.; Lotto, A.G.; Farrell, A.P. Effect of water temperature, timing, physiological condition, and lake thermal refugia on migrating adult Weaver Creek sockeye salmon (*Oncorhynchus nerka*). *Can. J. Fish. Aquat. Sci.* **2010**, *67*, 70–84. [CrossRef]

157. Breau, C.; Cunjak, R.A.; Peake, S.J. Behaviour during elevated water temperatures: Can physiology explain movement of juvenile Atlantic salmon to cool water? *J. Anim. Ecol.* **2011**, *80*, 844–853. [CrossRef]

158. Dugdale, S.J.; Bergeron, N.E.; St-Hilaire, A. Temporal variability of thermal refuges and water temperature patterns in Atlantic salmon rivers. *Remote Sens. Environ.* **2013**, *136*, 358–373. [CrossRef]

159. Wang, T.; Kelson, S.J.; Greer, G.; Tompson, S.E.; Carlson, S.M. Tributary confluences are dynamic thermal refuges for a juvenile salmonid in a warming river network. *River Res. Appl.* **2020**, *36*, 1076–1086. [CrossRef]

160. Caissie, D. The thermal regime of rivers: A review. *Freshw. Biol.* **2006**, *51*, 1389–1406. [CrossRef]

161. Hinch, S.G.; Rand, P.S. Optimal swimming speeds and forward-assisted propulsion: Energy-conserving behaviours of upriver-migrating adult salmon. *Can. J. Fish. Aquat. Sci.* **2000**, *57*, 2470–2478. [CrossRef]

162. McElroy, B.; DeLonay, A.; Jacobson, R. Optimum swimming pathways of fish spawning migrations in rivers. *Ecology* **2012**, *93*, 29–34. [CrossRef]

163. Armstrong, J.B.; Ward, E.J.; Schindler, D.E.; Lisi, P.J. Adaptive capacity at the northern front: Sockeye salmon behaviourally thermoregulate during novel exposure to warm temperatures. *Cons. Physiol.* **2016**, *4*, cow039. [CrossRef]

164. Fakhari, M.; Raymond, J.; Martel, R.; Dugdale, S.J.; Bergeron, N.E. Identification of thermal refuges and water temperature patterns in salmonid-bearing subarctic rivers of northern Quebec. *Geographies* **2022**, *2*, 528–548. [CrossRef]

165. Fenkes, M.; Shiels, H.A.; Fitzpatrick, J.L.; Nudds, R.L. The potential impacts of migratory difficulty, including warmer waters and altered flow conditions, on the reproductive success of salmonid fishes. *Comp. Biochem. Physiol. A* **2016**, *193*, 11–21. [CrossRef]

166. Videler, J.J.; Weihs, D. Energetic advantages of burst-and-coast swimming of fish at high speeds. *J. Exp. Biol.* **1082**, *97*, 169–178. [CrossRef]

167. Jonsson, B.; Gravem, F.R. Use of space and food by resident and migrant brown trout. *Environ. Biol. Fish.* **1985**, *14*, 281–293. [CrossRef]

168. Hedger, R.D.; Kjellman, M.; Thorstad, E.B.; Strøm, J.F.; Rikardsen, A.H. Diving and feeding of adult Atlantic salmon when migrating through the coastal zone in Norway. *Environ. Biol. Fish.* **2022**, *105*, 589–604. [CrossRef]

169. Jonsson, N.; Jonsson, B.; Hansen, L.P. Changes in proximate composition and estimates of energetic costs during upstream migration and spawning in Atlantic salmon *Salmo salar*. *J. Anim. Ecol.* **1997**, *66*, 425–436. [CrossRef]

170. Hendry, A.P.; Berg, O.K. Secondary sexual characters, energy use, senescence, and the cost of reproduction in sockeye salmon. *Can. J. Zool.* **1999**, *77*, 1663–1675. [CrossRef]

171. Glebe, B.D.; Leggett, W.C. Temporal, intra-population differences in energy allocation and use by American shad (*Alosa sapidissima*) during the spawning migration. *Can. J. Fish. Aquat. Sci.* **1981**, *38*, 795–805. [CrossRef]

172. Kallio-Nyberg, I.; Saloniemi, I.; Koljonen, M.L. Increasing temperature associated with increasing grilse proportion and smaller grilse size of Atlantic salmon. *J. Appl. Ichthyol.* **2020**, *36*, 288–297. [CrossRef]

173. Quinn, T.P.; Adams, D.J. Environmental changes affecting the migratory timing of American shad and sockeye salmon. *Ecology* **1996**, *77*, 1151–1162. [CrossRef]

174. Cooke, S.J.; Hinch, S.G.; Crossin, G.T.; Patterson, D.A.; English, K.K.; Healey, M.C.; Macdonald, J.S.; Shrimpton, J.M.; Young, J.L.; Lister, A.; et al. Physiological correlates of coastal arrival and river entry timing in late summer Fraser River sockeye salmon (*Oncorhynchus nerka*). *Behav. Ecol.* **2008**, *19*, 747–758. [CrossRef]

175. Robarts, M.D.; Quinn, T.P. The migratory timing of adult summer-run steelhead in the Columbia River over six decades of environmental change. *Trans. Am. Fish. Soc.* **2002**, *131*, 523–536. [CrossRef]

176. Berman, C.H.; Quinn, T.P. Behavioural thermoregulation and homing by spring chinook salmon, *Oncorhynchus tshawytscha* (Walbaum), in the Yakima River. *J. Fish Biol.* **1991**, *39*, 301–312. [CrossRef]

177. Goniea, T.M.; Keefer, M.L.; Bjornn, T.C.; Peery, C.A.; Bennett, D.H.; Stuehrenberg, L.C. Behavioral thermoregulation and slowed migration by adult fall Chinook salmon in response to high Columbia River water temperatures. *Trans. Am. Fish. Soc.* **2006**, *135*, 408–419. [CrossRef]

178. Hyatt, K.D.; Stockwell, M.M.; Rankin, D.P. Impact and adaptation responses of Okanagan River sockeye salmon (*Oncorhynchus nerka*) to climate variation and change effects during freshwater migration: Stock restoration and fisheries management implications. *Can. Water Res. J.* **2003**, *28*, 689–713. [CrossRef]

179. Jonsson, B.; Jonsson, N. Egg incubation temperature affects the timing of the Atlantic salmon *Salmo salar* homing migration. *J. Fish Biol.* **2018**, 1016–1020. [CrossRef]

180. Takatsu, K.; Selz, O.M.; Brodersen, J. Temperature regime during embryogenesis alters subsequent behavioural phenotypes of brown trout. *Biol. Lett.* **2022**, *18*, 20220369. [CrossRef]

181. Bærum, K.M.; Finstad, A.G.; Ulvan, E.M.; Haugen, T.O. Population consequences of climate change through effects on functional traits of lentic brown trout in the sub-Arctic. *Sci. Rep.* **2021**, *11*, 14246. [CrossRef]

182. Jonsson, B.; Greenberg, L. Egg incubation temperature influences population-specific outmigration rate of juvenile brown trout *Salmo trutta*. *J. Fish Biol.* **2022**, *100*, 909–917. [CrossRef]

183. Koteja, P. The evolution of concepts on the evolution of endothermy in birds and mammals. *Physiol. Biochem. Zool.* **2004**, *77*, 1043–1050. [CrossRef]

184. Brown, J.H.; Sibly, R.M. Life-history evolution under a production constrain. *Proc. Natl. Acad. Sci. USA* **2006**, *103*, 17595–17599. [CrossRef]
185. Korwin-Kossakowski, M. The influence of temperature during the embryonic period on larval growth and development in carp *Cyprinus carpio* L. and grass carp, *Ctenopharyngodon idella* (Val.): Theoretical and practical aspects. *Arch. Pol. Fish.* **2008**, *16*, 231–314. [CrossRef]
186. Martell, D.J.; Kieffer, J.D.; Trippel, E.A. Effects of temperature during early life history on embryonic and larval development and growth in haddock. *J. Fish Biol.* **2005**, *66*, 1558–1575. [CrossRef]
187. Carballo, C.; Firmino, J.; Anjos, L.; Santos, S.; Power, D.M.; Manchad, M. Short- and long-term effects on growth and expression patterns in response to incubation temperatures in Singalese sole. *Aquaculture* **2018**, *495*, 222–231. [CrossRef]
188. Jonsson, B.; Jonsson, N. Differences in growth between offspring of anadromous and freshwater brown trout *Salmo trutta*. *J. Fish Biol.* **2021**, *99*, 18–24. [CrossRef]
189. Mousseau, T.A.; Fox, C.W. The adaptive significance of maternal effects. *Trends Ecol. Evol.* **1998**, *13*, 403–407. [CrossRef]
190. Salinas, S.; Munch, S.B. Thermal legacies: Trans-generational effects of temperature on growth in avertebrate. *Ecol. Lett.* **2012**, *15*, 159–163. [CrossRef]
191. Stamps, J.A.; Luttbeg, B. Sensitive period diversity: Insight from evolutionary models. *Q. Rev. Biol.* **2022**, *97*, 243–295. [CrossRef]
192. Nettle, D.; Bateson, M. Adaptive developmental plasticity: What is it, how can we recognize it and when can it evolve? *Proc. R. Soc. Lond. B* **2015**, *282*, 20151005. [CrossRef]
193. Schlichting, C.D. The evolution of phenotypic plasticity in plants. *Annu. Rev. Ecol. Syst.* **1986**, *17*, 667–693. [CrossRef]
194. Kelly, S.A.; Panhuis, T.M.; Stoehr, A.M. Phenotypic plasticity: Molecular mechanisms and adaptive significance. *Comp. Physiol.* **2012**, *2*, 1417–1439.
195. Elphick, M.J.; Shine, R. Longterm effects of incubation temperature on the morphology and locomotor performance of hatchling lizards (*Bassiana duperreyi*, Scincidae). *Biol. J. Linn. Soc.* **1998**, *63*, 429–447. [CrossRef]
196. Richards, C.L.; Bossdorf, O.; Pigliucci, M. What role does heritable epigenetic variation play in phenotypic evolution. *BioScience* **2010**, *60*, 232–237. [CrossRef]
197. Rey, O.; Danchin, E.; Mirouze, M.; Loot, C.; Blanchet, S. Adaptation to global change: A transposable element-epigenetics perspective. *Trends Ecol. Evol.* **2016**, *31*, 514–526. [CrossRef]
198. Donelson, J.M.; Wong, M.; Booth, D.J.; Munday, P.L. Transgenerational plasticity of reproduction depends on rate of warming across generations. *Evol. Appl.* **2016**, *9*, 1072–1081. [CrossRef]
199. Hu, J.; Barrett, R.D.H. Epigenetics in natural animal populations. *J. Evol. Biol.* **2017**, *30*, 1612–1632. [CrossRef]
200. Tariel, J.; Luquet, É.; Plénet, S. Interactions between maternal, paternal, developmental, and immediate environmental effects on anti-predator behaviour of the snail *Physa acuta*. *Front. Ecol. Evol.* **2020**, *8*, 591074. [CrossRef]
201. Kawecki, T.J.; Ebert, D. Conceptual issues in local adaptation. *Ecol. Lett.* **2004**, *7*, 1225–1241. [CrossRef]
202. Bateson, P.; Gluckman, P.; Hanson, M. The biology of developmental plasticity and the Predictive Adaptive Response hypothesis. *J. Physiol.* **2014**, *592*, 2357–2368. [CrossRef]
203. Lea, A.J.; Altman, J.; Alberts, S.C.; Tung, J. Resource base influences genome-wide DNA Methylation levels in wild baboons (*Papio cynocephalus*). *Mol. Ecol.* **2016**, *25*, 1681–1696. [CrossRef]
204. Varriale, A.; Bernardi, G. DNA methylation and body temperature in fishes. *Gene* **2006**, *385*, 111–121. [CrossRef]
205. Lallias, D.; Bernard, M.; Ciobotaru, C.; Dechamp, N.; Labbé, L.; Goardon, L.; Le Calvez, J.-M.; Bideau, M.; Fricot, A.; Prézelin, A.; et al. Sources of variation of DNA methylation in rainbow trout: Combined effects of temperature and genetic background. *Epigenetics* **2021**, *16*, 1031–1052. [CrossRef]
206. Skjærven, K.H.; Hamre, K.; Penglase, S.; Finn, R.N.; Olsvik, P.A. Thermal stress alters expression of genes involved in one carbon and DNA methylation pathways in Atlantic cod embryos. *Comp. Biochem. Physiol. A Mol. Integr. Physiol.* **2014**, *173*, 17–27. [CrossRef]
207. Ryu, T.; Veilleux, H.D.; Munday, P.L.; Jung, I.; Donelson, J.M.; Ravasi, T. An Epigenetic signature for within-generational plasticity of a reef fish to ocean warming. *Front. Mar. Sci.* **2020**, *7*, 00284. [CrossRef]
208. Anastasiadi, D.; Diaz, N.; Piferrer, F. Small ocean temperature increases elicit stage-dependent changes in DNA methylation and gene expression in a fish, the European sea bass. *Sci. Rep.* **2017**, *7*, 12401. [CrossRef]
209. Beemelmanns, A.; Ribas, L.; Anastasiadi, D.; Moraleda-Prados, J.; Zanuzzo, F.S.; Rise, M.L.; Gamperl, A.K. DNA methylation dynamics in Atlantic salmon (*Salmo salar*) challenged with high temperature and moderate hypoxia. *Front. Mar. Sci.* **2021**, *7*, 604878. [CrossRef]
210. Finstad, A.G.; Forseth, T.; Jonsson, B.; Bellier, E.; Hesthagen, T.; Jensen, A.J.; Hessen, D.O.; Foldvik, A. Competition exclusion along climate gradients: Energy efficiency influences the distribution of two salmonid fishes. *Glob. Chang. Biol.* **2011**, *17*, 1703–1711. [CrossRef]

 *fishes*

Article

A Benchmark for Atlantic Salmon Conservation: Genetic Diversity and Structure in a Southern European Glacial Refuge before the Climate Changed

Ana Almodóvar *, Graciela G. Nicola, Daniel Ayllón, Sheila Leal, Daniel F. Marchán and Benigno Elvira

Department of Biodiversity, Ecology and Evolution, Faculty of Biological Sciences, Complutense University of Madrid (UCM), 28040 Madrid, Spain; gracielg@ucm.es (G.G.N.); daniel.ayllon@bio.ucm.es (D.A.); sheilaleal@bio.ucm.es (S.L.); danief01@ucm.es (D.F.M.); belvira@bio.ucm.es (B.E.)
* Correspondence: aalmodovar@bio.ucm.es

Abstract: Atlantic salmon *Salmo salar* supports highly valuable commercial and recreational fisheries in Europe, but its stocks are currently overexploited and threatened by climate change. Its southernmost populations (in northern Spain) play a key role in conserving the species' original genetic diversity, which is endangered due to decades-long (1970s to 1990s) massive stocking with non-native stocks. Their decline is well documented, but the effect of stock transfer and conservation efforts is unclear. Nine microsatellite loci were amplified from archival samples (scales from 1958–1959) from eight Spanish rivers to analyse the species' natural genetic dynamics before its decline started. Allelic richness was high in the historical populations (the 1950s) and above most contemporary estimates. Private alleles were found in most rivers, indicating high local uniqueness and relative isolation among river basins. Some alleles are regional markers since they are rare or absent from contemporary northern European populations. Effective population size suggested good conservation status, with higher values than those estimated for contemporary populations. Strong population structure and genetic differentiation between rivers were found, with limited gene flow, restricted to geographically close populations. Our estimates of historical genetic diversity and structure from southernmost salmon populations are a powerful benchmark to guide conservation programs.

Keywords: climate change; conservation genetics; effective population size; stocking; genetic variability; introgression; population genetics; Salmonidae; *Salmo salar*

Key Contribution: Genetic diversity and effective size of southernmost European Atlantic salmon populations have declined dramatically since the 1950s, coupled with a loss of the historical genetic structure. Human-induced pressures, including extensive stocking with non-native stocks and ongoing climate change, are responsible for the changed population genetic structure of these rear-edge populations.

Citation: Almodóvar, A.; Nicola, G.G.; Ayllón, D.; Leal, S.; Marchán, D.F.; Elvira, B. A Benchmark for Atlantic Salmon Conservation: Genetic Diversity and Structure in a Southern European Glacial Refuge before the Climate Changed. *Fishes* **2023**, *8*, 321. https://doi.org/10.3390/fishes8060321

Academic Editor: Bror Jonsson

Received: 25 May 2023
Revised: 13 June 2023
Accepted: 14 June 2023
Published: 16 June 2023

1. Introduction

Preservation of genetic diversity not only is paramount for the long-term conservation of species [1,2] but also enhances the capacity of ecosystems to withstand, recover, and adapt to environmental perturbations [3–5]. However, anthropogenic pressures on natural populations have long modified the genetic structure and gene flow patterns of fish species worldwide. These pressures hinder current attempts to identify which gene pools/stocks are genetically differentiated and which must be prioritised for conservation.

Atlantic salmon (*Salmo salar*) is an anadromous species with a very high socioeconomic status in Europe as it supports valuable commercial and recreational fisheries. However, since the second half of the 20th century, the species experienced alarming reductions in population size and genetic diversity due to overexploitation, anthropogenic impacts in fresh waters (i.e., water and habitat quality degradation, connectivity loss, and population admixing with non-native domesticated strains), and possible ecosystem change with reduced feeding opportunities for Atlantic salmon in the North Atlantic Ocean occurring in 2005. Production was reduced after that with smaller, slower growing, and later maturing salmon [6,7]. Consequently, the species was assessed as vulnerable (VU A2ace) in Europe [8]. These impacts have been more severe in southern European populations, where effective population sizes have been reported to be rather low [9] and negatively affected by current climate change [10–12]. Apart from the climate-driven ecosystem regime shifts leading to poorer trophic conditions that have occurred in their marine feeding habitats [11,12], southern populations have experienced the most serious impacts upon their local environmental conditions (mainly temperature and river flow) of its entire distribution [7,11,13]. These southern European salmon populations also have the singularity of following a westerly migration route towards Greenland instead of following the easterly branch of the North Atlantic current into the Norwegian Sea [14,15]. Better knowledge of these marine migration routes can help understand the causes of the ongoing salmon population decline.

Spanish populations have been profusely studied in terms of their genetic variability, population structure, and response to different management and conservation measures [16]. These studies were typically based on samples collected from the 1980s onwards, which corresponds to a period of drastic reduction in population sizes. This collapse was driven by the concurrence of an abrupt acceleration in the anthropogenic warming trend and the warm phase of the Atlantic Multidecadal Oscillation, which caused a major regime shift in biophysical conditions throughout the North Atlantic salmon feeding grounds [11]. Massive stocking of non-native salmon was performed in Spanish rivers between the 1970s and 1990s following the populations' collapse in an attempt to aid population recovery [17]. Salmon were mainly imported from Scotland and released (between 20,000 and 100,000) annually at similar densities in all rivers [12]. However, information regarding the genetic diversity and structure of Spanish populations before this time is scarce [12,18,19]. In addition, conclusions from previous studies are strikingly disparate. Some results indicate a strong loss of diversity and population structure, significant levels of introgression, and a general population homogenization across river basins [12,20,21]. In contrast, others show a low impact of admixture with non-native strains [22] and a reduction in the effective population size of native stocks, while maintaining relatively high levels of genetic variation [18]. Finally, others reveal different effects between western and eastern Spanish rivers [23].

In order to carefully assess the current conservation status of Spanish Atlantic salmon populations and to understand the effects of anthropogenic disturbances, climate-driven environmental changes, and conservation efforts, it would be desirable to have an overview of historical genetic diversity (prior to stocking and the severe impacts of climate change). This can provide a benchmark for restoration attempts [24]. To this end, salmon scales collected in 1958–1959 in eight river basins that cover most of its Spanish distribution were analysed for variation at nine microsatellite loci. Microsatellites can be superior markers to SNPs for studying genetic diversity in natural populations, so microsatellite DNA polymorphism analysis can help support the success of conservation and restoration projects as an economically advantageous research technique [25,26]. Microsatellites were chosen in the present study because they also allow comparison with the significant amount of data already available in the literature and databases.

Our main objectives were to (a) study the genetic diversity of each population; (b) infer connectivity, migration rates, and population structure across river basins; (c) obtain indicators of the historical conservation status of each population, such as effective population size and evidence of bottleneck events. This information will provide insight into the changes experienced by Spanish salmon populations in the last 70 years and a benchmark for comparisons with current populations in the context of ongoing climate change.

2. Materials and Methods

2.1. Study Area

A total of 374 archival samples of dried scales of Atlantic salmon from 8 northern Spanish rivers (Navia, Narcea, Sella, Deva-Cares, Nansa, Pas, Asón, and Bidasoa; Figure 1) were analysed. Archival samples were collected in 1958–1959 from returning salmon by the National Inland Fishing and Hunting Service of the Spanish Ministry of Agriculture (Madrid). Scales of salmon returning after two winters at sea were selected.

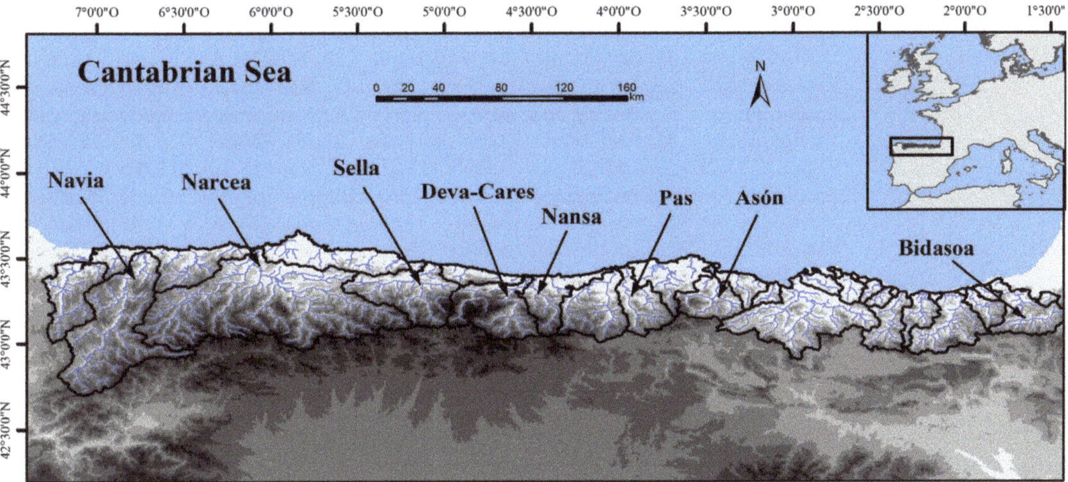

Figure 1. Location of the study area in Spain.

The Spanish rivers constitute the southernmost part of the Atlantic salmon distribution in Europe. In this area, the geomorphology and hydrogeology of the rivers are determined by the proximity of the Cantabrian Mountains to the sea, which produces very steep, V-shaped valleys that widen in their downstream sections. The main courses of the river basins are narrow and short, less than 100 km long, except for the River Navia, with a length of 159 km. In general, there is a dominance of fast-water habitats, and all are very well-preserved streams free from human activities causing water or habitat quality degradation.

However, the local hydroclimatic conditions in the study area have markedly changed over the last decades. Almodóvar et al. [11] detected a significant upward trend in regional temperature anomalies during 1950–2011, with an abrupt increase in the late 1980s. During this period, local climate warming concurred with a significant continuous increase in the frequency of low-flow events and a decrease in the minimum flow over seven or thirty consecutive days in the main rivers draining into the Cantabrian Sea and the Bay of Biscay. Such hydroclimatic changes were significantly linked to the abrupt decline observed in salmon captures in the study area since the 1970s, which was strongly accelerated from the late 1980s onwards [11]. In addition to the archival samples, reference samples of Atlantic salmon microsatellites for the eastern Atlantic (SALSEA) [27] were used for comparison.

2.2. Trends in Population Abundance

Atlantic salmon catch per unit effort (CPUE) data collected by the management authorities of Asturias, Cantabria, and Navarra in the study rivers for the last 73 years (1950–2022) were analysed. River catches were used as a proxy for abundance because this is the only measure of Atlantic salmon abundance consistently recorded in Spanish rivers over time. Until the early 2000s, the local authorities pooled salmon catches from the three rivers of Cantabria (Nansa, Pas, and Asón); so for consistency, we pooled their catches for the whole 73-year series. The salmon fishing season lasts from March to July, and the fishing regulations, number of fishing licences, and, thus, fishing effort have not changed significantly over the study period [11]. The nonparametric Mann–Kendall test, as modified by Yue and Wang [28] to account for serial correlation, was employed to detect significant annual upward or downward trends of salmon abundance in all the study rivers using the *modifiedmk* version 1.6 R package [29]. To detect regime shifts in the total salmon catch time series, a split moving-window boundary analysis [30] was used. This test computes the statistical contrast of the means of two halves of a window (width of 20 years here) as it moves along the time series through ANOVA tests ($\alpha = 0.01$).

2.3. DNA Extraction and Microsatellite Genotyping

The QIAamp DNA Mini Kit (QIAGEN, IZASA, Madrid, Spain) was used to extract genomic DNA from archival scales. The quality and concentration of DNA were determined using spectrophotometry and were verified by 0.8% agarose gel electrophoresis. The following nine microsatellite loci were analysed: *Ssa197*, *Ssa85* [31], *SSOSL417*, *SSOSL85*, *SSOLS311* [32], *SSOSL438* [33], *SsaF43* [34], *SSspG7*, and *SSsp2210* [35]. For each locus, a polymerase chain reaction (PCR) was performed in a volume of 25 μL. PCR reactions contained 1X Mg-free PCR buffer, 1.5–2.5 mM $MgCl_2$, 0.16 mM of each primer, 200 μM of dNTPs, 1–1.25 U of Biotools HotSplit DNA polymerase (Biotools, Madrid, Spain), and 50–100 ng genomic DNA template. Specific annealing temperatures for the loci were 50 °C (*SSOSL85*, *SSOSL311*), 53 °C (*SSOSL417*, *SSOSL438*), 55 °C (*SsaF43*, *SSspG7*), or 58 °C (*Ssa197*, *SSsp2210*, *Ssa85*).

The amplifications were performed in a PCR machine GeneAmp® PCR System 9700 (Applied Biosystems, Waltham, MA, USA) using the following conditions, with occasional modifications to adapt to specific primers and/or samples: 95 °C 5 min, 30–40 cycles of 95 °C 20–40 s, 50 °C up to 58 °C, 20–40 s, and 72 °C 20–40 s, with a final extension at 72 °C 10 min. An ABI PRISM 3730 sequencer (Applied Biosystems, Waltham, MA, USA) was used to visualise the PCR products, and allele scoring was performed manually with the Peak Scanner™ Software v1.0 (Applied Biosystems, Waltham, MA, USA). To avoid cross-contamination while working with the archival samples, the entire process of buffer preparation, DNA extraction, and PCR amplification were performed in a laminar flow hood. The working area and equipment were cleaned with UV light followed by swabbing with ethanol before each new step in the protocol, and strict cleaning procedures were respected by all staff. Negative and positive controls were applied in all steps of the extraction and genotyping process.

2.4. Genetic Diversity

MICRO-CHECKER v2.2.3 (University of Hull, Hull, UK) [36] was used to assess the frequency of null alleles and scoring errors due to stuttering or large allelic drop-out. The combined use of two or three methods has been suggested as the best strategy for minimizing the false-positive and false-negative rates [37]. For this reason, two different analysis methods were additionally used to test for the presence of null alleles: CERVUS v3.0.3 (Field Genetics, London, UK) [38] and ML-NullFreq (Montana State University, Bozeman, MT, USA) [39].

GENEPOP v4.1 (Université de Montpellier, Montpellier, France) [40] was used to perform tests for departure from Hardy–Weinberg equilibrium and linkage disequilibrium for each locus and sample. The statistical significance was evaluated using the Bonferroni corrections. GENETIX v4.05.2 (Université de Montpellier, Montpellier, France) [41] was used to estimate Wright's fixation index (F_{IS}) for samples' deviation from Hardy–Weinberg expectations for heterozygote disequilibrium following Weir and Cockerham [42]. Allele frequencies, number of alleles, and observed (H_o) and expected (H_e) heterozygosities were calculated at the population level with GENETIX v4.05.2 [41].

2.5. Connectivity and Population Structure

FSTAT v2.9.3 (Université de Lausanne, Lausanne, Switzerland) [43] was used to compute allelic richness (A_R) and the genetic differentiation (F_{ST}) between population pairs using sequential Bonferroni-corrected *p*-values (10,000 permutations). GENEALEX v6.501 (Australian National University, Acton, Australia) [44] was used to visualise linearised F_{ST} values $F_{ST}/(1 - F_{ST})$ among locations via principal coordinates analysis (PCoA).

The Bayesian clustering method implemented in STRUCTURE v2.3.4 (Stanford University, Stanford, CA, USA) [45] was applied to further explore population structure. Structure analyses were performed for 1 to 10 clusters (ΔK) with 10 replicates for each simulated cluster. Analyses were run using an admixture model with correlated allele frequencies, 1,000,000 MCMC generations, and a burn-in period of 250,000 steps. Optimal ΔK was determined using STRUCTURE HARVESTER v0.6.94 (University of California, Los Angeles, Los Angeles, CA, USA) [46,47]. Replicates were aggregated using CLUMP v1.1.2 (Stanford University, Stanford, CA, USA) [48] and graphically displayed using DISTRUCT v1.1 (Stanford University, Stanford, CA, USA) [49]. Since STRUCTURE can capture major structures in a dataset but overlook finer scale structure [46], a hierarchical approach was followed. STRUCTURE was further applied independently to the identified genetic clusters until either each river represented its own cluster or optimal ΔK for a group of populations approached 1.

Relationships between genetic differentiation and landscape characteristics were examined using two approaches. First, the Mantel test was used to assess the significance of the correlation between linearised F_{ST} [50] and pairwise geographic distance between locations (measured as the shortest sea distance between river mouths [51]). Mantel tests with 9999 permutations were conducted in the R v3.3.3. package *ade4* v1.7-11 [52,53]. Distance-based canonical redundancy analysis (dbRDA) of pairwise differentiation, implemented in the R package *vegan* v2.5-2 [54], was used as an additional method to study the correlation between $F_{ST}s$ and geographic distance.

Evidence for demographic bottlenecks was examined following two approaches. First, BOTTLENECK v1.2.02 (INRA-URLB, Montpellier, France) [55,56] was used to test for heterozygosity excess, assuming a two-phase mutation model (*TPM*) with 80% stepwise mutations (*SMM*) [57] and 10,000 iterations. The statistical significance of heterozygous excess was tested by one-tailed Wilcoxon's signed-rank test. In addition, we estimated the ratio of the number of alleles to the range of allele size (*M*-ratio) following Garza and Williamson [58]. The *M*-ratio was most likely to correctly detect a population size reduction if the bottleneck was more ancient, prolonged, and had a large Θ value ($\Theta = 4N_e\mu$) after the initial population decline [59]. The M_P_VAL software (NOAA Fisheries Santa Cruz Laboratory, University of California, Santa Cruz, CA, USA) [58] was used to estimate the *M*-ratio, which was compared with a critical value of M (Mc) from a theoretical population in mutation-drift equilibrium, implemented in the CRITICAL_M software (NOAA Fisheries Santa Cruz Laboratory, University of California, Santa Cruz, CA, USA) [58], assuming pre-bottleneck effective population size of 50, 100, 500 or 1000 and a mutation rate (μ) of 5×10^{-4}. According to the recommendations of Perry et al. [60], the proportion of one-step mutations (p_g) was set to 0.22, and the mean size of non-one-step mutations (Δ_g) to 3.1.

BAYESASS v3.0.3 (University of California, Davis, CA, USA) [61] was used to estimate recent migration rates (m) among rivers, using 2,000,000 burn-in and 20,000,000 iterations. Delta values for allele frequencies, inbreeding coefficients, and migration rates were set to 0.6, 0.7, and 1, respectively; these values provide adequate mixing within the ideal range of 20% and 60% [62].

The coalescent method implemented in MIGRATE-N v3.2.7 (Evolution and Genomics, Ballwin, MO, USA) [63,64] was chosen to explore migration rates between and within rivers. Estimations of mutation-scaled migration rates M ($M = m/\mu$) and Θ ($\Theta = 4N_e\mu$) were calculated using a Bayesian search strategy and a Brownian motion microsatellite model. Parameter space was searched using 10 short chains and 1 long chain with 3 replicates for 20,000,000 generations, an increment step of 20, and a burn-in of 250,000. Likewise, the parameter space was explored using 4 chains with an adaptive heating scheme (temperatures: 1.0, 1.2, 1.5, 3.0) to ensure that run results did not reflect local likelihood peaks. Finally, the Bayesian assignment procedure of Rannala and Mountain [65] implemented in GENECLASS v2 (INRA-URLB, Montpellier, France) [66] was used to identify putative first-generation migrants in Spanish rivers. The Paetkau et al. [67] resampling method was used to estimate the probability of each individual being a migrant based on 10,000 simulated individuals and a type 1 error threshold significance of 0.01.

2.6. Demographic Parameters

Estimates of census population size (N_c) were determined by the harmonic mean of ten years of annual river catches of returning adults collected by local management authorities. The effective number of breeders within a reproductive year (N_b) was calculated using the linkage disequilibrium (LD) method implemented in NeESTIMATOR v2.1 (Department of Agriculture and Fisheries, Queensland Government, Brisbane, Australia) [68]. A minimum allele frequency cutoff value of 0.02 was employed and 95% confidence intervals were obtained using the jack-knife method. This approach is based on LD between alleles from unlinked neutral loci within populations with random mating and implements a bias correction under a wide range of sample sizes.

Several studies [69–73] have shown that N_b (LD) estimates are robust to predict gene flow when migration rates are low and flow occurs between populations with weak genetic differentiation, even with small effective population sizes and missing data adjustment. This method appears to be the most suitable to estimate effective population size, showing consistent values across different demographic scenarios. In cases with overlapping generations, N_b estimates can be biased. For this reason, we applied the method developed by Waples et al. [74], in which two simple life-history traits are used to adjust genetic estimates of N_b to correct biases due to age structure. N_b was adjusted using the ratio between adult life span (AL) and age-at-maturity (α), following the equation:

$$N_{b(adj)} = N_b / (1.103 - 0.245 \mathrm{Log}\,(\mathrm{AL}/\alpha))$$

N_b is more easily quantifiable and thus can be a useful tool for managers, but remains less used than N_e [9]. Adjusted age-at-spawning was estimated between 2.65 to 3.04 years in Spanish rivers [18]; hence, an average age-at-maturity of 2.8 years was assumed for all rivers. The AL value was calculated as described by Waples et al. [74], using a maximum breeding age (ω) of 5 years for Atlantic salmon. The $N_{e(adj)}$ was calculated using the equation proposed by Waples et al. [74]:

$$N_{e(adj)} = N_{b(adj)} / (0.485 + 0.758 \mathrm{Log}\,(\mathrm{AL}/\alpha))$$

The effective size ratios $N_{b(adj)}/N_c$ and $N_{e(adj)}/N_c$ also were calculated following Perrier et al. [9].

3. Results

3.1. Long-Term Trends in Population Abundance

Total salmon abundance in the study area decreased significantly over the 1950–2022 period (Mann–Kendall test; $\tau = -0.493$, adjusted Kendall $p < 0.001$, Sen's slope = −13% change per decade). The same downward trend was observed in the individual rivers (Navia, $\tau = -0.611$, $p < 0.001$, Sen's slope = −3.8%; Narcea $\tau = -0.316$, $p < 0.001$, Sen's slope = −9%; Sella, $\tau = -0.494$, $p < 0.001$, Sen's slope = −11%; Deva-Cares, $\tau = -0.674$, $p < 0.001$, Sen's slope = −15.1%; Cantabria's rivers, $\tau = -0.708$, $p < 0.001$, Sen's slope = −15%; Bidasoa, $\tau = -0.210$, $p < 0.05$, Sen's slope = −6.1%) (Figure 2). There was a regime shift in CPUE in the early 1970s followed by an abrupt decline from 1988–1989 and a more recent shift in 2008–2009 (Figure 2). Total salmon abundance dropped by 45% from the 1950–1972 period (mean = 5519.4 ± 1770.9) to the 1973–1988 period (mean = 3020.6 ± 1207.9), decreasing from this period to 1989–2008 (mean = 1728.9 ± 563.1) by 43%. Finally, salmon populations were further reduced by 52% from 2009 onwards (mean = 820.6 ± 343.7). On the whole, total salmon abundance was reduced by 85% from 1950–1972 to 2009–2022.

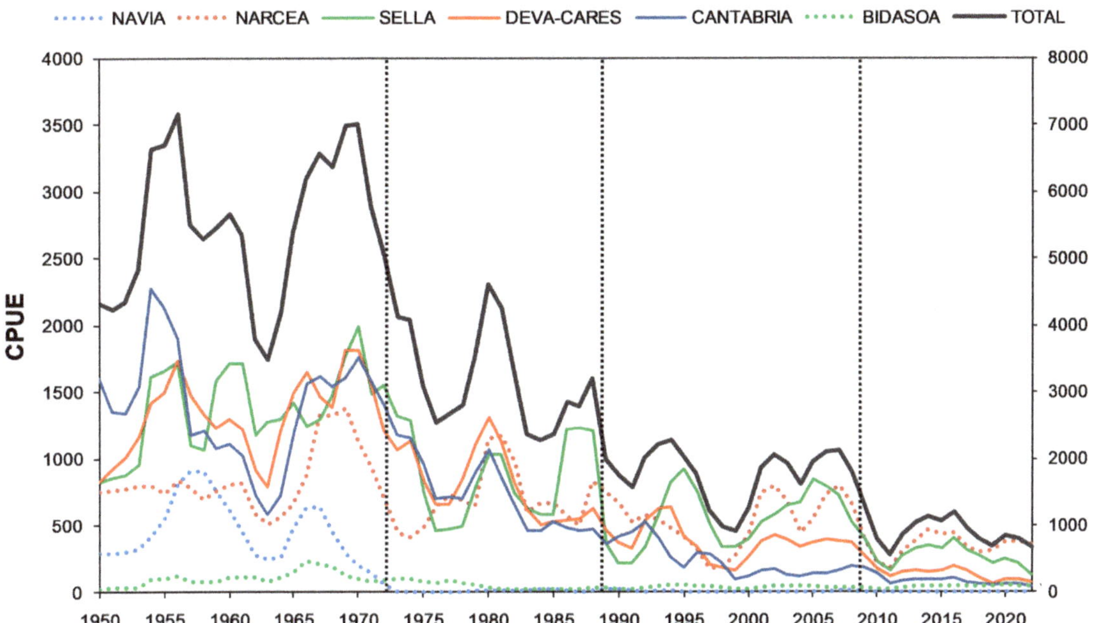

Figure 2. Long-term changes in catch per unit effort (CPUE) (3-year moving average) of Atlantic salmon in the study area (left *y*-axis: captures from study rivers; right *y*-axis: total captures) from 1950 to 2022. The dashed vertical lines indicate the detected significant shifts in the total capture trend (1972–1973; 1988–1989; 2008–2009). Cantabria includes rivers Nansa, Pas, and Asón.

3.2. Genetic Diversity

Variations at nine microsatellite loci were examined in archival samples (the 1950s) from eight Spanish rivers. The locus *Ssa85* was removed from the data set because it was difficult to amplify in some individuals and because of the occurrence of some alleles in the Navia, Narcea, and Asón rivers using MICRO-CHECKER (at the Bonferroni confidence level), CERVUS ($F_{(null)} \geq 0.20$) and ML-NullFreq ($F_{(null)} \geq 0.11$, $p < 0.05$) analyses.

A total of 110 alleles were found among all archival salmon samples. The mean number of alleles per locus ranged from 6.13 for the River Pas to 8.75 for the River Bidasoa (Table 1). The number of alleles varied from two at locus *SSOSL438* to 12 at locus *SSOSL311*. Allelic richness showed significant differences among rivers (ANOVA, $p < 0.05$). Higher values were observed in the rivers Sella, Deva-Cares, and Bidasoa (A_R mean = 8.2, range: 7.8–8.7), i.e., those with higher flows. Lower values were found in the rivers Navia, Narcea, Nansa, Pas, and Asón (A_R mean = 7.1, range: 6.0–7.5), i.e., those with lower flows. Observed heterozygosity ranged from 0.58 (the River Pas) to 0.75 (the River Deva-Cares). No statistically significant deviation from Hardy–Weinberg equilibrium or any evidence of linkage between pairs of loci was detected.

Table 1. Genetic diversity indices in archival samples (the 1950s) of Atlantic salmon from Spain: sample size (N), observed number of alleles (A), allelic richness (A_R), expected (H_e) and observed heterozygosity (H_o), F_{IS} values, and deviations from Hardy-Weinberg equilibrium (HWE). No deviation from HWE expectation was statistically significant after Bonferroni corrections.

River		Total	Locus							
			Ssa197	*SSOSL85*	*SSOSL311*	*SSOSL417*	*Sssp2210*	*SsspG7*	*SsaF43*	*SSOSL438*
Navia	N	48	48	48	48	48	48	48	48	47
	A	7.875	8	9	11	8	8	9	6	4
	A_R	7.191	7.706	8.178	9.756	7.098	7.458	8.454	5.387	3.489
	H_o	0.623	0.708	0.625	0.646	0.563	0.792	0.813	0.479	0.362
	H_e	0.643	0.808	0.659	0.717	0.635	0.806	0.786	0.412	0.317
	F_{IS}	0.040	0.133	0.062	0.11	0.125	0.029	−0.023	−0.154	−0.13
	HWE	0.081	0.047	0.268	0.132	0.112	0.391	0.448	0.115	0.209
Narcea	N	48	48	48	48	48	48	48	48	48
	A	7.875	10	7	9	9	9	10	5	4
	A_R	7.515	9.433	6.454	8.918	8.701	8.837	9.116	4.729	3.929
	H_o	0.677	0.729	0.708	0.708	0.688	0.792	0.813	0.604	0.375
	H_e	0.687	0.8	0.661	0.766	0.707	0.788	0.799	0.632	0.34
	F_{IS}	0.024	0.099	−0.061	0.086	0.038	0.005	−0.006	0.054	−0.092
	HWE	0.192	0.095	0.292	0.156	0.848	0.515	0.538	0.342	0.278
Sella	N	49	49	49	49	49	49	49	49	49
	A	8.75	12	11	9	10	9	11	5	3
	A_R	8.217	11.727	10.304	8.405	9.3	8.121	10.183	4.693	3
	H_o	0.722	0.776	0.776	0.776	0.837	0.878	0.694	0.571	0.469
	H_e	0.731	0.853	0.814	0.758	0.761	0.794	0.779	0.544	0.545
	F_{IS}	0.023	0.101	0.057	−0.012	−0.09	−0.095	0.119	−0.041	0.149
	HWE	0.194	0.061	0.228	0.541	0.141	0.111	0.048	0.416	0.149
Deva-Cares	N	46	46	46	46	46	46	46	46	46
	A	8.5	10	8	12	12	9	9	4	4
	A_R	7.811	9.886	6.988	10.749	10.51	8.865	7.792	3.988	3.706
	H_o	0.745	0.804	0.717	0.935	0.739	0.826	0.674	0.717	0.544
	H_e	0.697	0.844	0.688	0.822	0.729	0.762	0.642	0.604	0.488
	F_{IS}	−0.057	0.058	−0.032	−0.127	−0.004	−0.073	−0.039	−0.177	−0.102
	HWE	0.031	0.213	0.435	0.029	0.572	0.195	0.412	0.063	0.283
Nansa	N	48	44	47	48	47	46	48	48	47
	A	7	10	11	11	6	6	5	4	3
	A_R	6.702	9.748	10.347	10.368	5.984	5.509	4.724	4	2.937
	H_o	0.678	0.841	0.745	0.813	0.809	0.565	0.646	0.729	0.277
	H_e	0.667	0.866	0.77	0.855	0.715	0.599	0.584	0.673	0.274
	F_{IS}	−0.006	0.041	0.043	0.06	−0.12	0.067	−0.096	−0.074	0.002
	HWE	0.472	0.297	0.31	0.218	0.105	0.288	0.238	0.272	0.638
Pas	N	47	46	47	47	45	47	47	47	46
	A	6.125	11	7	8	6	5	5	5	2
	A_R	5.96	10.633	6.615	7.741	5.998	4.984	4.969	4.744	1.997
	H_o	0.583	0.783	0.617	0.766	0.733	0.532	0.553	0.596	0.087
	H_e	0.619	0.839	0.694	0.825	0.745	0.614	0.564	0.585	0.083
	F_{IS}	0.068	0.078	0.121	0.082	0.027	0.144	0.03	−0.007	−0.034
	HWE	0.017	0.14	0.102	0.154	0.412	0.089	0.452	0.555	1
Asón	N	48	48	48	48	48	48	48	48	48
	A	7.625	9	7	10	7	9	9	5	5
	A_R	7.191	8.378	6.71	9.116	6.706	8.417	8.688	4.928	4.587
	H_o	0.703	0.792	0.813	0.667	0.854	0.688	0.75	0.521	0.542
	H_e	0.683	0.774	0.724	0.736	0.708	0.725	0.815	0.469	0.516
	F_{IS}	−0.019	−0.012	−0.112	0.104	−0.196	0.062	0.09	−0.101	−0.04
	HWE	0.289	0.539	0.12	0.131	0.013	0.297	0.131	0.225	0.441

Table 1. *Cont.*

River		Total	Locus								
			Ssa197	*SSOSL85*	*SSOSL311*	*SSOSL417*	*Sssp2210*	*SsspG7*	*SsaF43*	*SSOSL438*	
Bidasoa	N	40	38	37	40	35	40	40	40	36	
	A	8.75	8	9	12	10	8	11	7	5	
	A_R	8.67	7.989	8.89	11.845	10	7.861	10.971	6.861	4.944	
	H_o	0.716	0.868	0.73	0.825	0.714	0.7	0.9	0.575	0.417	
	H_e	0.728	0.79	0.721	0.873	0.709	0.787	0.875	0.63	0.437	
	F_{IS}	0.029	−0.086	0.002	0.067	0.008	0.123	−0.016	0.099	0.061	
	HWE	0.138	0.194	0.574	0.173	0.556	0.087	0.541	0.172	0.391	

Results from the comparison of the genetic diversity parameters (A_R, H_e, and H_o) and effective population size N_e between the archival (the 1950s) and contemporary (the 1990s and 2000s) samples of Spanish Atlantic salmon are shown in Table 2. Average allelic richness A_R for the historical (the 1950s) samples was higher than contemporary averages. However, the average expected He and observed Ho heterozygosity in archival samples were not apparently different from those of contemporary populations, with the exception of the lower 1990s values for the rivers Narcea and Sella. Finally, the average effective population size N_e for the archival 1950s samples was always higher than contemporary values, showing an evident decline.

Table 2. Comparison of average allelic richness (A_R), and expected heterozygosity (H_e), observed heterozygosity (H_o), and effective population size (N_e) between archival (the 1950s) and contemporary (the 1990s and 2000s) samples of Atlantic salmon from Spain.

River	Period	Av. A_R	Av. He	Av. Ho	Av. Ne	Sources
Navia	1950s	7.19	0.643	0.623	667	Present study
	1990s	-	-	-	-	
Narcea	1950s	6.68	0.617	0.607	325	Present study
	1990s	6.32	0.536	0.487	114	[75]
Sella	1950s	8.22	0.731	0.722	1088	Present study
	1990s	6.65	0.532	0.487	107	[75]
Deva-Cares	1950s	7.81	0.697	0.745	729	Present study
	1990s	5.97	0.689	0.564	-	[18,19,75,76]
	2000s	4.68	0.660	0.690	-	[19]
Nansa	1950s	6.70	0.667	0.678	197	Present study
	1990s	5.34	0.772	0.620	68	[18,19]
	2000s	4.88	0.680	0.710	-	[19]
Pas	1950s	5.96	0.619	0.583	180	Present study
	1990s	5.15	0.769	0.650	-	[18,19]
	2000s	4.48	0.670	0.670	-	[19]
Asón	1950s	7.19	0.683	0.703	1051	Present study
	1990s	5.20	0.773	0.650	42	[18,19]
	2000s	4.99	0.700	0.740	-	[19]
Bidasoa	1950s	8.67	0.728	0.716	689	Present study
	2000s	-	0.850	0.750	-	[77]

All study rivers contained private alleles, ranging in number from 1 to 8 (for the River Bidasoa) (Table 3). All the microsatellite loci showed private alleles, with *SSOSL311* and *SSOSL417* showing the highest number of allelic variants across rivers (7 in both cases). Frequencies for these alleles ranged from 1.04% to 7.10%, the highest value corresponding to *SSspG7*110* from the River Sella.

Table 3. Private alleles in archival samples (the 1950s) of Atlantic salmon from Spain after comparison with the SALSEA baseline. For each microsatellite locus, the first column indicates the allele and the second its frequency (in percentage) in the river. The number of private alleles in each river and locus (#) are also included.

River	# Private Alleles per River	*Ssa197*		*SSOSL85*		*SSOSL311*		*SSOSL417*		*Sssp2210*		*SSspG7*		*SsaF43*		*SSOSL438*	
Navia	3	-	-	*214	1.0%	*173	1.0%	*209	1.0%	-	-	-	-	-	-	-	-
Narcea	1	-	-	-	-	-	-	*185	1.0%	-	-	-	-	-	-	-	-
Sella	5	*151	2.0%	-	-	-	-	*159	1.0%	-	-	*106 *110 *138	2.0% 7.1% 1.0%	-	-	-	-
Deva-Cares	6	-	-	-	-	*123 *171	1.1% 1.1%	*189 *199	2.2% 1.1%	*158	3.3%	- -	- -	- -	- -	*128	1.1%
Nansa	1	-	-	*216	3.2%	-	-	-	-	-	-	-	-	-	-	-	-
Pas	1	-	-	-	-	-	-	-	-	-	-	-	-	*128	6.4%	-	-
Asón	3	-	-	*204	1.0%	*127	1.0%	-	-	*138	3.1%	-	-	-	-	-	-
Bidasoa	8	-	-	-	-	*137 *149 *163	2.5% 6.3% 3.8%	*175 *183	2.9% 5.7%	*170	1.3%	-	-	*112	1.3%	*142	1.4%
# Private alleles per locus		1		3		7		7		3		3		2		2	
Total alleles		14		15		19		18		13		15		9		7	

After comparison with the SALSEA baseline for the coincident microsatellite loci (*Ssa197**, *SSsp2210**, *SspG7**, and *SsaF43**), a few of these private alleles were found exclusively in Spanish rivers and were absent or extremely rare in northern European samples. For example, *Ssa197*151* and *SSspG7*110* were only found in the historical samples from the River Sella and the 2000s samples from the rivers Ulla and Eo; *SSspG7*106* and *SsaF43*128* were found in historical samples from the rivers Sella and Pas, and only (at a very low frequency) in three isolated populations from Ireland and Norway and Iceland, respectively.

3.3. Connectivity and Population Structure

All pairwise F_{ST} values were statistically significant after the Bonferroni corrections (Table 4). They ranged from 0.008 to 0.098 (average F_{ST} = 0.054). The lowest level of genetic differentiation was observed between the rivers Sella and Deva-Cares (F_{ST} = 0.008, $p < 0.010$) and rivers Navia and Narcea (F_{ST} = 0.019, $p < 0.01$), whereas the highest level of differentiation was observed between the rivers Sella and Pas (F_{ST} = 0.098, $p < 0.01$).

Table 4. Pairwise F_{ST} values in archival samples (the 1950s) of Atlantic salmon from Spain. All values were significant after the Bonferroni correction.

	Navia	Narcea	Sella	Deva-Cares	Nansa	Pas	Asón
Narcea	0.019						
Sella	0.054	0.039					
Deva-Cares	0.059	0.040	0.008				
Nansa	0.064	0.049	0.049	0.035			
Pas	0.085	0.079	0.098	0.082	0.083		
Asón	0.052	0.048	0.036	0.044	0.052	0.085	
Bidasoa	0.033	0.037	0.043	0.052	0.071	0.067	0.052

Principal coordinate analysis (PCoA) based on linearised F_{ST}s showed the separation of three groups in the studied rivers (Figure 3). The first axis, which accounted for 54.5% of molecular variance, separated the River Pas population from the rest of the rivers. The second axis, which explained 25.6% of the variance, separated the rivers Narcea, Navia, and Bidasoa from the Sella, Deva-Cares, Nansa, and Asón group, while the River Pas occupied an intermediate position.

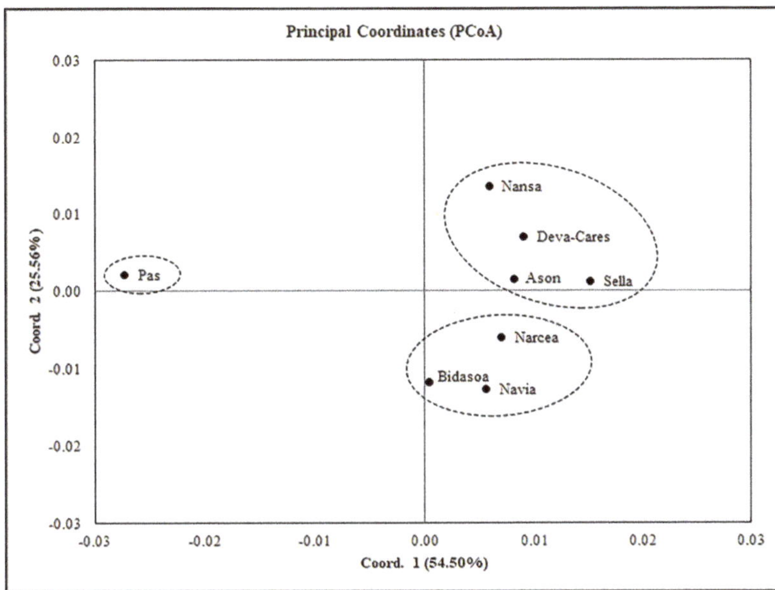

Figure 3. Principal coordinate analysis of linearised pairwise F_{ST} values in archival samples (the 1950s) of Atlantic salmon populations from Spain.

The results of the Evanno test suggested that $\Delta K = 3$ was the optimal cluster number for the first hierarchical level for the STRUCTURE analysis. Furthermore, the most significant increase in the LnP(D) values was detected at $\Delta K = 3$. This Bayesian approach confirmed the same three genetic units found in the PCoA analysis (Figure 4). The River Asón population was assigned to the Sella, Deva-Cares, and Nansa clusters, to which it had the greatest percentage of membership (Q value = 62.0%). At the second and third hierarchical levels, all rivers were genetically different, except for Deva-Cares and Sella populations, which were indistinguishable from each other and presented a very low genetic divergence ($F_{ST} = 0.008$), so they were considered together for further analyses.

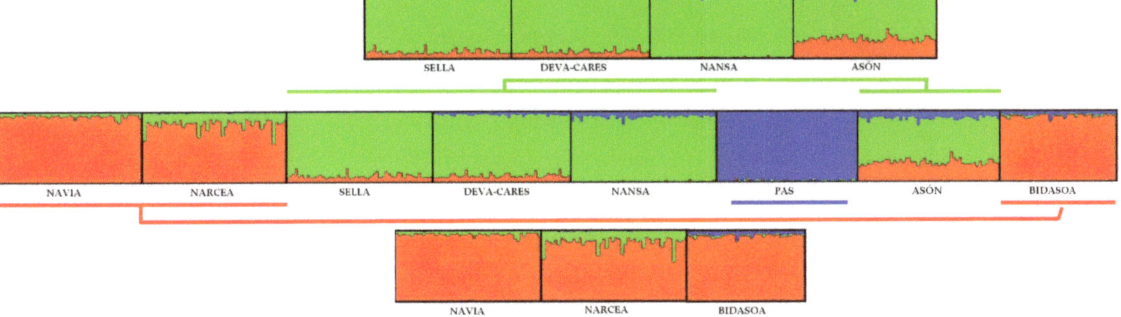

Figure 4. Estimation of the percentage of membership to each of three clusters inferred by STRUCTURE in archival samples (the 1950s) of Atlantic salmon from Spain, based on variation at nine microsatellite loci. Each vertical bar represents one individual and each black square delimits a population corresponding to a river. The graphs above and below the main graph represent the first hierarchical level (Evanno test, $\Delta K = 3$), with brackets indicating the correspondence between the populations. Each cluster is represented by one colour: red, green, or blue.

Mantel and dbRDA failed to detect isolation by distance among the analysed populations and significant correlations between genetic and geographic distances (Mantel: $r = -0.083$, $p > 0.05$; dbRDA: $r = 0.801$, $p > 0.05$). The results of the bottleneck tests did not show any such effect on the studied populations (Table 5). The BOTTLENECK test did not detect significant excess heterozygosity for the *TPM* model with 80% stepwise mutations (*SMM*) ($p > 0.05$, Table 5). Similarly, *M*-ratio values were not significantly lower than the simulated critical value of *M* (*Mc*) under mutation-drift equilibrium for the lower and upper values of pre-bottleneck effective population size values ($p > 0.05$, Table 5). Both tests did not show old or recent bottlenecks in the archival samples of Spanish salmon populations.

Table 5. BOTTLENECK tests of archival samples (the 1950s) of Atlantic salmon from Spain. The expected recent bottleneck is presented as *p*-values from Wilcoxon's signed-rank test, assuming a two-phase mutation model (*TPM*) with 80% stepwise mutations (*SMM*). M-ratio and critical value of *M* (*Mc*) assuming pre-bottleneck effective population size (*Ne*) of 50, 100, 500, and 1000, and a mutation rate (μ) of 5×10^{-4}.

River	*SMM*	*M*-ratio	*Mc*			
			Ne = 50	*Ne* = 100	*Ne* = 500	*Ne* = 1000
Navia	$p = 0.981$	0.771	0.754	0.744	0.701	0.685
Narcea	$p = 0.680$	0.781	0.754	0.744	0.701	0.685
Sella	$p = 0.422$	0.777	0.758	0.746	0.701	0.683
Deva-Cares	$p = 0.875$	0.782	0.756	0.748	0.703	0.680
Nansa	$p = 0.273$	0.761	0.754	0.744	0.701	0.685
Pas	$p = 0.273$	0.802	0.758	0.749	0.703	0.799
Asón	$p = 0.809$	0.776	0.754	0.744	0.701	0.685
Bidasoa	$p = 0.727$	0.765	0.756	0.749	0.700	0.677

Recent migration rates between most isolated populations were low and asymmetric ($m = 0.020$–0.100) or very low ($m < 0.020$) (Figure 5). Higher migration rates were detected between the Navia (source) and Narcea (destination) populations ($m = 0.119$, range 0.099–0.141); Narcea and Navia ($m = 0.245$, range 0.181–0.309), and the Nansa and Sella + Deva-Cares populations ($m = 0.169$, range 0.074–0.264). Migration rates estimated by MIGRATE were very low (0.002–0.022) and showed no clear patterns. In addition, several of the chains did not reach convergence.

Finally, GENECLASS analysis identified nine individuals as migrants ($p < 0.01$) (Table 6). The migrants' populations of origin were in accordance with their percentage of membership estimated from STRUCTURE analysis and their population migration rates estimated by BAYESASS, except for one migrant found in the River Pas. This individual was identified as being from the River Navia but showed a high percentage of membership (*Q* value = 93.5%) to the Pas cluster, and overall, there was a very low migration rate between rivers (average $m = 0.013$).

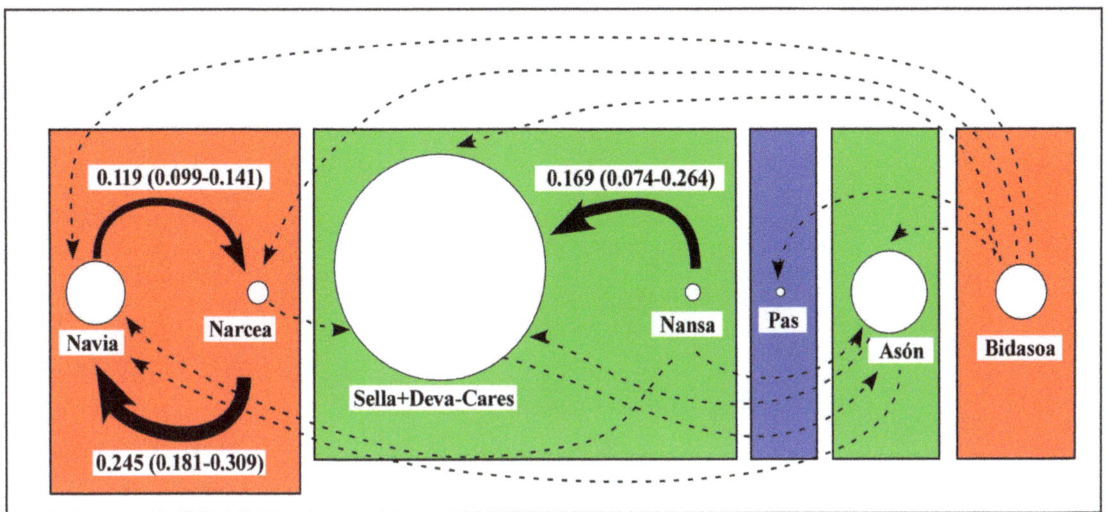

Figure 5. Graphical representation of migration estimates using BAYESASS in archival samples (the 1950s) of Atlantic salmon from Spain. The circle size for each population is proportional to the adjusted effective size (*Ne(adj)*). Coloured rectangles correspond to the 1st-level STRUCTURE clusters. Values over 0.1 are represented as thick black arrows (proportional to migration values) and shown above or below them. Values between 0.02 and 0.1 are shown as dashed arrows.

Table 6. Estimates of the census size (N_c), the effective number of breeders (N_b, [CI 95%]), the adjusted number of breeders ($N_{b(adj)}$), the adjusted effective size ($N_{e(adj)}$), and the effective size ratios $N_{b(adj)}/N_c$, and $N_{e(adj)}/N_c$ in archival samples (the 1950s) of Atlantic salmon from Spain. Populations in the Deva-Cares and Sella were pooled together after considering the 1st-level STRUCTURE clusters.

River	N_c	Migrants Removed	N_b [CI 95%]	$N_{b(adj)}$ [CI 95%]	$N_{e(adj)}$ [CI 95%]	$N_{b(adj)}/N_c$	$N_{e(adj)}/N_c$
Navia	701	1	320 [162–∞]	285 [145–∞]	667 [338–∞]	0.407	0.951
Narcea	1298	2	156 [80–∞]	139 [71–∞]	325 [166–∞]	0.107	0.250
Deva-Cares + Sella	2238	2	1167 [161–∞]	1040 [144–∞]	2431 [336–∞]	0.465	1.086
Nansa	124	0	95 [54–275]	85 [48–245]	197 [111–572]	0.681	1.592
Pas	136	1	86 [48–259]	77 [43–230]	180 [100–538]	0.566	1.323
Asón	1051	2	645 [197–∞]	575 [175–∞]	1344 [410–∞]	0.547	1.278
Bidasoa	102	1	331 [83–∞]	295 [74–∞]	689 [173–∞]	2.890	6.753

3.4. Demographic Parameters

Estimates of the census size, the effective number of breeders, the adjusted number of breeders, the adjusted effective size, and the effective size ratios are shown in Table 6. $N_{b(adj)}$ was lowest for the rivers Pas and Nansa (77 and 85 respectively) and highest for the Deva-Cares and Sella populations (1040), with the remaining values being in the 139–575 range. For $N_{e(adj)}$, a similar trend was observed, with the lowest values found for rivers Pas and Nansa (180, 197) and the highest values for the Deva-Cares and Sella populations (2431), while the rest of the populations varied between 325 and 1344. The River Narcea had the lowest $N_{b(adj)}/N_c$ value (0.107), while the rest of the populations showed higher values (0.497–0.681). The River Bidasoa presented an unusually high value (2.890). The $N_{e(adj)}/N_c$ values ranged between 0.250 (Narcea) and 1.592 (Nansa), with the Bidasoa again showing a very high value (6.753).

4. Discussion

4.1. Genetic Diversity

The historical (the 1950s) allelic richness of Spanish salmon populations was comparable to those found in some northern European populations [78,79] and was markedly higher than those observed in the 1990s Spanish samples [18]. These findings indicate a healthy level of genetic diversity prior to a severe loss in the following four decades. This genetic diversity loss contrasts with the increase of allelic richness observed in other salmonid populations after stocking [80] but could be explained by the loss of private alleles, an increase in the fishing effort [76], and the population collapse caused by climate-driven environmental changes in their breeding and feeding areas [10,11], leading to genetic drift in the resulting small populations. The increase in allelic richness in the Spanish 2000s samples [76], surpassing historical pre-stocking values, could be explained as a result of the genetic introgression caused by the stock transfer from northern Europe [12].

The absence of significant deviations from Hardy–Weinberg equilibrium (*HWE*) is consistent with other wild populations presenting low levels of disturbance (i.e., Icelandic populations [81]). In contrast, Consuegra et al. [18] found significant deviations from *HWE* (heterozygote deficit) in the rivers Nansa (samples from 1996 and 1997) and Deva (samples from 1993 and 1996), while Griffiths et al. [82] found that genotype frequencies in samples from the rivers Narcea, Asón, and Sella from 2004 significantly deviated from *HWE*. According to Griffiths et al. [82], these deviations from *HWE* expectations may be attributed to the decline in salmon population sizes in the Spanish river basins, combined with the effects of historical stocking and supportive breeding measures.

The number of private alleles found in historical samples and their absence in current populations offers further evidence of the loss of local genetic diversity. The existence of this native genetic diversity provides evidence of the relative historical isolation of salmon populations in these rivers before climate change and the effects of stocking on affected populations.

4.2. Connectivity and Population Structure

Historical Spanish populations showed a strong and significant genetic structuring very similar to that found in the River Teno complex (Norway and Finland) salmon population (F_{ST} = 0.065) [83]. Likewise, a pattern of spatial genetic differentiation was also described in the Spanish rivers, with an overall F_{ST} value of 0.019 [75]. Separating the pairwise comparisons by year (1993 vs. 1999), an increase in genetic differentiation (from 0.016 to 0.023) was found [75]. Even then, these values are far from those obtained here for the historical (the 1950s) populations (0.054). An overall F_{ST} value of 0.053 for Spanish rivers sampled in 1988, which dropped to 0.017 in 1996 and recovered to 0.033 in 2007, was reported [21]. These comparisons support the view of original marked genetic differentiation between populations, which was disrupted by human-mediated pressures, such as stock transfer and possible alterations of homing behaviour due to climate change, and a slow later recovery.

The results from the STRUCTURE and PCoA analyses confirmed a significant level of genetic structure in the region, which would correspond to the situation existing before the effects of climate change and foreign stocking intensified in the late 1980s. Indeed, right after the end of stock transfer practices (the 1990s), some Spanish populations (from which the rivers Sella, Narcea, and Cares overlap with our study) could be considered a single genetic unit [21]. In 2007, after several years of supportive breeding, the genetic structure changed into three separate units. This suggests that the removal of human-mediated transfer practices allowed the recovery of the natural genetic structure of salmon populations from Spanish rivers, even if the allocation to genetic units did not exactly match the clusters found in the 1950s.

While most of the allocations to genetic clusters are consistent with geographic location (with neighbouring rivers being the most closely related), other allocations are surprising and warrant more complex explanations. The River Pas population seems to be very isolated genetically, but it is geographically located between the other river basins. On the other hand, the population from the River Bidasoa (the easternmost river basin) was included in the same cluster as the rivers Narcea and Navia (the westernmost river basins). This cluster could be explained by the existence of undocumented pre-1960s human-mediated transfers, which could have homogenised these distant populations. Another compatible scenario is the existence in the past of a widespread lineage, linked by frequent migration, corresponding to the Narcea-Navia-Bidasoa cluster, from which smaller populations (Bidasoa) would have eventually become isolated, diverging into distinct genetic units. Therefore, larger populations (Narcea-Navia) could have retained more of the original signal of panmixia. However, the separation of these populations into different clusters in the lower hierarchical level appears to be more consistent with the geographical location of the populations and should reflect their connectivity and spatial genetic structure more accurately.

A STRUCTURE analysis including samples from the rivers Deva, Asón, Pas, and Nansa, and samples pooled by decades since the 1950s was recently carried out by Ciborowski et al. [19]. As temporal variation added heterogeneity to the dataset and the microsatellites used were not all the same as in the present study, it is difficult to compare the results of both analyses. There were some similarities and some discrepancies. As in the second hierarchical level, each river was assigned to its own cluster. However, populations from the rivers Nansa and Pas could be assigned to the same cluster considering the 1960s data in isolation, while in our study, the Pas population is clearly isolated from the rest at the first hierarchical level.

Significant migration only occurred between contiguous rivers, which were recovered in the same PCoA and STRUCTURE (1st-level) clusters. Estimates obtained by MIGRATE were generally very low and differed from the BAYESASS migration rates in magnitude and direction. Estimates from MIGRATE tend to be comparatively low and do not necessarily reflect connectivity inferred by BAYESASS (commonly interpreted as contemporary connectivity) [24]. In this light, we decided to focus on BAYESASS migration rates for interpretation.

Estimated migration rates for Atlantic salmon have been rarely reported for the Spanish rivers. An earlier study in the 1990s [18], using microtagging-recapture and Bayesian assignment tests based on microsatellites to estimate the proportion of salmon migrants among the rivers Asón, Pas, and Nansa, obtained a range from 0.07 to 0.31. Migration rates obtained for most population pairs in our study are within or below this range (even though some of them are far from each other), indicating infrequent departures from homing behaviour among historical populations.

More studies on Atlantic salmon migration rates have been carried out in the rest of its geographical distribution. Several tagging studies have shown that straying in salmonids is rare, typically ranging between 1 and 10% in geographically close rivers [24]. In the River Teno (Norway and Finland, samples from 1979 to 2001), recent migration rates (using BAYESASS) varied between 0.003–0.260 [84]. In Canadian rivers from Newfoundland and Labrador, a similar range was obtained (0.011–0.303) using contemporary samples and the same methodology [24]. Both studies considered such migration rates as low. Once again, estimates from our historical populations seem to overlap with those previous rates, but with lower maximum values. It is remarkable that migration from larger to smaller populations and vice versa were detected. Some authors [18] suggested that larger, more stable populations should act as a sink attracting strays from neighbouring rivers, while others [85–87] found that larger populations act as a source. Empirical studies show that larger populations can behave as sinks and that patterns can switch over evolutionary time scales [24]. In any case, we found that gene flow from smaller to larger populations was quantitatively prevailing. On the other hand, warmer and drier local environmental conditions seem to cause increased straying between neighbouring rivers in southern populations of Atlantic salmon, resulting in a higher gene flow between close populations [77]. Thus, ongoing climate change might increase straying rates in the future. Values obtained in current populations from Spain could be compared with migration rates from historical populations to extrapolate and test this hypothesis across the whole Cantabrian watershed.

The absence of isolation by distance is not uncommon in Atlantic salmon populations at large spatial scales [24,88]. Considering the strong homing behaviour of the species, it seems intuitive that genetic distances between populations from different river basins have an important geographical component. However, it seems more likely that the spatial genetic structure in the Spanish populations may be determined by the geological and ecological characteristics of the rivers (isolation by the environment [89]). Thus, the studied rivers can be classified into Atlantic siliceous rivers (Navia, Narcea), Atlantic calcareous rivers (Sella, Deva-Cares, Nansa, Pas, Asón), and Pyrenean rivers (Bidasoa). Salmon are more likely to stray into rivers that are ecologically similar to their river of origin [90], which is consistent with the correspondence between the 1st-level genetic clusters and the grouping of the rivers on the former classification. Other environmental features of the rivers (temperature, water discharge, etc.) could explain further subdivision in 2nd-level clusters and isolation of the River Pas, the closer relationship between the Sella and Deva-Cares samples, or the exclusion of Nansa (geographically closer to Sella and Deva-Cares).

4.3. Demographic Parameters

The population abundance trends estimated in this study are consistent with previous studies in the area [11], showing a widespread and strong population decline linked to climate change from the 1980s. Consequently, the estimates of the census size, the effective and adjusted number of breeders, the adjusted effective size, and the effective size ratios in historical samples were generally closer to values found in well-preserved, high-latitude salmon populations than to current Spanish populations, which experienced a strong decline linked to climate change from the 1980s [10,11]. For example, the average $N_{b(adj)}$ from our study was 356, the average $N_{b(adj)}$ estimate from Canadian (Quebec) populations was 211 [9], and the average N_b estimated for some Spanish rivers was 107.5 [75]. The same applies to $N_{e(adj)}$: our estimates from historical populations ranged from 180 to 2431 (average 833), while estimates for the 1990s and 2000s populations ranged from 37 to 96 (average 60 [76]), 64–260 [17,75] or 38–175 [18]. These results are consistent with the general declines in Ne observed in salmon populations worldwide [91].

As evidence of the reasonably good conservation status of the historical populations, all had N_e above the minimum threshold ($N_e = 95$) for retaining 90% of genetic diversity over 100 years [76], and several showed N_e values close to or well over the threshold ($N_e = 1000$), below which long-term maintenance of evolutionary potential is uncertain [92]. Considering other N_e estimates from the study area [76], which included three of the studied rivers (Cares, Sella, and Narcea), it is established that the decline in effective population size between the 1990s and 2007–2008 already started between the 1950s and the 1990s, probably due to marine overfishing and climate-driven environmental changes in both freshwater and marine habitats. Contrary to Consuegra et al. [18], who found evidence of bottlenecks in the more recent (1993–1998) samples of the populations from the rivers Deva, Nansa, Pas, and Asón, there was no signal of bottlenecking in the populations sampled in the 1950s. This is other evidence of the good conservation status of the historical populations that highlights the importance of their decline in the last decades of the 20th century.

5. Conclusions

Historical (the 1950s) Atlantic salmon populations from Spain showed a remarkable native allelic uniqueness, different indicators of good conservation status (not dissimilar from current Canadian and Scandinavian populations), and a marked population structure with limited migration between close river basins. Since then, study populations have experienced a continuous decline that was accelerated in the 1980s, which could be attributed to intense climate-driven biophysical changes in their marine and freshwater habitats. This decline contributed to a reduction in genetic diversity since the 1990s, which was reversed due to massive stocking with non-native stocks, a practice that could not prevent a loss of native alleles and the population collapse. Furthermore, the synergistic effects of a warmer climate, drier rivers, and the homogenizing effect of human-mediated introgression have driven the genetic erosion of native population structure.

Despite efforts to recover targeted stocks, the current situation is alarming and merits urgent measures to prevent local extinction. Stocking practices have been highly detrimental and their effects, combined with those of climate change, can be considered the main anthropogenic impact affecting salmon survival and the integrity of the species' genome regionally. Comparisons of current populations with the benchmark provided in our study could help identify which factors have affected salmon populations the most in this area in the last 70 years, in order to guide conservation programs and evaluate the adaptive potential of the species under ongoing climate change. Efforts should target recovering the historical structure of populations and preserving their unique gene pools.

Author Contributions: A.A. planned, supervised, and coordinated the work. A.A., S.L. and D.F.M. performed laboratory work and data analyses. A.A. wrote the paper with major contributions from B.E., G.G.N. and D.A. A.A. and B.E. gathered the funding and coordinated the project administration. All authors have read and agreed to the published version of the manuscript.

Funding: This research was funded by the Spanish Ministry of Science and Innovation through research project CGL2012-36049 and by the Spanish Ministry of Economy, Industry and Competitiveness through research project CGL2017-84269-P.

Institutional Review Board Statement: All laboratory procedures complied with current Spanish and European conservation legislation and strictly adhered to regulations on the handling of wild animals (Directive 2010/63/EU).

Data Availability Statement: The data presented in this study are available upon request from the corresponding author (A.A.: aalmodovar@bio.ucm.es).

Acknowledgments: We are very grateful to Jerónimo de la Hoz (Environmental Agency, Regional Government of Asturias, Spain) for his help. We are also very grateful to John Gilbey, who kindly provided us with the microsatellite baseline generated by the SALSEA-Merge Project (Available online: www.nasco.int/sas/salseamerge.htm, accessed on 15 May 2022).

Conflicts of Interest: The authors declare no conflict of interest.

References

1. Schindler, D.E.; Hilborn, R.; Chasco, B.; Boatright, C.P.; Quinn, T.P.; Rogers, L.A.; Webster, M.S. Population diversity and the portfolio effect in an exploited species. *Nature* **2010**, *465*, 609–612. [CrossRef] [PubMed]
2. Hoban, S.; Archer, F.I.; Bertola, L.D.; Bragg, J.G.; Breed, M.F.; Bruford, M.W.; Hunter, M.E. Global genetic diversity status and trends: Towards a suite of Essential Biodiversity Variables (EBVs) for genetic composition. *Biol. Rev.* **2022**, *97*, 1511–1538. [CrossRef]
3. Luck, G.W.; Daily, G.C.; Ehrlich, P.R. Population diversity and ecosystem services. *Trends Ecol. Evol.* **2003**, *18*, 331–336. [CrossRef]
4. Frankham, R. Genetics and extinction. *Biol. Conserv.* **2005**, *126*, 131–140. [CrossRef]
5. Reusch, T.B.H.; Ehlers, A.; Hämmerli, A.; Worm, B. Ecosystem recovery after climatic extremes enhanced by genotypic diversity. *Proc. Natl. Acad. Sci. USA* **2005**, *102*, 2826–2831. [CrossRef] [PubMed]
6. Dadswell, M.; Spares, A.; Reader, J.; McLean, M.; McDermott, T.; Samways, K.; Lilly, J. The decline and impending collapse of the Atlantic Salmon (*Salmo salar*) population in the North Atlantic Ocean: A review of possible causes. *Rev. Fish. Sci. Aquac.* **2021**, *30*, 215–258. [CrossRef]

7. Thorstad, E.B.; Bliss, D.; Breau, C.; Damon-Randall, K.; Sundt-Hansen, L.E.; Hatfield, E.M.; Sutton, S.G. Atlantic salmon in a rapidly changing environment. Facing the challenges of reduced marine survival and climate change. *Aquat. Conserv. Mar. Freshw. Ecosyst.* **2021**, *31*, 2654–2665. [CrossRef]

8. Nieto, A.; Ralph, G.M.; Comeros-Raynal, M.T.; Kemp, J.; García-Criado, M.; Allen, D.J.; Williams, J.T. *European Red List of Marine Fishes*; Publications Office of the European Union: Luxembourg, 2015.

9. Perrier, C.; April, J.; Cote, G.; Bernatchez, L.; Dionne, M. Effective number of breeders in relation to census size as management tools for Atlantic salmon conservation in a context of stocked populations. *Conserv. Genet.* **2016**, *17*, 31–44. [CrossRef]

10. Nicola, G.G.; Elvira, B.; Jonsson, B.; Ayllón, D.; Almodóvar, A. Local and global climatic drivers of Atlantic salmon decline in southern Europe. *Fish Res.* **2018**, *198*, 78–85. [CrossRef]

11. Almodóvar, A.; Ayllón, D.; Nicola, G.G.; Jonsson, B.; Elvira, B. Climate-driven bio-physical changes in feeding and breeding environments explain the decline of southernmost European Atlantic salmon populations. *Can. J. Fish. Aquat. Sci.* **2019**, *76*, 1581–1595. [CrossRef]

12. Almodóvar, A.; Leal, S.; Nicola, G.G.; Hórreo, J.L.; García-Vázquez, E.; Elvira, B. Long-term stocking practices threaten the original genetic diversity of the southernmost European populations of Atlantic salmon *Salmo salar*. *Endanger. Species Res.* **2020**, *41*, 303–317. [CrossRef]

13. Gallagher, B.K.; Geargeoura, S.; Fraser, D.J. Effects of climate on salmonid productivity: A global meta-analysis across freshwater ecosystems. *Glob. Chang. Biol.* **2022**, *28*, 7250–7269. [CrossRef]

14. Almodóvar, A.; Nicola, G.G.; Ayllón, D.; Trueman, C.N.; Davidson, I.; Kennedy, R.; Elvira, B. Stable isotopes suggest the location of marine feeding grounds of south European Atlantic salmon in Greenland. *ICES J. Mar. Sci.* **2020**, *77*, 593–603. [CrossRef]

15. Rikardsen, A.H.; Righton, D.; Strøm, J.F.; Thorstad, E.B.; Gargan, P.; Sheehan, T.; Aarestrup, K. Redefining the oceanic distribution of Atlantic salmon. *Sci. Rep.* **2021**, *11*, 12266. [CrossRef] [PubMed]

16. Hórreo, J.L.; de la Hoz, J.; Machado-Schiaffino, G.; Pola, I.G.; García-Vázquez, E. Restoration and enhancement of Atlantic salmon populations: What we have learned from North Iberian rivers. *Knowl. Manag. Aquat. Ecosyst.* **2011**, *402*, 23. [CrossRef]

17. Ribeiro, A.; Morán, P.; Caballero, A. Genetic diversity and effective size of the Atlantic salmon *Salmo salar* L. inhabiting the River Eo (Spain) following a stock collapse. *J. Fish Biol.* **2008**, *72*, 1933–1944. [CrossRef]

18. Consuegra, S.; Verspoor, E.; Knox, D.; García de Leániz, C. Asymmetric gene flow and the evolutionary maintenance of genetic diversity in small, peripheral Atlantic salmon populations. *Conserv. Genet.* **2005**, *6*, 823–842. [CrossRef]

19. Ciborowski, K.; Jordan, W.C.; García de Leániz, C.; Consuegra, S. Temporal and spatial instability in neutral and adaptive (MHC) genetic variation in marginal salmon populations. *Sci. Rep.* **2017**, *7*, 42416. [CrossRef] [PubMed]

20. Ayllón, F.; Martínez, J.L.; García-Vázquez, E. Loss of regional population structure in Atlantic salmon, *Salmo salar* L. following stocking. *ICES J. Mar. Sci.* **2006**, *63*, 1269–1273. [CrossRef]

21. Hórreo, J.L.; Machado-Schiaffino, G.; Ayllón, F.; Griffiths, A.M.; Bright, D.; Stevens, J.R.; García-Vázquez, E. Impact of climate change and human-mediated introgression on southern European Atlantic salmon populations. *Glob. Chang. Biol.* **2011**, *17*, 1778–1787. [CrossRef]

22. Blanco, G.; Ramos, M.D.; Vázquez, E.; Sánchez, J.A. Assessing temporal and spatial variation in wild populations of Atlantic salmon with particular reference to Asturias (Northern Spain) rivers. *J. Fish Biol.* **2005**, *67*, 169–184. [CrossRef]

23. Campos, J.L.; Posada, D.; Morán, P. Introgression and genetic structure in northern Spanish Atlantic salmon (*Salmo salar* L.) populations according to mtDNA data. *Conserv. Genet.* **2008**, *9*, 157–169. [CrossRef]

24. Palstra, F.P.; O'Connell, M.F.; Ruzzante, D.E. Population structure and gene flow reversals in Atlantic salmon (*Salmo salar*) over contemporary and long-term temporal scales: Effects of population size and life history. *Mol. Ecol.* **2007**, *16*, 4504–4522. [CrossRef]

25. Kaczmarczyk, D. Techniques based on the polymorphism of microsatellite DNA as tools for conservation of endangered populations. *Appl. Ecol. Environ. Res.* **2019**, *17*, 1599–1615. [CrossRef]

26. Wenne, R. Microsatellites as molecular markers with applications in exploitation and conservation of aquatic animal populations. *Genes* **2023**, *14*, 808. [CrossRef]

27. Gilbey, J.; Coughlan, J.; Wennevik, V.; Prodöhl, P.; Stevens, J.R.; García de Leániz, C.; Verspoor, E. A microsatellite baseline for genetic stock identification of European Atlantic salmon (*Salmo salar* L.). *ICES J. Mar. Sci.* **2018**, *75*, 662–674. [CrossRef]

28. Yue, S.; Wang, C.Y. The Mann-Kendall test modified by effective sample size to detect trend in serially correlated hydrological series. *Water Resour. Manag.* **2004**, *18*, 201–218. [CrossRef]

29. Patakamuri, S.K.; O'Brien, N. Modifiedmk: Modified Versions of Mann Kendall and Spearman's Rho Trend Tests, R Package Version 1.6. 2022. Available online: https://cran.r-project.org/web/packages/modifiedmk/ (accessed on 9 May 2023).

30. Webster, R. Automatic soil-boundary location from transect data. *J. Int. Assoc. Math. Geol.* **1973**, *5*, 27–37. [CrossRef]

31. O'Reilly, P.T.; Hamilton, L.C.; McConnell, S.K.; Wright, J.M. Rapid analysis of genetic variation in Atlantic salmon (*Salmo salar*) by PCR multiplexing of dinucleotide and tetranucleotide microsatellites. *Can. J. Fish. Aquat. Sci.* **1996**, *53*, 2292–2298. [CrossRef]

32. Slettan, A.; Olsaker, I.; Lie, Ø. Atlantic salmon, *Salmo salar*, microsatellites at the SSOSL25, SSOSL85, SSOSL311, SSOSL417 loci. *Anim. Genet.* **1995**, *26*, 281–282. [CrossRef] [PubMed]

33. Slettan, A.; Olsaker, I.; Lie, Ø. Polymorphic Atlantic salmon, *Salmo salar* L. microsatellites at the SSOSL438, SSOSL439 and SSOSL444 loci. *Anim. Genet.* **1996**, *27*, 57–58. [CrossRef] [PubMed]

34. Sánchez, J.A.; Clabby, C.; Ramos, D.; Blanco, G.; Flavin, F.; Vázquez, E.; Powell, R. Protein and microsatellite single locus variability in *Salmo salar* L. (*Atlantic salmon*). *Heredity* **1996**, *77*, 423–432. [CrossRef] [PubMed]

35. Paterson, S.; Piertney, S.B.; Knox, D.; Gilbey, J.; Verspoor, E. Characterization and PCR multiplexing of novel highly variable tetranucleotide Atlantic salmon (*Salmo salar* L.) microsatellites. *Mol. Ecol. Notes* **2004**, *4*, 160–162. [CrossRef]
36. Van Oosterhout, C.; Hutchinson, W.F.; Wills, D.P.; Shipley, P. MICRO-CHECKER: Software for identifying and correcting genotyping errors in microsatellite data. *Mol. Ecol. Notes* **2004**, *4*, 535–538. [CrossRef]
37. Dabrowski, M.J.; Pilot, M.; Kruczyk, M.; Zmihorski, M.; Umer, H.M.; Gliwicz, J. Reliability assessment of null allele detection: Inconsistencies between and within different methods. *Mol. Ecol.* **2014**, *14*, 361–373. [CrossRef]
38. Kalinowski, S.T.; Taper, M.L.; Marshall, T.C. Revising how the computer program CERVUS accommodates genotyping error increases success in paternity assignment. *Mol. Ecol.* **2007**, *16*, 1099–1106. [CrossRef]
39. Kalinowski, S.T.; Taper, M.L. Maximum likelihood estimation of the frequency of null alleles at microsatellite loci. *Conserv. Genet.* **2006**, *7*, 991–995. [CrossRef]
40. Rousset, F. GENEPOP'007: A complete reimplementation of the GENEPOP software for Windows and Linux. *Mol. Ecol. Resour.* **2008**, *8*, 103–106. [CrossRef] [PubMed]
41. Belkhir, K.; Borsa, P.; Chikhi, L.; Raufaste, N.; Bonhomme, F. GENETIX 4.05, logiciel sous Windows TM pour la génétique des populations. In *Laboratoire Génome, Populations, Interactions, CNRS UMR 5171*; Université de Montpellier II: Montpellier, France, 2004.
42. Weir, B.S.; Cockerham, C.C. Estimating *F*-statistics for the analysis of population structure. *Evolution* **1984**, *38*, 1358–1370. [CrossRef]
43. Goudet, J. FSTAT (Version 1.2): A computer program to calculate *F*-statistics. *J. Hered.* **1995**, *86*, 485–486. [CrossRef]
44. Peakall, R.; Smouse, P.E. GENALEX 6: Genetic analysis in Excel. Population genetic software for teaching and research. *Mol. Ecol. Notes* **2006**, *6*, 288–295. [CrossRef]
45. Pritchard, J.K.; Stephens, M.; Donnelly, P. Inference of population structure using multilocus genotype data. *Genetics* **2000**, *155*, 945–959. [CrossRef]
46. Evanno, G.; Regnaut, S.; Goudet, J. Detecting the number of clusters of individuals using the software STRUCTURE: A simulation study. *Mol. Ecol.* **2005**, *14*, 2611–2620. [CrossRef]
47. Earl, D.A.; von Holdt, B.M. STRUCTURE HARVESTER: A website and program for visualizing STRUCTURE output and implementing the Evanno method. *Conserv. Genet. Resour.* **2012**, *4*, 359–361. [CrossRef]
48. Jakobsson, M.; Rosenberg, N.A. CLUMPP: A cluster matching and permutation program for dealing with label switching and multimodality in analysis of population structure. *Bioinformatics* **2007**, *23*, 1801–1806. [CrossRef]
49. Rosenberg, N.A. DISTRUCT: A program for the graphical display of population structure. *Mol. Ecol. Notes* **2004**, *4*, 137–138. [CrossRef]
50. Rousset, F. Genetic differentiation and estimation of gene flow from *F*-statistics under isolation by distance. *Genetics* **1997**, *145*, 1219–1228. [CrossRef]
51. King, T.L.; Kalinowski, S.T.; Schill, W.B.; Spidle, A.P.; Lubinski, B.A. Population structure of Atlantic salmon (*Salmo salar* L.): A range-wide perspective from microsatellite DNA variation. *Mol. Ecol.* **2001**, *10*, 807–821. [CrossRef]
52. Thioulouse, J.; Chessel, D.; Dolédec, S.; Olivier, J. ADE-4: A multivariate analysis and graphical display software. *Stat. Comput.* **1997**, *7*, 75–83. [CrossRef]
53. Chessel, D.; DuFour, A.B.; Thioulouse, J. The ade4 Package—I: One-table Methods. *R News* **2004**, *4*, 5–10.
54. Oksanen, J.; Blanchet, F.; Guillaume, F.M.; Kindt, R.; Legendre, P.; McGlinn, D.; Wagner, H. Vegan: Community Ecology Package. R Package (Version 2.5-2). 2008. Available online: https://CRAN.R-project.org/package=vegan (accessed on 5 May 2023).
55. Cornuet, J.M.; Luikart, G. Description and power analysis of two tests for detecting recent population bottlenecks from allele frequency data. *Genetics* **1996**, *144*, 2001–2014. [CrossRef] [PubMed]
56. Piry, S.; Luikart, G.; Cornuet, J.M. Computer note. BOTTLENECK: A computer program for detecting recent reductions in the effective size using allele frequency data. *J. Hered.* **1999**, *90*, 502–503. [CrossRef]
57. Di Rienzo, A.; Peterson, A.C.; Garza, J.C.; Valdes, A.M.; Slatkin, M.; Freimer, N.B. Mutational processes of simple-sequence repeat loci in human populations. *Proc. Natl. Acad. Sci. USA* **1994**, *91*, 3166–3170. [CrossRef] [PubMed]
58. Garza, J.C.; Williamson, E. Detection of reduction in population size using data from microsatellite loci. *Mol. Ecol.* **2001**, *10*, 305–318. [CrossRef]
59. Williamson-Natesan, E.G. Comparison of methods for detecting bottlenecks from microsatellite loci. *Conserv. Genet.* **2005**, *6*, 551–562. [CrossRef]
60. Perry, M.Z.; Kirby, R.; Reid, B.N.; Stoelting, R.; Doucet-Bëer, E.; Robinson, S.; Palsbøll, P.J. Reliability of genetic bottleneck tests for detecting recent population declines. *Mol. Ecol.* **2012**, *21*, 3403–3418. [CrossRef] [PubMed]
61. Wilson, G.A.; Rannala, B. Bayesian inference of recent migration rates using multilocus genotypes. *Genetics* **2003**, *163*, 1177–1191. [CrossRef] [PubMed]
62. Rannala, B. *BayesAss Edition 3.0 User's Manual*; University of California: Davis, CA, USA, 2011; Available online: https://www.rannala.org/?page_id=245 (accessed on 5 May 2023).
63. Beerli, P. Comparison of Bayesian and maximum likelihood inference of population genetic parameters. *Bioinformatics* **2006**, *22*, 341–345. [CrossRef]
64. Beerli, P.; Palczewski, M. Unified framework to evaluate panmixia and migration direction among multiple sampling locations. *Genetics* **2010**, *185*, 313–326. [CrossRef]

65. Rannala, B.; Mountain, J.L. Detecting immigration by using multilocus genotypes. *Proc. Natl. Acad. Sci. USA* **1997**, *94*, 9197–9201. [CrossRef]

66. Piry, S.; Alapetite, A.; Cornuet, J.M.; Paetkau, D.; Baudouin, L.; Estoup, A. GENECLASS2: A software for genetic assignment and first-generation migrant detection. *J. Hered.* **2004**, *95*, 536–539. [CrossRef] [PubMed]

67. Paetkau, D.; Slade, R.; Burden, M.; Estoup, A. Genetic assignment methods for the direct, real-time estimation of migration rate: A simulation-based exploration of accuracy and power. *Mol. Ecol.* **2004**, *13*, 55–65. [CrossRef] [PubMed]

68. Do, C.; Waples, R.S.; Peel, D.; Macbeth, G.M.; Tillett, B.J.; Ovenden, J.R. NeEstimator v2: Re-implementation of software for the estimation of contemporary effective population size (N_e) from genetic data. *Mol. Ecol. Resour.* **2014**, *14*, 209–214. [CrossRef] [PubMed]

69. Gómez-Uchida, D.; Knight, T.W.; Ruzzante, D.E. Interaction of landscape and life history attributes on genetic diversity, neutral divergence and gene flow in a pristine community of salmonids. *Mol. Ecol.* **2009**, *18*, 4854–4869. [CrossRef] [PubMed]

70. Waples, R.S.; England, P.R. Estimating contemporary effective population size on the basis of linkage disequilibrium in the face of migration. *Genetics* **2011**, *189*, 633–644. [CrossRef] [PubMed]

71. Gilbert, K.J.; Whitlock, M.C. Evaluating methods for estimating local effective population size with and without migration. *Evolution* **2015**, *69*, 2154–2166. [CrossRef] [PubMed]

72. Whiteley, A.R.; Coombs, J.A.; Donnell, M.J.O.; Nislow, K.H.; Letcher, B.H. Keeping things local: Subpopulation N_b and N_e in a stream network with partial barriers to fish migration. *Evol. Appl.* **2017**, *10*, 348–365. [CrossRef]

73. Bernos, T.A.; Yates, M.C.; Fraser, D.J. Fine-scale differences in genetic and census population size ratios between two stream fishes. *Conserv. Genet.* **2018**, *9*, 265–274. [CrossRef]

74. Waples, R.S.; Antao, T.; Luikart, G. Effects of overlapping generations on linkage disequilibrium estimates of effective population size. *Genetics* **2014**, *197*, 769–780. [CrossRef]

75. Borrell, Y.J.; Bernardo, D.; Blanco, G.; Vázquez, E.; Sánchez, J.A. Spatial and temporal variation of genetic diversity and estimation of effective population sizes in Atlantic salmon (*Salmo salar* L.) populations from Asturias (Northern Spain) using microsatellites. *Conserv. Genet.* **2008**, *9*, 807–819. [CrossRef]

76. Hórreo, J.L.; Machado-Schiaffino, G.; Griffiths, A.M.; Bright, D.; Stevens, J.R.; García-Vázquez, E. Atlantic salmon at risk: Apparent rapid declines in effective population size in southern European populations. *Trans. Am. Fish. Soc.* **2011**, *140*, 605–610. [CrossRef]

77. Valiente, A.G.; Beall, E.; García-Vázquez, E. Population genetics of south European Atlantic salmon under global change. *Glob. Chang. Biol.* **2010**, *16*, 36–47. [CrossRef]

78. Säisä, M.; Koljonen, M.L.; Gross, R.; Nilsson, J.; Tähtinen, J.; Koskiniemi, J.; Vasemägi, A. Population genetic structure and postglacial colonization of Atlantic salmon (*Salmo salar*) in the Baltic Sea area based on microsatellite DNA variation. *Can. J. Fish. Aquat. Sci.* **2005**, *62*, 1887–1904. [CrossRef]

79. Finnegan, A.K.; Griffiths, A.M.; King, R.A.; Machado-Schiaffino, G.; Porcher, J.P.; García-Vázquez, E.; Stevens, J.R. Use of multiple markers demonstrates a cryptic western refugium and postglacial colonisation routes of Atlantic salmon (*Salmo salar* L.) in northwest Europe. *Heredity* **2013**, *111*, 34–43. [CrossRef] [PubMed]

80. Valiquette, E.; Perrier, C.; Thibault, I.; Bernatchez, L. Loss of genetic integrity in wild lake trout populations following stocking: Insights from an exhaustive study of 72 lakes from Québec, Canada. *Evol. Appl.* **2014**, *7*, 625–644. [CrossRef] [PubMed]

81. Gudmundsson, L.A.; Gudjónsson, S.; Marteinsdóttir, G.; Scarnecchia, D.L.; Daníelsdóttir, A.K.; Pampoulie, C. Spatio-temporal effects of stray hatchery-reared Atlantic salmon *Salmo salar* on population genetic structure within a 21 km-long Icelandic river system. *Conserv. Genet.* **2013**, *14*, 1217–1231. [CrossRef]

82. Griffiths, A.M.; Machado-Schiaffino, G.; Dillane, E.; Coughlan, J.; Hórreo, J.L.; Bowkett, A.E.; McGinnity, P. Genetic stock identification of Atlantic salmon (*Salmo salar*) populations in the southern part of the European range. *BMC Genet.* **2010**, *11*, 31. [CrossRef] [PubMed]

83. Vähä, J.P.; Erkinaro, J.; Falkegård, M.; Orell, P.; Niemelä, E. Genetic stock identification of Atlantic salmon and its evaluation in a large population complex. *Can. J. Fish. Aquat. Sci.* **2016**, *74*, 327–338. [CrossRef]

84. Vähä, J.P.; Erkinaro, J.; Niemelä, E.; Primmer, C.R. Temporally stable genetic structure and low migration in an Atlantic salmon population complex: Implications for conservation and management. *Evol. Appl.* **2008**, *1*, 137–154. [CrossRef]

85. Fraser, D.J.; Lippe, C.; Bernatchez, L. Consequences of unequal population size, asymmetric gene flow and sex-biased dispersal on population structure in brook charr (*Salvelinus fontinalis*). *Mol. Ecol.* **2004**, *13*, 67–80. [CrossRef]

86. Manier, M.K.; Arnold, S.J. Population genetic analysis identifies source-sink dynamics for two sympatric garter snake species (*Thamnophis elegans* and *Thamnophis sirtalis*). *Mol. Ecol.* **2005**, *14*, 3965–3976. [CrossRef]

87. Hansen, M.M.; Skaala, O.; Jensen, L.F.; Bekkevold, D.; Mensberg, K.L.D. Gene flow, effective population size and selection at major histocompatibility complex genes: Brown trout in the Hardanger Fjord, Norway. *Mol. Ecol.* **2007**, *16*, 1413–1425. [CrossRef]

88. Bradbury, I.R.; Hamilton, L.C.; Robertson, M.J.; Bourgeois, C.E.; Mansour, A.; Dempson, J.B. (2014). Landscape structure and climatic variation determine Atlantic salmon genetic connectivity in the Northwest Atlantic. *Can. J. Fish. Aquat. Sci.* **2014**, *71*, 246–258. [CrossRef]

89. Sexton, J.P.; Hangartner, S.B.; Hoffman, A.A. Genetic isolation by environment or distance: Which pattern of gene flow is most common? *Evolution* **2013**, *68*, 1–15. [CrossRef]

90. Bowlby, H.D.; Fleming, I.A.; Gibson, A.J.F. Applying landscape genetics to evaluate threats affecting endangered Atlantic salmon populations. *Conserv. Genet.* **2016**, *17*, 823–838. [CrossRef]

91. Lehnert, S.J.; Kess, T.; Bentzen, P.; Kent, M.P.; Lien, S.; Gilbey, J.; Clément, M.; Jeffery, N.W.; Waples, R.S.; Bradbury, I.R. Genomic signatures and correlates of widespread population declines in salmon. *Nat. Commun.* **2019**, *10*, 2996. [CrossRef] [PubMed]
92. Frankham, R.; Bradshaw, C.J.A.; Brook, B.W. Genetics in conservation management: Revised recommendations for the 50/500 rules, red list criteria and population viability analyses. *Biol. Conserv.* **2014**, *170*, 56–63. [CrossRef]

 *fishes*

Article

High-Temperature Stress Induces Autophagy in Rainbow Trout Skeletal Muscle

Alfredo Molina [1,2,3], Phillip Dettleff [4], Valentina Valenzuela-Muñoz [2,5], Cristian Gallardo-Escarate [2,5] and Juan Antonio Valdés [1,2,3,*]

[1] Laboratorio de Biotecnología Molecular, Facultad de Ciencias de la Vida, Universidad Andres Bello, Santiago 8370035, Chile; amolina@unab.cl

[2] Interdisciplinary Center for Aquaculture Research (INCAR), Universidad de Concepción, Concepción 4030000, Chile; valevalenzuela@udec.cl (V.V.-M.); crisgallardo@udec.cl (C.G.-E.)

[3] Centro de Investigación Marina Quintay (CIMARQ), Universidad Andres Bello, Quintay 2340000, Chile

[4] Escuela de Medicina Veterinaria, Facultad de Agronomía e Ingeniería Forestal, Facultad de Ciencias Biológicas y Facultad de Medicina, Pontificia Universidad Católica de Chile, Santiago 7820436, Chile; phillip.dettleff@uc.cl

[5] Laboratory of Biotechnology and Aquatic Genomics, Department of Oceanography, University of Concepción, Concepción 4030000, Chile

* Correspondence: jvaldes@unab.cl; Tel.: +56-2-26615815

Abstract: Ectothermic animals, such as teleosts, have increasingly been exposed to stressful high-temperature events due to global warming. Currently, the effects of thermal stress on skeletal muscle, a key tissue for fish growth, are unknown. This study examined the impact of high-temperature stress on the skeletal muscle transcriptome of rainbow trout (*Oncorhynchus mykiss*) in control (15 °C) and high-temperature (20 °C) conditions. Additionally, we examined the plasmatic levels of cortisol, glucose, and creatine kinase activity, and examined oxidative damage and autophagy activation in skeletal muscle. High-temperature stress induced significant increases in cortisol and glucose plasmatic levels. Nevertheless, no changes were observed in creatine kinase activity in plasma and skeletal muscle oxidation. Skeletal muscle RNA was isolated and sequenced using the HiSeq Illumina platform. A total of 383,796,290 reads were mapped onto the reference rainbow trout genome. The transcriptomic analysis showed that 293 genes were upregulated in the high-temperature group, mainly associated with autophagosome assembly, amino acid transport, and the glutamine metabolic process. On the other hand, 119 genes were downregulated in the high-temperature group, mainly associated with digestion, proteolysis, and the muscle contraction process. In addition, RT-qPCR of differentially expressed representative genes and Western blot analysis of LC3-II/LC3-I levels confirmed skeletal muscle autophagy induced by high temperature. This study sheds light on intriguing facets of the adaptive response of rainbow trout skeletal muscle to high-temperature stress and provides significant insights into the physiology of autophagy in teleosts.

Keywords: cortisol; high-temperature stress; autophagy; RNA-Seq; skeletal muscle

Key Contribution: This study evaluates the effects of high-temperature stress on the skeletal muscle transcriptome of rainbow trout (*O. mykiss*) and contributes to a better understanding of the potential role of autophagy as a negative regulator of skeletal muscle atrophy.

Citation: Molina, A.; Dettleff, P.; Valenzuela-Muñoz, V.; Gallardo-Escarate, C.; Valdés, J.A. High-Temperature Stress Induces Autophagy in Rainbow Trout Skeletal Muscle. *Fishes* **2023**, *8*, 303. https://doi.org/10.3390/fishes8060303

Academic Editor: Qiuning Liu

Received: 2 May 2023
Revised: 28 May 2023
Accepted: 2 June 2023
Published: 6 June 2023

1. Introduction

Global warming refers to the ongoing rise in the average global temperature of the Earth [1]. This phenomenon is exerting an impact on animal populations across the globe, primarily through chronic temperature increases and a heightened incidence of heat-waves [2]. Increases in atmospheric temperatures are reflected in higher temperatures of marine and freshwater habitats, home to a great diversity of living organisms [3]. It is projected that, by the end of this century, the mean temperature of the ocean will have risen by 1–4 °C, which will exert negative effects on the physiology of aquatic organisms [3]. Additionally, changes in ocean temperatures will influence the intensity and frequency of phenomena on the Pacific coast known as El Niño–Southern Oscillation (ENSO) [4]. Given that fishes are ectothermic organisms, and their body temperature is equivalent to that of the surrounding water environment, teleosts are particularly vulnerable to the effects of global warming [5].

The physiological mechanisms that are essential for mounting an adequate response to temperature stress entail the activation of a well-coordinated network of neuroendocrine pathways, including the brain–sympathetic–chromaffin (BSC) axis and the hypothalamic–pituitary–interrenal (HPI) axis, which are responsible for the production of catecholamines and cortisol, respectively [6]. The secretion of these hormones plays a critical role in regulating energy metabolism and maintaining an organism's homeostasis [7]. Specifically, the cortisol-mediated response in teleosts plays a fundamental role in facilitating energetic adaptation to temperature stress by enabling long-term glucose metabolism via gluconeogenesis and protein catabolism [8].

Skeletal muscle is considered to be one of the primary tissues impacted by the stress induced by rising temperatures [8]. Skeletal muscles make up approximately 60% of the body mass in teleosts, playing a crucial role in their locomotion, metabolism, and growth [9]. The growth of skeletal muscle is a multifaceted process that is coregulated by mechanisms associated with myoblast proliferation (hyperplasia), increase in muscular fiber size through protein synthesis (hypertrophy), and muscle protein degradation (atrophy) [10]. Although several studies have described the physiological and molecular effects of temperature stress in fish tissues, very few have focused on skeletal muscles [11]. In a recent study conducted on a marine teleost species, red cusk-eel (*Genypterus chilensis*), it was found that thermal stress leads to skeletal muscle oxidation and atrophy [12]. Additionally, RNA-Seq assays revealed that high-temperature stress induced the expression of various autophagy-associated genes in the liver of the species [13]. Autophagy is a fundamental cellular mechanism that plays an essential role in energetic catabolism and the lysosome-mediated degradation of cell components. It is a highly conserved process that allows the cell to recycle damaged or unwanted organelles and proteins [14]. In mammalian skeletal muscle, autophagy has been described as an essential mechanism for maintaining the structure and proper functioning of this tissue [15]. Skeletal muscle with an impaired autophagy process can be affected by myopathy disorder, which is related to excessive protein accumulation in the muscle cells [16]. However, very little is known about autophagy in teleost skeletal muscles and its regulation by environmental variables such as temperature [17]. In this study, we evaluated the effects of high-temperature stress on global gene expression in the skeletal muscle of rainbow trout to comprehend the impact of temperature stress on skeletal muscle and predict the influence of global warming on the adaptive capability of fish during their development. The transcriptomic information obtained here enabled us to identify potential genes and signaling pathways associated with the high-temperature stress response in rainbow trout.

2. Materials and Methods

2.1. Experimental Thermal Stress Protocol

Juvenile rainbow trout (*Oncorhynchus mykiss*) (1 year old; 13.22 g $\pm$ 1.34; total n = 20) were obtained from Pisciculture Rio Blanco (V region, Chile). Fish were maintained under natural temperatures and light:dark photoperiod conditions (15 °C $\pm$ 1 °C and L:D 12:12), in aerated dechlorinated water, with water turnover of 0.5 L min^{-1} and fed daily with commercial pellets containing 45% protein, 22% lipids, 16% carbohydrates, and 17% other components. Fish were acclimatized for 1 week before the trial and exposed to a thermal stress protocol previously described [12]. Briefly, this protocol consists of increasing the temperature over 24 h at a rate of 1 °C in 5 h. During the remaining 4 days of the trial, the control temperature was maintained at 15 $\pm$ 1 °C (control group; n = 5) and the stress temperature at $20 \pm 1\ ^\circ$C (stress group; n = 5). Biological replicates of the trials were included. On the fifth day, five individuals per group were sampled. For plasma isolation, blood samples were collected and centrifuged at $5000\times g$ for 10 min and stored at $-80\ ^\circ$C. Finally, sample individuals were euthanized by an overdose of anesthetic (benzocaine, 300 mg/L). Skeletal muscle was collected, frozen in liquid nitrogen, and maintained at $-80\ ^\circ$C until further analysis.

2.2. Cortisol, Creatine Kinase Activity, and Glucose Quantification in Blood Plasma

The cortisol plasmatic concentration was quantified using the Cayman Cortisol ELISA Kit (Cayman Chemical, Ann Arbor, MI, USA; catalog number 500360). The plasmatic activity of creatine kinase (CK) was quantified using the Abcam Creatine Kinase Activity Assay Kit (Abcam, Cambridge, UK; catalog number 155901). The glucose plasmatic concentration was quantified using the Abcam Glucose Uptake Assay Kit (Abcam, Cambridge, UK; catalog number 136955). The use of these kits in rainbow trout has been previously verified [18].

2.3. DNA and Protein Oxidative Damage in Skeletal Muscle

DNA oxidative damage and protein carbonylation were determined using the commercially available kits OxiSelect Oxidative DNA Damage Quantification (catalog number STA-320) and OxiSelect Protein Carbonyl Spectrophotometric Assay (catalog number STA-310) (Cell Biolabs, San Diego, CA, USA), respectively, following the manufacturer's instructions. For further details, see Rivas-Aravena et al. [18].

2.4. Skeletal Muscle RNA Extraction and Sequencing

Total RNA was extracted from the skeletal muscles of both the control and stress groups using the EZNA Total RNA Kit II (OMEGA Bio-Tek Inc., Norcross, GA, USA). The RNA concentration was measured using a Qubit 2.0 Fluorometer (Life Technology, Carlsbad, CA, USA), and RNA integrity was confirmed by capillary electrophoresis using a Fragment Analyzer Automated CE System (Advanced Analytical Technologies, Inc., Ankeny, IA, USA). The ratio of absorbance at 260 nm and 280 nm was used to assess the purity of RNA. Samples with RQN values of at least 9 were selected for further analysis. The construction of cDNA libraries was carried out with 1 µg of RNA using the TruSeq RNA Sample Preparation Kit v2 (Illumina, San Diego, CA, USA). Libraries were sequenced with the Hiseq technology (Illumina) at Macrogen (Seoul, Republic of Korea) using a paired-end technique (2×150 bp).

2.5. RNA-Seq and GO Analysis

The sequencing reads were processed to remove sequences of low quality (Q20) and those less than 30 bp in length. To detect differentially expressed genes (DEGs), the reads were mapped to the last version of rainbow trout (*O. mykiss*) reference genome by CLC Genomics Workbench 9.0 (Qiagen, Germantown, MD USA), using default parameters. Gene expression levels were estimated using the RPKM value (reads per kilobase per million mapped reads). Genes exhibiting a fold-change value greater than 2.0 and a false discovery rate (FDR) *p*-value less than 0.05 were considered as differentially expressed. The identification of DAVID GO and KEGG enrichment analysis of DEGs was previously described [19].

2.6. RNA-Seq Validation by Real-Time RT-qPCR

All qPCR assays followed MIQE guidelines [20]. Preserved skeletal muscles from each sampled fish were homogenized, and total RNA was extracted using TRIzol reagent (Invitrogen, Carlsbad, CA, USA). Isolated RNAs with A260/280 ratios between 1.9 and 2.0 were selected for further processing. Next, 1 µg of RNA from each sample was reverse transcribed into cDNA using the ImProm-II Reverse Transcription System (Promega, WI, USA). Real-time qPCR was performed using a Stratagene MX3000P qPCR system (Stratagene, La Jolla, CA, USA) following the procedure described by Rivas-Aravena et al. [18]. The list of primers used in this study is provided in Supplementary Table S1. The housekeeping genes used for normalization were β-actin (*actβ*) and 40S ribosomal protein S30 (*fau*). These genes were previously obtained by using the geNorm program, which obtained the normalization factor and subsequent relative expression levels [18].

2.7. LC3 Western Blot Analysis

To validate the effects of high-temperature stress on the induction of autophagy, we analyzed the levels of the microtubule-associated proteins 1A/1B light chain 3B (LC3) by Western blot. To extract skeletal muscle proteins, 0.1 g of tissue was homogenized in 1 mL of lysis buffer containing 50 mM Tris-HCl (pH 7.4), 150 mM NaCl, 1 mM EDTA, 1% NP-40, and a protease inhibitor cocktail (Calbiochem, Billerica, MA, USA). Pierce BCA Protein Assay Kit (Thermo Scientific, Rockford, IL, USA) was used for protein concentration measurement. Then, 50 g of proteins was resolved in SDS-PAGE and analyzed using the Western blot procedure described by Rivas-Aravena et al. [18]. Antibodies against LC3 (catalog number 12741; dilution 1:2000) and β-actin (catalog number 4967; dilution 1:5000) were obtained from Cell Signaling Technology (Danvers, MA, USA). After incubation for 1 h with HRP-conjugated secondary antibodies (dilution 1:2000), membranes were developed by enhanced chemiluminescence (Amersham Biosciences, Amersham, UK). The films were scanned and densitometric analysis was carried out with ImageJ [21].

2.8. Statistical Analysis

Based on the raw data, the mean and standard error of the mean (±SEM) were calculated for each indicator. Differences in means among the groups were assessed using one-way ANOVA, followed by Bonferroni's post hoc test for multiple comparisons. All statistical analyses were performed using GraphPad Prism v.8.0 software (GraphPad Software Inc., La Jolla, CA, USA).

3. Results

3.1. Cortisol, Glucose, and Creatine Kinase Activity Quantification in Plasma, and Oxidation in Skeletal Muscle Tissue

Blood plasma cortisol and glucose levels significantly increased after five days of exposure to high temperature in the stressed group (Figure 1a,b). No significant differences in plasma creatine kinase activity were observed between the control and stressed groups (Figure 1c).

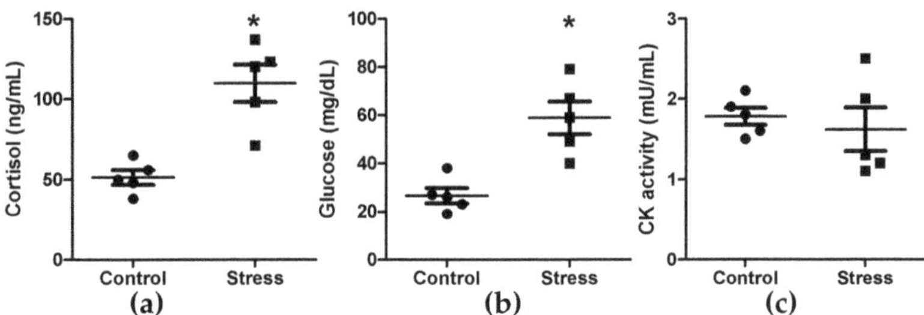

Figure 1. Levels of cortisol, glucose, and creatine kinase activity in plasma. (**a**) cortisol (**b**) glucose (**c**) creatine kinase in blood plasma were assessed in juvenile rainbow trout kept under high-temperature (20 °C) stress and optimal temperature (15 °C) regime. The results are expressed as mean and standard error of the mean (±SEM, n = 5 per treatment). Differences between control and stress groups are shown by * $p < 0.05$.

To evaluate the skeletal muscle oxidation induced by high temperature, we measured protein and DNA oxidative damage. Thermal stress did not induce protein carbonylation and DNA oxidation as compared with the control group (Figure 2a,b).

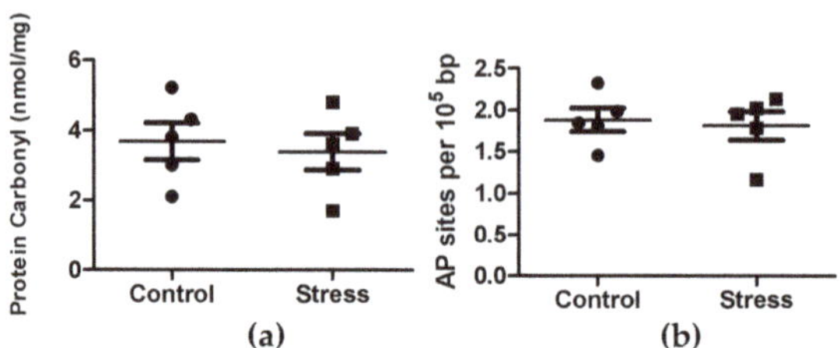

Figure 2. Oxidative damage quantification in skeletal muscle. (**a**) Protein carbonylation (**b**) DNA oxidative damage in juvenile rainbow trout kept under high-temperature (20 °C) stress and optimal temperature (15 °C) regime. The results are expressed as mean and standard error of the mean (±SEM, n = 5 per treatment). No statistical differences between groups were detected.

3.2. Transcriptomic Analysis and Pathway Enrichment Analysis

To analyze the effect of high-temperature stress on global gene expression, we performed RNA-Seq analysis. A total of 383,796,290 trimmed reads were mapped in the rainbow trout reference genome, covering 85.9% of its size. Principal component analysis (PCA) revealed a high similarity in the biological replicates (Supplementary Figure S1). Differential expression analysis revealed that 293 genes were upregulated and 119 genes were downregulated under high-temperature stress (Supplementary Table S2). To analyze the biological role of the DEGs, GO term enrichment analysis was performed with the DAVID database. The upregulated DEGs were enriched in biological processes (BPs), such as autophagosome assembly, amino acid transport, and glutamine metabolic process (Figure 3). Gene Ontology (GO) terms for upregulated genes were assigned to RNA binding and nucleoplasm for molecular function (MF) and cellular component (CC), respectively (Supplementary Tables S3 and S4). Among KEGG pathways, mitophagy—animal, autophagy—animal, and spinocerebellar ataxia were over-represented (Table 1).

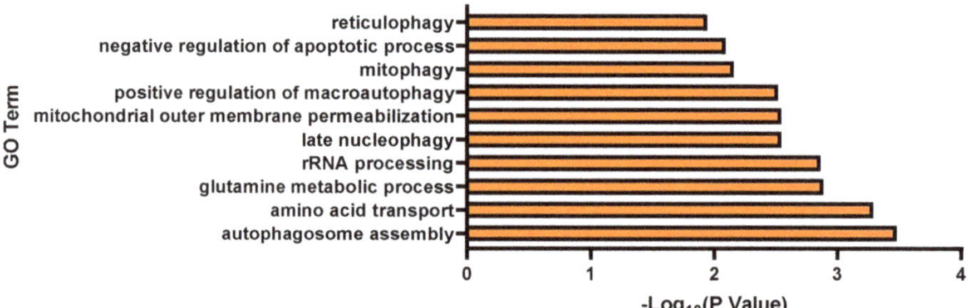

Figure 3. Gene enrichment analysis of biological processes (BPs). The graph indicates the −log10(*p*-value) enriched BPs of upregulated genes between the control and stress groups with *p*-values < 0.05.

The downregulated differentially expressed genes (DEGs) were enriched in biological processes (BPs) such as digestion, proteolysis, and muscle contraction (Figure 4). The Gene Ontology (GO) terms for downregulated genes were assigned to serine-type endopeptidase activity and cytosol for molecular function (MF) and cellular component (CC), respectively (Supplementary Tables S5 and S6). The KEGG pathways assigned to the differentially expressed genes (DEGs) included pancreatic secretion, protein digestion and absorption, and adrenergic signaling in cardiomyocytes (Table 2).

Table 1. Enriched KEGG pathways of upregulated DEGs in rainbow trout skeletal muscles.

KEGG Pathway	p-Value	Upregulated Genes
Mitophagy—animal	1.45×10^{-4}	*mapk10, bnip3l, bnip3, ubc, atg9a, ulk1, rab7a*
Autophagy—animal	9.89×10^{-4}	*mapk10, bnip3, atg9a, ulk1, raf1, rab7a, atg4d, atg2b*
Spinocerebellar ataxia	1.07×10^{-3}	*mapk10, por, psmd11, psmd2, psmd3, atp2a2, ulk1, atg2b*
Protein processing in ER	2.99×10^{-3}	*mapk10, hsp90ab1, hspa1l, canx, cul1, plaa, cryaa*
Alzheimer's disease	1.05×10^{-2}	*mapk10, gsk3b, por, psmd11, cdk5, psmd2, psmd3, atp2a2, ulk1, raf1, atg2b*
Antigen processing	1.12×10^{-2}	*hsp90ab1, hspa1l, hspa4, canx, rfxap*
ErbB signaling pathway	1.50×10^{-2}	*mapk10, map2k4, gsk3b, myc, raf1*
Pathways of neurodegeneration	1.70×10^{-2}	*mapk10, gsk3b, por, psmd11, cdk5, psmd2, ubc, psmd3, atp2a2, ulk1, raf1, atg2b*
Legionellosis	2.58×10^{-2}	*hspa1l, rab1b, bnip3, eef1a2*
mTOR signaling pathway	3.05×10^{-2}	*gsk3b, cab39, ulk1, raf1, lpin1, wdr24*

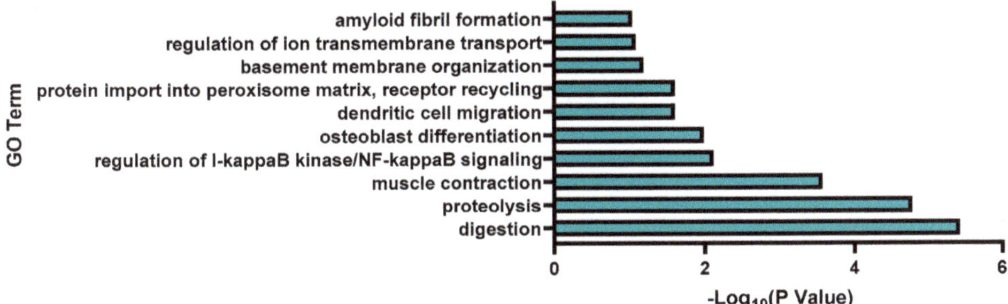

Figure 4. Gene enrichment analysis of biological processes (BPs). The graph indicates the $-\log_{10}(p\text{-value})$ enriched BPs of downregulated genes between the control and stress groups with p-values of <0.05.

Table 2. Enriched KEGG pathways of downregulated DEGs in rainbow trout skeletal muscle.

KEGG Pathway	p-Value	Downregulated Genes
Pancreatic secretion	1.10×10^{-10}	*prss1, cela2a, cpb1, ctrb2, ctrb1, amy1c, atp2a1, cel, prss3, prss2*
Protein digestion and absorption	2.35×10^{-6}	*prss1, cela2a, cpb1, ctrb2, ctrb1, prss3, prss2*
Adrenergic signaling in cardiomyocytes	2.59×10^{-3}	*cacnb1, tpm3, atp2a1, scn1b, myh7*
cGMP-PKG signaling pathway	3.82×10^{-3}	*atp2a1, vdac1, raf1, prkg1, myh7*
Influenza A	4.15×10^{-3}	*prss1, vdac1, raf1, prss3, prss2*
Cardiac muscle contraction	4.57×10^{-3}	*cacnb1, tpm3, atp2a1, myh7*
Hypertrophic cardiomyopathy	5.03×10^{-3}	*cacnb1, tpm3, atp2a1, myh7*
Dilated cardiomyopathy	6.02×10^{-3}	*cacnb1, tpm3, atp2a1, myh7*
Diabetic cardiomyopathy	4.37×10^{-2}	*atp5f1b, atp2a1, vdac1, sdha*
Chemical carcinogenesis—ROS	5.51×10^{-2}	*atp5f1b, vdac1, sdha, raf1*

3.3. RNA-Seq Result Validation by Real-Time RT-qPCR and Western Blot

For real-time RT-qPCR validation, we selected four upregulated genes associated with autophagy (*mapk10, bnip3, atg9a,* and *raf1*) and four downregulated genes related to protein digestion (*prss1, cela1, ctrb2,* and *prss2*) (Supplementary Figure S2). The gene expression fold-changes measured by these two methods (RNA-Seq and RT-qPCR) were highly correlated, with a significant R^2 value of 0.86 (p-value, 0.0001). Finally, to verify the presence of autophagy in skeletal muscle, we analyzed the LC3-II/LC3-I levels. We found that LC3-II protein was significantly upregulated in the stressed group (Figure 5a), measuring 2.1-fold that of control conditions (Figure 5b).

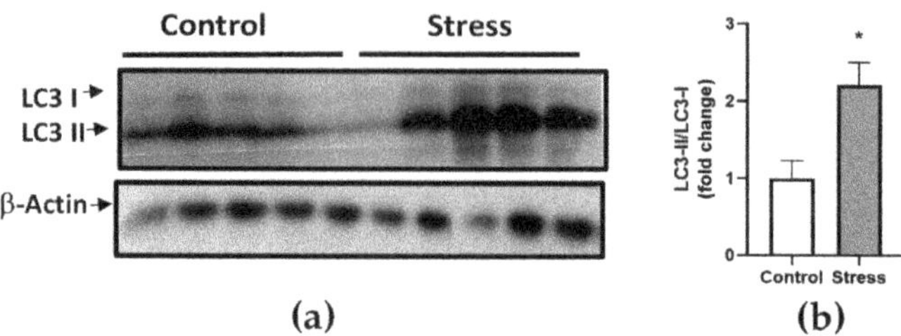

(a)   (b)

Figure 5. High-temperature stress induces autophagy in rainbow skeletal muscle. (**a**) Western blot of LC3 I/II and H2B. (**b**) Densitometric analysis of the Western blot showing an LC3 II/LC3 I ratio. The results are expressed as mean and standard error of the mean ($\pm$SEM, n = 5 per treatment). Differences between control and stress groups are shown by * $p < 0.05$.

4. Discussion

In the present study, we investigated the effects of high-temperature stress on the global gene expression response in the skeletal muscle of rainbow trout (*O. mykiss*). The effect of temperature stress on fish physiology is an important research focus in the face of ongoing global warming and its potential impact on freshwater and marine aquaculture [22]. The ability of fish to cope with temperature variations is related to their phenotypic plasticity, that is, different phenotypes triggered by variable environmental conditions. In this sense, the acclimatization process mediated by transcriptional changes is essential for the implementation of an ad hoc metabolic, neuroendocrine, and immunological response to climate change [23]. In the present study, we found that the applied stress protocol resulted in a significant increase in the blood plasma levels of cortisol and glucose, reaching concentrations of approximately 110 ng/mL and 59 mg/dL, respectively. These results are consistent with those previously reported to date in multiple other teleost fishes [24–28]. For instance, studies conducted on adult rainbow trout (*O. mykiss*) have shown that sudden increases in temperature can induce a similar rise in glucose and blood plasma cortisol levels [26–28]. This contrasts with studies where the temperature was gradually increased [29]. In a recently published article, it was determined that a similar protocol of gradually increasing temperature induced an increment in cortisol and plasma glucose levels in the red cusk-eel (*G. chilensis*), a marine teleost [12]. Considering that cortisol has been described as a potent catabolic hormone, inducing muscle atrophy in vertebrates [19,30], we quantified plasma creatine kinase and oxidative damage in muscle tissue as indicators of skeletal muscle atrophy. Interestingly, our results showed that high-temperature stress did not elevate the levels of blood plasma creatine kinase and oxidative damage markers in skeletal muscle tissue. This observation contrasts with the findings of thermal stress studies on red cusk-eel (*G. chilensis*), which reported an increase in skeletal muscle oxidative damage and an upregulation in the expression of several genes associated with muscle atrophy [12]. Similarly, in the African sharptooth catfish (*Clarias gariepinus*), it was reported that the long-term exposure to high-temperature stress has a negative consequence on skeletal muscle growth performance and structural integrity [31]. In Atlantic salmon (*Salmo salar*), thermal stress has been shown to cause skeletal muscle discoloration and loss of skeletal muscle integrity [32], while in gilt-head seabream (*Sparus aurata*), exposure to high temperatures induces an inflammatory and oxidative response in red muscle [33]. To gain further insights into the molecular mechanisms underlying these responses, we conducted RNA-Seq analysis in the present study.

Differential expression analysis revealed that upregulated genes were mainly associated with autophagosome assembly and mitophagy. The obtained results were validated by RT-qPCR analysis of selected genes and LC3 I/II Western blot, confirming that temperature stress induces autophagy in rainbow trout skeletal muscle. Autophagy is a catabolic mechanism that is fundamental for physiological balance and responsible for the delivery of cytoplasmic components to the lysosomes for digestion [34]. Autophagy is triggered by various stimuli, including lack of nutrients, reactive oxygen species (ROS), endoplasmic reticulum stress, and the presence of microorganisms [35]. Although various reports in teleosts have investigated autophagy in processes such as reproduction [36–38], hepatic metabolism [13,39,40], and immune response to infections [41,42], there is limited evidence of the role of autophagy in skeletal muscle catabolism. Studies of fine flounder (*Paralichthys adspersus*) have shown that stress induced by high-density farming can upregulate the expression of genes involved in the autophagy process, which is believed to be a protective mechanism against apoptosis regulated by the ubiquitin–proteasome pathway [43]. Similar findings were reported in red cusk-eel (*G. chilensis*), where it was demonstrated that exposure to handling stress increased the expression of genes associated with autophagy and the ubiquitin–proteasome pathway [44]. In rainbow trout (*O. mykiss*), it was determined that infectious pancreatic necrosis virus (IPNV) infection induced a dynamic response between autophagy and the proteasomal pathways in skeletal muscle [45]. Furthermore, autophagy-mediated skeletal muscle atrophy has been found to be induced by intensive exercise in zebrafish (*Danio rerio*) [46]. Among the genes with differential expression validated by RT-qPCR are *mapk10*, *bnip3*, *atg9a*, and *raf1*. *Mapk10*, also known as Jun Kinase 3 (JNK3), encodes for a serine/threonine-protein kinase involved in various processes such as cell differentiation, apoptosis, and autophagy [47]. In fish, its expression has been related to stress due to hypoxia and salinity in Asian seabass (*Lateolabrax maculatus*) [48]. *Bnip3*, also known as BCL2/adenovirus E1B protein-interacting protein 3, regulates apoptosis, modulating the permeability of the outer mitochondrial membrane [49]. It has also been described that *bnip3* is also a potent inducer of autophagy in many tissues [49]. Its overexpression in skeletal muscle induced by starvation was reported in the fine flounder (*P. adspersus*) [50]. *Atg9a*, also known as autophagy-related protein 9A, encodes for a lipid scramblase involved in autophagosomal membrane expansion, directly regulating autophagy [51]. Gene expression analysis showed that cadmium presence in water induced *atg9a* gene expression, positively modulating autophagy in the liver tissue of Chinese ink carp (*Procypris merus*) [40]. *Raf1* encodes for the RAF proto-oncogene serine/threonine-protein kinase, which acts as a critical regulator of autophagy and the link between the membrane-associated Ras GTPases and the MAPK/ERK signaling pathway [52]. In Chinese rare minnows (*Gobiocypris rarus*), the chemical compound Carbamazepine increases its expression, inducing DNA damage and apoptosis in the liver [53]. To our knowledge, there are no previous reports linking the expression of these genes with high-temperature stress and autophagy in lower vertebrates.

Interestingly, temperature-induced autophagy is a phenomenon that has been reported in mammalian skeletal muscle models. In wild boar (*Sus scrofa*), it has been reported that short-term temperature stress results in apoptosis and autophagy in skeletal muscle [54]. Similarly, it has been reported that in rat (*Rattus norvegicus*) skeletal muscle, the autophagy pathway is activated in response to temperature stress, as a compensatory mechanism for muscle atrophy induced by tendon cutting [55]. In mice (*Mus musculus*), temperature stress treatment rescues denervation-induced mitophagy (autophagy in mitochondria) and the consequent skeletal muscle atrophy [56]. Further, in cell models of mammalian skeletal muscle (C2C12 myotubes), it was reported that acute heat exposure induced autophagy resulting in an elevation in AMPK, Beclin-1, and LC3 II levels, similar to our observations [57]. In recent years, evidence has emerged showing that autophagy is a fundamental mechanism of the cellular redox balance, acting in the molecular responses to reactive oxygen species (ROS) [58]. Reactive oxygen species (ROS) are produced during mitochondrial ATP production, which can cause mitochondrial damage, triggering further cell apoptosis [59]. To prevent apoptosis, dysfunctional mitochondria are eliminated through mitophagy, a selective process of autophagy that targets mitochondria [60]. Surprisingly, we did not detect oxidative damage in the muscle tissue of fish subjected to high-temperature stress. However, we did observe an over-representation of biological processes related to mitophagy and negative regulation of apoptosis, indicating that autophagy in teleost skeletal muscle is a protective mechanism against oxidative damage induced by temperature stress.

5. Conclusions

In the present study, we demonstrated that temperature stress induces autophagy in the skeletal muscle of *O. mykiss*. Considering the absence of protein carbonylation and DNA oxidation in skeletal muscle and the absence of skeletal muscle atrophy markers in plasma, we propose that autophagy acts as a negative regulator of damage in this tissue, inducing the mitophagy process and inhibiting cell apoptosis.

Supplementary Materials: The following supporting information can be downloaded at https://www.mdpi.com/article/10.3390/fishes8060303/s1. Figure S1: Principal component analyses (PCAs) of cDNA libraries; Figure S2: RT-qPCR validation of selected differentially expressed genes in the skeletal muscle rainbow trout in response to high-temperature stress. Table S1: Primer sequences used in qPCR analysis; Table S2: List of DEGs in control vs. stress groups; Table S3: Molecular function (MF) of differentially upregulated genes; Table S4: Cellular component (CC) of differentially upregulated genes; Table S5: Molecular function (MF) of differentially downregulated genes; Table S6: Cellular component (CC) of differentially downregulated genes.

Author Contributions: Conceptualization, A.M. and J.A.V.; methodology, P.D., V.V.-M. and J.A.V.; software, J.A.V.; validation, P.D.; formal analysis, C.G.-E. and J.A V.; investigation, J.A.V.; resources, A.M. and J.A.V.; data curation, J.A.V.; writing—original draft preparation, J.A.V.; writing—review and editing, J.A.V.; visualization, A.M. and J.A.V.; supervision, J.A.V.; project administration, A.M. and J.A.V.; funding acquisition, J.A.V. All authors have read and agreed to the published version of the manuscript.

Funding: This research was funded by the Fondo Nacional de Desarrollo Científico y Tecnológico (FONDECYT) under grant numbers 1230794 (to Juan Antonio Valdés), 11230153 (to Phillip Dettleff), 11220307 (to Valentina Valenzuela-Muñoz), and 1210852 (to Cristian Gallardo-Escarate); Fondo de Financiamiento de Centros de Investigación en Áreas Prioritarias (FONDAP) INCAR 15110027 grant; and the Concurso de Apoyo a Centros de Excelencia en Investigación FONDAP 2022 1522A0004.

Institutional Review Board Statement: The study adhered to animal welfare procedures, and the protocol was approved by the bioethical committee of Universidad Andres Bello and the National Commission for Scientific and Technological Research of the Chilean government (protocol code 010-2023).

Informed Consent Statement: Not applicable.

Data Availability Statement: The raw read sequences obtained from sequencing were deposited in the Sequence Read Archive (SRA) under BioProject accession number PRJNA930332 (SRR23318096, SRR23318099). The datasets generated and analyzed during the current study are not publicly available owing to privacy or ethical restrictions but are available from the corresponding author upon reasonable request.

Conflicts of Interest: The authors declare no conflict of interest. The funders had no role in the design of the study; in the collection, analyses, or interpretation of data; in the writing of the manuscript; or in the decision to publish the results.

References

1. Abbass, K.; Qasim, M.Z.; Song, H.; Murshed, M.; Mahmood, H.; Younis, I. A Review of the Global Climate Change Impacts, Adaptation, and Sustainable Mitigation Measures. *Environ. Sci. Pollut. Res.* **2022**, *29*, 42539–42559. [CrossRef] [PubMed]
2. Chophel, Y. Global Warming and Climate Change (GWCC) Realities. In *The Nature, Causes, Effects and Mitigation of Climate Change on the Environment*; Harris, S.A., Ed.; IntechOpen: Rijeka, Croatia, 2022; ISBN 978-1-83968-611-5.
3. Garcia-Soto, C.; Cheng, L.; Caesar, L.; Schmidtko, S.; Jewett, E.B.; Cheripka, A.; Rigor, I.; Caballero, A.; Chiba, S.; Báez, J.C.; et al. An Overview of Ocean Climate Change Indicators: Sea Surface Temperature, Ocean Heat Content, Ocean PH, Dissolved Oxygen Concentration, Arctic Sea Ice Extent, Thickness and Volume, Sea Level and Strength of the AMOC (Atlantic Meridional Overturning Circulation). *Front. Mar. Sci.* **2021**, *8*, 642372. [CrossRef]
4. Wang, B.; Luo, X.; Yang, Y.-M.; Sun, W.; Cane, M.A.; Cai, W.; Yeh, S.-W.; Liu, J. Historical Change of El Niño Properties Sheds Light on Future Changes of Extreme El Niño. *Proc. Natl. Acad. Sci. USA* **2019**, *116*, 22512–22517. [CrossRef] [PubMed]
5. Alfonso, S.; Gesto, M.; Sadoul, B. Temperature Increase and Its Effects on Fish Stress Physiology in the Context of Global Warming. *J. Fish. Biol.* **2021**, *98*, 1496–1508. [CrossRef] [PubMed]
6. Pankhurst, N.W. The Endocrinology of Stress in Fish: An Environmental Perspective. *General. Comp. Endocrinol.* **2011**, *170*, 265–275. [CrossRef]

7.   Chen, S.; Yong, Y.; Ju, X. Effect of Heat Stress on Growth and Production Performance of Livestock and Poultry: Mechanism to Prevention. *J. Therm. Biol.* **2021**, *99*, 103019. [CrossRef]
8.   Faught, E.; Vijayan, M.M. Mechanisms of Cortisol Action in Fish Hepatocytes. *Comp. Biochem. Physiol. Part B Biochem. Mol. Biol.* **2016**, *199*, 136–145. [CrossRef]
9.   Manneken, J.D.; Dauer, M.V.P.; Currie, P.D. Dynamics of Muscle Growth and Regeneration: Lessons from the Teleost. *Exp. Cell Res.* **2022**, *411*, 112991. [CrossRef]
10.   Johnston, I.A.; Bower, N.I.; Macqueen, D.J. Growth and the Regulation of Myotomal Muscle Mass in Teleost Fish. *J. Exp. Biol.* **2011**, *214*, 1617–1628. [CrossRef]
11.   Sadoul, B.; Vijayan, M.M. Stress and Growth. In *Fish Physiology*; Elsevier: Amsterdam, The Netherlands, 2016; Volume 35, pp. 167–205. ISBN 978-0-12-802728-8.
12.   Dettleff, P.; Zuloaga, R.; Fuentes, M.; Gonzalez, P.; Aedo, J.; Estrada, J.M.; Molina, A.; Valdés, J.A. Physiological and Molecular Responses to Thermal Stress in Red Cusk-Eel (*Genypterus chilensis*) Juveniles Reveals Atrophy and Oxidative Damage in Skeletal Muscle. *J. Therm. Biol.* **2020**, *94*, 102750. [CrossRef]
13.   Dettleff, P.; Zuloaga, R.; Fuentes, M.; Gonzalez, P.; Aedo, J.; Estrada, J.M.; Molina, A.; Valdés, J.A. High-Temperature Stress Effect on the Red Cusk-Eel (*Geypterus chilensis*) Liver: Transcriptional Modulation and Oxidative Stress Damage. *Biology* **2022**, *11*, 990. [CrossRef] [PubMed]
14.   Chen, W.; Chen, Y.; Liu, Y.; Wang, X. Autophagy in Muscle Regeneration: Potential Therapies for Myopathies. *J. Cachexia Sarcopenia Muscle* **2022**, *13*, 1673–1685. [CrossRef] [PubMed]
15.   Han, X.; Goh, K.Y.; Lee, W.X.; Choy, S.M.; Tang, H.-W. The Importance of MTORC1-Autophagy Axis for Skeletal Muscle Diseases. *Int. J. Mol. Sci.* **2022**, *24*, 297. [CrossRef]
16.   Chatzinikita, E.; Maridaki, M.; Palikaras, K.; Koutsilieris, M.; Philippou, A. The Role of Mitophagy in Skeletal Muscle Damage and Regeneration. *Cells* **2023**, *12*, 716. [CrossRef] [PubMed]
17.   Zhou, Z.; He, Y.; Wang, S.; Wang, Y.; Shan, P.; Li, P. Autophagy Regulation in Teleost Fish: A Double-Edged Sword. *Aquaculture* **2022**, *558*, 738369. [CrossRef]
18.   Rivas-Aravena, A.; Fuentes-Valenzuela, M.; Escobar-Aguirre, S.; Gallardo-Escarate, C.; Molina, A.; Valdés, J.A. Transcriptomic Response of Rainbow Trout (*Oncorhynchus mykiss*) Skeletal Muscle to Flavobacterium Psychrophilum. *Comp. Biochem. Physiol. Part D Genom. Proteom.* **2019**, *31*, 100596. [CrossRef]
19.   Aedo, J.E.; Zuloaga, R.; Bastías-Molina, M.; Meneses, C.; Boltaña, S.; Molina, A.; Valdés, J.A. Early Transcriptomic Responses Associated with the Membrane-Initiated Action of Cortisol in the Skeletal Muscle of Rainbow Trout (*Oncorhynchus mykiss*). *Physiol. Genom.* **2019**, *51*, 596–606. [CrossRef]
20.   Bustin, S.A.; Benes, V.; Garson, J.A.; Hellemans, J.; Huggett, J.; Kubista, M.; Mueller, R.; Nolan, T.; Pfaffl, M.W.; Shipley, G.L.; et al. The MIQE Guidelines: Minimum Information for Publication of Quantitative Real-Time PCR Experiments. *Clin. Chem.* **2009**, *55*, 611–622. [CrossRef]
21.   Schneider, C.A.; Rasband, W.S.; Eliceiri, K.W. NIH Image to ImageJ: 25 Years of Image Analysis. *Nat. Methods* **2012**, *9*, 671–675. [CrossRef]
22.   Gizińska, J.; Sojka, M. How Climate Change Affects River and Lake Water Temperature in Central-West Poland—A Case Study of the Warta River Catchment. *Atmosphere* **2023**, *14*, 330. [CrossRef]
23.   Yuan, D.; Wang, H.; Liu, X.; Wang, S.; Shi, J.; Cheng, X.; Gu, H.; Xiao, S.; Wang, Z. High Temperature Induced Metabolic Reprogramming and Lipid Remodeling in a High-Altitude Fish Species, Triplophysa Bleekeri. *Front. Mar. Sci.* **2022**, *9*, 1017142. [CrossRef]
24.   LeBlanc, S.; Middleton, S.; Gilmour, K.M.; Currie, S. Chronic Social Stress Impairs Thermal Tolerance in the Rainbow Trout (*Oncorhynchus mykiss*). *J. Exp. Biol.* **2011**, *214*, 1721–1731. [CrossRef]
25.   Basu, N.; Nakano, T.; Grau, E.G.; Iwama, G.K. The Effects of Cortisol on Heat Shock Protein 70 Levels in Two Fish Species. *Gen. Comp. Endocrinol.* **2001**, *124*, 97–105. [CrossRef] [PubMed]
26.   Kumar, P.; Pal, A.K.; Sahu, N.P.; Jha, A.K.; Priya, P. Biochemical and Physiological Stress Responses to Heat Shock and Their Recovery in Labeo Rohita Fingerlings. *Proc. Natl. Acad. Sci. India Sect. B Biol. Sci.* **2015**, *85*, 485–490. [CrossRef]
27.   Uren Webster, T.M.; Rodriguez-Barreto, D.; Martin, S.A.M.; Van Oosterhout, C.; Orozco-terWengel, P.; Cable, J.; Hamilton, A.; Garcia De Leaniz, C.; Consuegra, S. Contrasting Effects of Acute and Chronic Stress on the Transcriptome, Epigenome, and Immune Response of Atlantic Salmon. *Epigenetics* **2018**, *13*, 1191–1207. [CrossRef]
28.   Pérez-Casanova, J.C.; Afonso, L.O.B.; Johnson, S.C.; Currie, S.; Gamperl, A.K. The Stress and Metabolic Responses of Juvenile Atlantic Cod *Gadus morhua* L. to an Acute Thermal Challenge. *J. Fish Biol.* **2008**, *72*, 899–916. [CrossRef]
29.   Chadwick, J.G.; Nislow, K.H.; McCormick, S.D. Thermal Onset of Cellular and Endocrine Stress Responses Correspond to Ecological Limits in Brook Trout, an Iconic Cold-Water Fish. *Conserv. Physiol.* **2015**, *3*, cov017. [CrossRef]
30.   Aedo, J.E.; Zuloaga, R.; Aravena-Canales, D.; Molina, A.; Valdés, J.A. Role of Glucocorticoid and Mineralocorticoid Receptors in Rainbow Trout (*Oncorhynchus mykiss*) Skeletal Muscle: A Transcriptomic Perspective of Cortisol Action. *Front. Physiol.* **2023**, *13*, 1048008. [CrossRef]

31. Khieokhajonkhet, A.; Sangphrom, S.; Aeksiri, N.; Tatsapong, P.; Wuthijaree, K.; Kaneko, G. Effects of Long-Term Exposure to High Temperature on Growth Performance, Chemical Composition, Hematological and Histological Changes, and Physiological Responses in Hybrid Catfish [♂*Clarias Gariepinus* (Burchell, 1822) ×♀*C. Macrocephalus* (Günther, 1864)]. *J. Therm. Biol.* **2022**, *105*, 103226. [CrossRef]

32. Vo, T.T.M.; Amoroso, G.; Ventura, T.; Elizur, A. Histological and Transcriptomic Analysis of Muscular Atrophy Associated with Depleted Flesh Pigmentation in Atlantic Salmon (*Salmo salar*) Exposed to Elevated Seawater Temperatures. *Sci. Rep.* **2023**, *13*, 4218. [CrossRef]

33. Feidantsis, K.; Georgoulis, I.; Zachariou, A.; Campaz, B.; Christoforou, M.; Pörtner, H.O.; Michaelidis, B. Energetic, Antioxidant, Inflammatory and Cell Death Responses in the Red Muscle of Thermally Stressed Sparus Aurata. *J. Comp. Physiol. B* **2020**, *190*, 403–418. [CrossRef] [PubMed]

34. Galluzzi, L.; Baehrecke, E.H.; Ballabio, A.; Boya, P.; Bravo-San Pedro, J.M.; Cecconi, F.; Choi, A.M.; Chu, C.T.; Codogno, P.; Colombo, M.I.; et al. Molecular Definitions of Autophagy and Related Processes. *EMBO J.* **2017**, *36*, 1811–1836. [CrossRef] [PubMed]

35. Khandia, R.; Dadar, M.; Munjal, A.; Dhama, K.; Karthik, K.; Tiwari, R.; Yatoo, M.I.; Iqbal, H.M.N.; Singh, K.P.; Joshi, S.K.; et al. A Comprehensive Review of Autophagy and Its Various Roles in Infectious, Non-Infectious, and Lifestyle Diseases: Current Knowledge and Prospects for Disease Prevention, Novel Drug Design, and Therapy. *Cells* **2019**, *8*, 674. [CrossRef] [PubMed]

36. Cassel, M.; de Paiva Camargo, M.; Oliveira de Jesus, L.W.; Borella, M.I. Involution Processes of Follicular Atresia and Post-Ovulatory Complex in a Characid Fish Ovary: A Study of Apoptosis and Autophagy Pathways. *J. Mol. Hist.* **2017**, *48*, 243–257. [CrossRef] [PubMed]

37. Godoi, F.G.A.; Forner-Piquer, I.; Randazzo, B.; Habibi, H.R.; Lo Nostro, F.L.; Moreira, R.G.; Carnevali, O. Effects of Di-Isononyl Phthalate (DiNP) on Follicular Atresia in Zebrafish Ovary. *Front. Endocrinol.* **2021**, *12*, 677853. [CrossRef]

38. Cheng, Y.; Lai, F.; Wang, X.; Shang, D.; Zou, J.; Luo, M.; Xia, X.; Cheng, H.; Zhou, R. Srag Regulates Autophagy via Integrating into a Preexisting Autophagy Pathway in Testis. *Mol. Biol. Evol.* **2021**, *38*, 128–141. [CrossRef]

39. Seiliez, I.; Belghit, I.; Gao, Y.; Skiba-Cassy, S.; Dias, K.; Cluzeaud, M.; Rémond, D.; Hafnaoui, N.; Salin, B.; Camougrand, N.; et al. Looking at the Metabolic Consequences of the Colchicine-Based in Vivo Autophagic Flux Assay. *Autophagy* **2016**, *12*, 343–356. [CrossRef]

40. Dai, Z.; Cheng, J.; Bao, L.; Zhu, X.; Li, H.; Chen, X.; Zhang, Y.; Zhang, J.; Chu, W.; Pan, Y.; et al. Exposure to Waterborne Cadmium Induce Oxidative Stress, Autophagy and Mitochondrial Dysfunction in the Liver of Procypris Merus. *Ecotoxicol. Environ. Saf.* **2020**, *204*, 111051. [CrossRef]

41. Masud, S.; Prajsnar, T.K.; Torraca, V.; Lamers, G.E.M.; Benning, M.; Van Der Vaart, M.; Meijer, A.H. Macrophages Target Salmonella by Lc3-Associated Phagocytosis in a Systemic Infection Model. *Autophagy* **2019**, *15*, 796–812. [CrossRef]

42. Yang, M.; Lu, Z.; Li, F.; Shi, F.; Zhan, F.; Zhao, L.; Li, Y.; Li, J.; Lin, L.; Qin, Z. *Escherichia coli* Induced Ferroptosis in Red Blood Cells of *grass carp* (*Ctenopharyngodon Idella*). *Fish Shellfish Immunol.* **2021**, *112*, 159–167. [CrossRef]

43. Valenzuela, C.A.; Zuloaga, R.; Mercado, L.; Einarsdottir, I.E.; Björnsson, B.T.; Valdés, J.A.; Molina, A. Chronic Stress Inhibits Growth and Induces Proteolytic Mechanisms through Two Different Nonoverlapping Pathways in the Skeletal Muscle of a Teleost Fish. *Am. J. Physiol.-Regul. Integr. Comp. Physiol.* **2018**, *314*, R102–R113. [CrossRef] [PubMed]

44. Aedo, J.E.; Maldonado, J.; Aballai, V.; Estrada, J.M.; Bastias-Molina, M.; Meneses, C.; Gallardo-Escarate, C.; Silva, H.; Molina, A.; Valdés, J.A. MRNA-Seq Reveals Skeletal Muscle Atrophy in Response to Handling Stress in a Marine Teleost, the Red Cusk-Eel (*Genypterus chilensis*). *BMC Genom.* **2015**, *16*, 1024. [CrossRef] [PubMed]

45. Aedo, J.E.; Aravena-Canales, D.; Dettleff, P.; Fuentes-Valenzuela, M.; Zuloaga, R.; Rivas-Aravena, A.; Molina, A.; Valdés, J.A. RNA-Seq Analysis Reveals the Dynamic Regulation of Proteasomal and Autophagic Degradation Systems of Rainbow Trout (*Oncorhynchus mykiss*) Skeletal Muscle Challenged with Infectious Pancreatic Necrosis Virus (IPNV). *Aquaculture* **2022**, *552*, 738000. [CrossRef]

46. Sun, C.-C.; Zhou, Z.-Q.; Chen, Z.-L.; Zhu, R.-K.; Yang, D.; Peng, X.-Y.; Zheng, L.; Tang, C.-F. Identification of Potentially Related Genes and Mechanisms Involved in Skeletal Muscle Atrophy Induced by Excessive Exercise in Zebrafish. *Biology* **2021**, *10*, 761. [CrossRef] [PubMed]

47. Dhanasekaran, D.N.; Reddy, E.P. JNK-Signaling: A Multiplexing Hub in Programmed Cell Death. *Genes. Cancer* **2017**, *8*, 682–694. [CrossRef]

48. Tian, Y.; Wen, H.; Qi, X.; Zhang, X.; Li, Y. Identification of Mapk Gene Family in Lateolabrax Maculatus and Their Expression Profiles in Response to Hypoxia and Salinity Challenges. *Gene* **2019**, *684*, 20–29. [CrossRef]

49. Gao, A.; Jiang, J.; Xie, F.; Chen, L. Bnip3 in Mitophagy: Novel Insights and Potential Therapeutic Target for Diseases of Secondary Mitochondrial Dysfunction. *Clin. Chim. Acta* **2020**, *506*, 72–83. [CrossRef]

50. Valenzuela, C.A.; Escobar, D.; Perez, L.; Zuloaga, R.; Estrada, J.M.; Mercado, L.; Valdés, J.A.; Molina, A. Transcriptional Dynamics of Immune, Growth and Stress Related Genes in Skeletal Muscle of the Fine Flounder (Paralichthys Adpersus) during Different Nutritional Statuses. *Dev. Comp. Immunol.* **2015**, *53*, 145–157. [CrossRef]

51. Jin, M.; Klionsky, D.J. Transcriptional Regulation of ATG9 by the Pho23-Rpd3 Complex Modulates the Frequency of Autophagosome Formation. *Autophagy* **2014**, *10*, 1681–1682. [CrossRef]

52. Donninger, H.; Schmidt, M.L.; Mezzanotte, J.; Barnoud, T.; Clark, G.J. Ras Signaling through RASSF Proteins. *Semin. Cell Dev. Biol.* **2016**, *58*, 86–95. [CrossRef]

53. Yan, S.; Chen, R.; Wang, M.; Zha, J. Carbamazepine at Environmentally Relevant Concentrations Caused DNA Damage and Apoptosis in the Liver of Chinese Rare Minnows (*Gobiocypris rarus*) by the Ras/Raf/ERK/P53 Signaling Pathway. *Environ. Pollut.* **2021**, *270*, 116245. [CrossRef] [PubMed]

54. Ganesan, S.; Pearce, S.C.; Gabler, N.K.; Baumgard, L.H.; Rhoads, R.P.; Selsby, J.T. Short-Term Heat Stress Results in Increased Apoptotic Signaling and Autophagy in Oxidative Skeletal Muscle in Sus Scrofa. *J. Therm. Biol.* **2018**, *72*, 73–80. [CrossRef] [PubMed]

55. Hirunsai, M.; Srikuea, R. Autophagy-Lysosomal Signaling Responses to Heat Stress in Tenotomy-Induced Rat Skeletal Muscle Atrophy. *Life Sci.* **2021**, *275*, 119352. [CrossRef] [PubMed]

56. Tamura, Y.; Kitaoka, Y.; Matsunaga, Y.; Hoshino, D.; Hatta, H. Daily Heat Stress Treatment Rescues Denervation-Activated Mitochondrial Clearance and Atrophy in Skeletal Muscle: Heat Stress Treatment in Denervated Skeletal Muscle. *J. Physiol.* **2015**, *593*, 2707–2720. [CrossRef]

57. Summers, C.M.; Valentine, R.J. Acute Heat Exposure Alters Autophagy Signaling in C2C12 Myotubes. *Front. Physiol.* **2020**, *10*, 1521. [CrossRef]

58. Yun, H.R.; Jo, Y.H.; Kim, J.; Shin, Y.; Kim, S.S.; Choi, T.G. Roles of Autophagy in Oxidative Stress. *Int. J. Mol. Sci.* **2020**, *21*, 3289. [CrossRef]

59. Bolisetty, S.; Jaimes, E. Mitochondria and Reactive Oxygen Species: Physiology and Pathophysiology. *Int. J. Mol. Sci.* **2013**, *14*, 6306–6344. [CrossRef]

60. Wen, X.; Tang, L.; Zhong, R.; Liu, L.; Chen, L.; Zhang, H. Role of Mitophagy in Regulating Intestinal Oxidative Damage. *Antioxidants* **2023**, *12*, 480. [CrossRef]

Review

Potential Impact of Climate Change on Salmonid Smolt Ecology

Teppo Vehanen *, Tapio Sutela and Ari Huusko

Natural Resources Institute Finland, Latokartanonkaari 9, 00790 Helsinki, Finland; tapio.sutela@luke.fi (T.S.); ari.huusko@luke.fi (A.H.)
* Correspondence: teppo.vehanen@luke.fi

Abstract: The migratory life history of anadromous salmonids requires successful migration between nursery, feeding, and spawning habitats. Smolting is the major transformation anadromous salmonids undergo before migration to feeding areas. It prepares juvenile fish for downstream migration and their entry to seawater. We reviewed the effects of climate change on smolt ecology from the growth of juveniles in fresh water to early post-smolts in the sea to identify the potential effects of climate change on migratory salmonid populations during this period in their life history. The focus was especially on Atlantic salmon. The shift in suitable thermal conditions caused by climate change results in Atlantic salmon expanding their range northward, while at the southern edge of their distribution, populations struggle with high temperatures and occasional droughts. Climatic conditions, particularly warmer temperatures, affect growth during the freshwater river phase. Better growth in northern latitudes leads to earlier smolting. Thermal refuges, the areas of cooler water in the river, are important for salmonids impacted by climate change. Restoring and maintaining connectivity and a suitably diverse mosaic habitat in rivers are important for survival and growth throughout the range. The start of the smolt migration has shifted earlier as a response to rising water temperatures, which has led to concerns about a mismatch with optimal conditions for post-smolts in the sea, decreasing their survival. A wide smolt window allowing all migrating phenotypes from early to late migrants' safe access to the sea is important in changing environmental conditions. This is also true for regulated rivers, where flow regulation practices cause selection pressures on migrating salmonid phenotypes. The freshwater life history also affects marine survival, and better collaboration across life stages and habitats is necessary among researchers and managers to boost smolt production in rivers. Proactive measures are recommended against population declines, including sustainable land use in the catchment, maintaining a diverse mosaic of habitats for salmonids, restoring flow and connectivity, and conserving key habitats.

Keywords: climate change; salmonids; *Salmo*; rivers; fresh water; migration

Key Contribution: This paper makes a valuable contribution to understanding how climate change affects the key life stage of salmonids, smolting, and smolt migration. Smolting prepares entry from fresh water to salt water. Salmonids are a group of species with huge cultural and economic significance.

Citation: Vehanen, T.; Sutela, T.; Huusko, A. Potential Impact of Climate Change on Salmonid Smolt Ecology. *Fishes* **2023**, *8*, 382. https://doi.org/10.3390/fishes8070382

Academic Editor: Bror Jonsson

Received: 25 May 2023
Revised: 11 July 2023
Accepted: 12 July 2023
Published: 24 July 2023

1. Introduction

Anadromous and potamodromous salmonids migrate from their natal river to a feeding environment before returning for reproduction [1–3]. Migration enables fish to exploit many temporally productive and spatially discrete habitats with various fitness benefits (e.g., growth, reproduction, predator avoidance) [4]. Migratory life history requires unrestricted migration routes between nursery, feeding, and spawning habitats [5]. During each life stage, salmonids utilize the habitat that is advantageous for them. Migration between habitats thus clearly has an adaptive value [6]. Nursery and feeding habitats differ in environmental characteristics, and migrations precede adaptive physiological transformations and changes in the phenotype and behavior to be better suited for the new environment.

Smolting is the major transformation anadromous salmonids undergo before migration to feeding areas. Smolting prepares fish for downstream migration and entry to seawater. Atlantic salmon *Salmo salar* L., 1758 juveniles can stay in their natal river habitat to grow for 1–8 years before migrating [7,8]. Of the Pacific salmon, pink salmon *Oncorhynchus gorbuscha* (Walbaum, 1792) and chum salmon *Oncorhynchus keta* (Walbaum, 1792) can move almost directly after the emergence at the fry stage into seawater, while the others (masu salmon *Oncorhynchus masou* (Brevoort, 1856), *O. tshawytscha* (Walbaum, 1792), *O. nerka* (Walbaum, 1792), and steelhead (rainbow) trout *Oncorhynchus mykiss* (Walbaum, 1792)) spend one or more years in fresh water before migrating to the sea for feeding [7]. When smolting, the phenotype of fish changes as follows: the coloration of smolts becomes silvery, and the body shape becomes more streamlined [2]. This, with darkened fins, a dark back, and a white abdomen, camouflage the fish in the pelagic environment [5]. Behavioral changes include a loss of rheotaxis, and juveniles become more pelagic. Their tendency to group also increases [9]. Several physiological changes occur, for example, increased salinity tolerance, increased metabolism, and olfactory imprinting helping the fish locate their home stream on their return migration [10]. Environmental cues, e.g., photoperiod, temperature, and waterflow, regulate physiological changes and initiate migration [2,5]. Lake-living Atlantic salmon appear to smolts similarly to anadromous conspecifics ([11], but see [12]). This is apparently an inherited trait [13], even for salmon spending their entire life in fresh water, such as the Atlantic salmon residing in Lake Vänern, Sweden.

Both Atlantic and Pacific salmon populations have been in decline throughout their habitat ranges [13–15]. To reverse this trend, it is important to understand the role of different environmental and anthropogenic factors in the decline [16]. Numerous factors may impact population abundances negatively, and with the complex life history of migrating salmonids, the reasons are obviously multiple and difficult to unravel [17,18], although the ongoing climate warming appears particularly important, especially for low latitude populations. Anthropogenic activities have a long history of altering salmonid populations and, thus, smolt development and smolt migration. Smolts are sensitive to external impacts and behavior and survival during migrations [2]. Several anthropogenic activities may affect smolt development, behavior, and survival during migration, such as hydropower developments, land use, pollution, fish farming, and parasites like sea lice *Lepeophtheirus salmonis* (Krøyer, 1837) [2,5]. Temperature and flow interact with the other anthropogenic pressures to affect smolting and smolt migration.

Salmonids are a cold-water species. Global warming will generally have a major impact on their success. Historically, climatic variability has affected the patterns of abundance in Atlantic salmon and Pacific salmon populations [19–22]. Although estuarine and marine mortalities have been found to be important determinants of survival, marine mortality depends on factors acting in fresh water and during smolt migration [23]. Thorstad et al. [24] argue that the best strategy to mitigate the changing environmental conditions should be to ensure that the greatest number of wild smolts in the best condition migrate from rivers and coastal areas to feeding areas because mortality at sea is found to be density independent [25]. Survival at sea depends on the size of the smolts and environmental conditions when the smolts begin their sea sojourn. [18,23,26]. In research, it is important to address the links between river habitat conditions and the physiological requirements of salmonids during their juvenile life stages in freshwater habitats [27]. Climate change will continue to affect not only smolting and migration but also instream habitats across all seasons [27].

In this paper, we review climate change effects on (1) in-river habitat conditions in preparation for smolting, (2) the smolting process, (3) smolt migration, and (4) early post-smolt survival. Our focus is on *Salmo* spp., but when relevant, we also refer to the fish species in the Pacific salmon genus (*Onchorhynchus* spp.).

2. Climate Change and Salmonid Distribution

Human activities are estimated to have induced approximately 1.0 °C of global warming above pre-industrial levels (between 1880 and 2017), with a likely range of 0.8 °C to 1.2 °C. Global warming is likely to reach 1.5 °C in about 2030 if temperatures continue to increase at the current rate [28]. For example, a higher winter discharge, earlier snowmelt, and earlier onset of summer low flow periods are predicted throughout the range of Atlantic salmon [29,30].

Increasing global surface temperatures are very likely to lead to changes in precipitation and atmospheric moisture because of changes in atmospheric circulation, a more active hydrological cycle, and increases in the water-holding capacity throughout the atmosphere. Overall, global land precipitation has increased by about 2% since the beginning of the 20th century. There have been marked increases in precipitation in the latter part of the 20th century over northern Europe, though with a general decrease southward to the Mediterranean. Dry wintertime conditions over southern Europe and the Mediterranean and wetter-than-normal conditions over many parts of northern Europe and Scandinavia [31] are linked to the strong positive values of the North Atlantic Oscillation (NAO), with more anticyclonic conditions over southern Europe and stronger westerly winds over northern Europe (Ref. [32] conducted a review).

Northern Eurasia (north of approximately 40 °N) showed widespread and statistically significant increases in winter precipitation between 1921 and 2015, with values exceeding 1.2–1.6 mm mo^{-1} per decade west of the Ural Mountains and along the east coast, while southern Europe exhibits coherent yet weaker amplitude drying trends that attain statistical significance over the eastern Mediterranean. These precipitation trends occur in the context of changes in the large-scale atmospheric circulation, with negative SLP (Sea Level Pressure) trends over northern Eurasia and positive SLP trends over the central North Atlantic extending into southwestern Europe [33].

The magnitude of climate change is considered to depend on the atmospheric load of the two most important greenhouse gases: carbon dioxide (CO_2) and methane (CH_4). The terrestrial biosphere plays an important role in the global carbon balance. In boreal zones, forests and peatlands are an essential part of the global carbon cycle. Recent temperature increases have been associated with increasing forest fire activity in Canada since about 1970 and exceptionally warm summer conditions in Russia during the 2010 fire season reviewed by [34].

Atlantic salmon is distributed from northern Portugal (42 °N) to the River Kara in northern Russia in Europe [35], and West Atlantic salmon is distributed from the Connecticut River to the Ungava region of northern Quebec. Southern Atlantic salmon populations have declined dramatically and face the highest risk of extinction as global warming moves its thermal niche northward [36]. The suitable thermal habitat for salmon is expected to extend northward with the invasion of new spawning, nursery, and feeding areas north of the species' present distributional range but with the loss of the most southern populations [37–40]. Indeed, salmon are already responding to warmer temperatures by expanding their range northward into the Arctic Ocean [41,42] and disappearing from the southern edge of their distribution area [7,40,43–45]. The population complex of Atlantic salmon in Europe has experienced a multidecadal decline in recruitment, resulting in the lowest population abundances observed since 1970 [46]. Atlantic salmon abundance and productivity show similar patterns of decline across six widespread regions of North America [47]. Abundance declined in the late 1980s and early 1990s, after which it remained stable at low levels. Climate-driven environmental factors such as changes in plankton communities and prey availability at warmer ocean temperatures were linked to the low productivity of North Atlantic salmon populations [47]. Landlocked European populations of salmonids are found in Norway, Sweden, Finland, and Russian Karelia [48–51]. The landlocked populations of salmon have declined throughout their distribution range [51,52]. Brown trout (*Salmo trutta* L. 1758) is native to Europe and Asia, where anadromous populations are found from Portugal to the White Sea [7]. It must be noted that the taxonomic status of the brown trout species complex is challenging, and the high morphological and ecological diversity has led to the morphological description of populations belonging to species other than *S. trutta* in Europe [53]. In the future, the living conditions for trout will probably deteriorate in the southern part of the current distribution. In the northern part of their current distribution, global warming may improve feeding opportunities, growth, and survival conditions [7]. According to Filipe et al. [54], future brown trout distribution will become progressively and dramatically reduced in European watercourses. Their forecasts indicate that the greatest losses in suitable habitats will take place in southern Europe.

3. In-River Habitat Conditions in Preparation for Smolting

The most important climate-change-driven habitat changes that influence salmonid juveniles in rivers are changes in thermal and hydrological regimes [55–57]. These changes will affect how juveniles use their physical habitat and affect growth and survival.

Water temperature has various effects on the biology of salmonids. Thermal optima allow salmon to maximize growth; temperatures above thermal optima can stress fish and ultimately lead to mortality [58,59]. On a larger scale, northern populations are predicted to do better than southern populations under global warming [38,60,61], but even in the same river, the effects on different populations can vary [59]. Some northern populations can have an increase in parr recruitment and smolt production [61]. However, some Arctic salmonids are also already experiencing warm (>21 °C), physiologically challenging, migratory river conditions [62], and an increase in river water temperatures has already been observed in several rivers [63–65]. In general, high-latitude ecosystems are facing rapid warming, and cold-water fish will eventually be displaced by fish adapted to warmer water [66]. The range of temperatures at which fish survive or grow differs between development stages and salmonid species (for a review, see [7]). Atlantic salmon eggs survive between 0 and 16 °C, and alevins can develop normally up to 22 °C [67]. Growth takes place in temperatures between 6 °C and 22.5 °C, with maximum growth at around 16 °C, and the upper lethal temperature is 29.5 °C for parr but depends on the acclimatization temperature and the length of the acclimatization period. With the warming of surface waters, the risk of local extinctions will increase [68,69].

Smolt's characteristics are influenced by their earlier life in fresh water [23,70]. For example, the incubation of eggs at higher temperatures has resulted in fry with a reduced swimming performance or later returning adults [71,72]. According to Thompson and Beauchamp [73], the survival of steelhead trout in the marine environment can be driven by an overall higher growth rate established early in life in fresh water, which results in a larger size at smolt migration. Climate-induced instream thermal conditions affect parr size and the age of departure from the river [74]. For salmonid populations facing increased water temperatures, thermal heterogeneity in the river plays an important role in survival and growth [56,75]. The density of juveniles in thermal refuges has been found to increase after high-temperature effects [56]. Maintaining and restoring a diverse mixture of habitats and thermal refugia is important for salmonids impacted by climate change [76]. Thermal topology can also influence fish growth. Fish in the least complex network grew faster and were ready to smolt earlier than fish in the more spatially complex temperature network, i.e., in a river environment where the thermal diversity was higher [77]. Climate-induced high water temperatures can also interact with parr density, while in chinook salmon at low parr density, the effect of temperature on growth was positive, and at high densities, the relationship proved to be negative [75].

Especially in the southern margins of the salmonid distribution ranges, the availability of suitable cold-water environments becomes more important as the temperatures rise [78]. The temporal variability of these cold refuges is high; the most stable ones are typically groundwater seeps and cold-water tributaries [79]. For cold-water species like salmonids, headwater streams may become more important structural and thermal refuges. Head-waters are often less impacted by humans than the main streams (T. Vehanen, personal communication). On the other hand, high-elevation streams, especially those above snow-lines, can be especially vulnerable to climate change because they are likely to experience the greatest snow–rain transition [80]. Stream size is a limiting factor for some salmonid species, but for species like coho salmon, differently sized streams can provide an alternative rearing habitat [81]. For brown trout, small streams are important spawning and nursery habitats [82,83]. Brown trout are well adapted and influenced by habitat variables associated with the size of small streams, especially with flow variations [82,84], and the population traits of anadromous brown trout from a small stream differ from those in larger rivers [72].

Seasonal flow is another key element impacted by climate change, contributing to the habitat quality of salmonid juveniles [85]. Climate change has already altered the hydrological regimes of rivers. The changes for Atlantic salmon and brown trout include frequent periods with extreme weather, i.e., low- and high-flow events, precipitation falling as rain and less as snow, and a decrease in the ice-covered period [7,86]. These changes can have a negative impact on freshwater salmon's instream habitat [80,87]. Extremes in waterflow can decrease recruitment and survival. Generally, the early life stages, i.e., the eggs, emerging alevins, fry, and young juveniles, experience the highest mortalities [88,89]. High-flow events during the emergence of fry from the gravel can cause the flushing of fry to unsuitable habitats. The preferences for physical habitat parameters like water velocity and depth vary seasonally [90]. Climate change-induced high or low flows cause variation in this habitat suitable for salmonid parr. Low flow conditions are also often associated with an extended duration of high water temperatures [87]. The minimum levels of river flow have been found to be regulators for parr survival and, hence, for smolt production in Atlantic salmon and brown trout [61,84]. It is also predicted that stream hydrology will change when winters get warmer, and increased fluctuations in winter discharge and temperatures may lead to repeated ice formation and breakup [91–93]. As winters get warmer, there is less snow, more rain, and higher winter discharges. These changes can negatively affect the growth and survival of juvenile salmonids during the winter [94]. The ability of the young salmonids to swim against strong currents is poor at low temperatures [7,95], and salmonid parr prefers relatively slow flow rates in the winter [96]. Increased rain on snow with a high flow can lead to the ice scouring of the streambed, which results in higher egg mortality [97]. The mortality of salmonid eggs and fry may become higher with climate change in northern rivers.

Water temperature and flow variation, the two important aspects of climate change, are interacting with anthropogenic activities, such as land use in the catchment to affect the fish community in rivers [98]. Anthropogenic activities have long altered migratory fish by closing pathways and creating challenging migration conditions for smolt. Climate change can further strengthen these human-caused effects. Climate change will intensify precipitation and flood events in all climate regions [99], but the difference at the regional scale can be high [100]. Increased precipitation intensity enhances suspended solid and nutrient loadings in rivers, especially in human-altered catchments [101]. Increased rainfall with land use (i.e., forestry, agriculture) will intensify the brownification of surface waters due to the increased loading of dissolved organic carbon from the catchments [102,103]. This widespread phenomenon, especially in the boreal region, will deteriorate the habitat quality of salmonid juveniles habituated to good freshwater quality. A reduced freshwater habitat connectivity can decrease the growth of juveniles and may have deleterious impacts on later marine life stages [104]. Flow regulation typically creates flow and temperature conditions for fish species that prefer warm- and slow-water habitats and can thus favor invasive species. The physically challenging migratory conditions caused by flow regulation combined with large diurnal temperature fluctuations can restrict the migration of salmonids by limiting their ability to recover from fatiguing exercise [62]. A rapid temperature rise also has a negative effect on the osmoregulatory performance of Atlantic salmon smolts [105].

4. Smolting

In salmonids, genetic diversity combined with developmental flexibility leads to numerous pathways to residency, migration, or maturation [106], and especially among Pacific salmon, there are also other phenotypes than smolts out-migrating rivers [107,108]. Anadromous salmonid juveniles transform from parr to smolts to prepare for downstream migration and entry into seawater. Physiological and behavioral changes take place in the spring when juvenile salmonids undergo smolting. Smolting and smolt migration are considered critical life-history stages essential for survival [5]. While still in fresh water, fish undergo a preparatory smolting process involving morphological changes as they become silvery and streamlined [2] (Figure 1). Behavioral changes include decreasing rheotactic and optomotor sensitivities and fish's station-holding abilities [9]. The photoperiod and temperature regulate physiological changes via their impact on the neuroendocrine system [2]. Thus, because the photoperiod remains the same at the same date and site each year, the temperature will be critical in determining responses to future climate change. Within the same river system, the distance to the sea does not seem to play a role; populations are closer or further from the sea smolts at the same time [109,110]. Waterflow and its variability as another major environmental factor can act more as a timer to initiate migration [2,7], for example.

Figure 1. When smolting Atlantic salmon smolts become silvery and body-streamlined (Photo: River Tornionjoki (Finland) Atlantic salmon smolt, Ville Vähä).

Smolting varies depending on several factors like temperature, latitude, age and size, growth rate, and a combination of these factors. Hence, climate change with rising water temperatures obviously has an impact on the smolting process. Size and growth potentially affect the timing of migration and the survival of smolts [26,111–114]. Smolt age depends on growth rate. For instance, fast-growing parr smolts are younger and smaller than slow-growing parrs [115]. Warmer river temperatures increase the growth of parr and the share of fish smolting at an earlier age [74]. Temperature naturally correlates with latitude and is a strong predictor of migration timing in Atlantic salmon [116].

The migration decisions to smolt or not are decided between internal and external factors [114,117]. The important internal factors are the growth rate and the energetic status of individual fish [5]. Differences in the smolting rate between naturally anadromous and more resident populations have an inherited component [117]. A high growth rate in the late summer and the early fall of the year before migration can predict smolting [5]. However, high growth, especially during the winter, may induce the maturation of the parr [118].

Growth and energetics do not solely depend on temperature but on other factors like food availability. For example, it appears that smolting may be switched off via poor nutritional conditions preceding smolting [117,119]. Better growth conditions caused by an increase in the river temperature can increase the proportion of sexually mature male parr, which have a lower probability of migrating [120]. How climate change will affect individual growth rates and energetics in salmonid populations will depend intimately on how it affects the ecological status of rivers, particularly food availability.

Climate change may strengthen or weaken the effects of anthropogenic activities on water quality important for smolting salmonids. Pollutants, acidity, and sedimentation can adversely affect smolt development, which can have negative consequences on their readiness for life at sea [10,23]. Especially in northern temperate coastal regions, which will receive higher winter rainfall, phosphorus loading from land to streams is expected to increase, whereas a decline in warm temperate and arid climates is expected [121]. In the northern region, increasing precipitation will increase nutrient leaching, especially from areas affected by human alteration: agriculture, forestry, and other land use [122]. For example, acid leaks from the catchment are expected to increase. Increased acidity will have a major impact on the fish community, especially on acid-sensitive salmonids [123]. Even a short moderate exposure to acidity may require more than two weeks for the recovery of Atlantic salmon smolts [124]. Freshwater ecosystems are sensitive to anthropogenic flow regime alteration, which may cause temperature fluctuations. Close to its southernmost distribution, warming with low flows threaten coho salmon in California, and environmental flow protection is needed to support Pacific salmon in a changing climate [125]. Rapid temperature shifts have a negative impact on the hypo-osmoregulatory capacities of Atlantic salmon smolts [126]. There is an interaction of salinity and elevated temperature in the osmoregulatory performance of salmon smolt, and rapid temperature fluctuations above the threshold temperature (20 °C) have been found to cause iono-regulatory failure.

5. Smolt Migrations

Smolts start their downstream migration during a "period of readiness", a smolt window when they are physiologically prepared to meet the conditions in their marine feeding area [2,127]. In Atlantic salmon smolt, migration typically takes place during the spring and early summer at a length of 12–25 cm [128]. Temperature and flow are environmental cues for smolt migration. Migration times differ between years and rivers; the temperature can be a good predictor of the timing [129–131]. Warmer temperatures result in earlier migrations [130]. Typically, a correlation between the onset of the smolt run and the water temperature has been found [132]. Temperature experience, an accumulated temperature, or a combination of a temperature increase and temperature level in the river during the spring are the cues to initiate migration rather than any threshold temperature [130,133,134]. The initiation of smolt migration was positively associated with freshwater temperatures of up to about 10 °C and leveling off at higher values [18]. Another major environmental clue that plays an important role in initiating smolt migration is river flow. During the smolt window, increased waterflow initiates smolt migration [128,132,135,136], but high flows have also been found to have an opposite influence by depressing migration [130,132]. Depending on the conditions, the relative influence of water temperature and flow in initiating migration can differ across years [137]. Other environmental cues, like the photoperiod, have been found to control the initiation of downstream migration [138], but temperature and flow are the key environmental factors to be considered in response to climate change.

When ready, smolts lose their willingness to maintain station in a flow and start migrating downstream with the aid of the current. The speed of the current influences the downstream travel time, but smolts actively swim, typically following the mainstream in the surface water layer [7,120,139]. Smolts predominantly migrate at night, but this may change later in the migration period [2,5,140]. Smolts migrate downstream in schools of varying sizes. Relatively little is known about the formation of these groups. Some results indicate solitary movement from natal streams, followed by schooling further downstream [141]. A genetic component is involved as Atlantic salmon smolts migrate more in kin-structured groups than with unrelated individuals [142]. Some environmental factors like light and dark variations can influence schooling [143].

The timing of migrations has been adapted via evolution to avoid unfavorable conditions and arrive when environmental conditions are suitable for survival and growth [4]. Mismatched timing would lead to decreased fitness, depleted food sources, and/or increased predation. As described above, the environment has an effect on migration timing, but it is also influenced by inheritance [144,145]. The relative contribution of genetic differences remains uncertain [146]. Under climate-induced environmental changes, different migrating phenological traits may be important for the fitness of individuals [128]. It is obvious that Atlantic salmon migration timing is already responding to warming temperatures: The initiation of a smolt's seaward migration has occurred approximately 2.5 days earlier per decade throughout the basin of the North Atlantic [18]. Accordingly, the long time series analysis (1978–2008) of the timing of the smolt migration of Atlantic salmon in the River Bush, Northern Ireland, revealed that earlier downstream migration periods were evident across the time series [147]. Kastl et al. [125] found that an increase from 10.2 to 12.8 °C in mean seasonal water temperature accelerated the migration window by three weeks in coho salmon living near its southern distribution range in California, USA.

The earlier migration timing has given rise to growing concerns about smolts potentially missing the optimal environmental migration "window" [23]. Global warming also affects the receiving marine ecosystem by increasing surface seawater temperatures, and the results of this mismatch are difficult to predict. Climate change affects how and when species interact, potentially decoupling species interactions, combining others, and reconstructing predator–prey interactions [148]. Some of these mismatches may lead to increased predation on smolt production or cause starvation; some may have no effect. A better understanding of how these interactions work is crucial to predict vulnerability to the effects of climate change. Monitoring the timing and number of migrating smolts is important for revealing the effects of the changing climate on the smolt run. The quantification of migrating smolts to produce assessments of possible changes in natural reproduction, rates of survival, and patterns of migration, for example, by smolt trapping, is important for management (Figure 2).

The changed timing of smolt migration may lead to long-term changes to the migratory phenotypes of salmonids, e.g., [4]. A wide migration window with a diversity of phenotypes can act as a safeguard against uncertainty in resource availability, buffering the variability in predator pressure or thermal mismatch. The survival of phenotypes can depend on seasonally fluctuating conditions, such as thermal or hydrological circumstances affecting food availability, either directly or indirectly [149]. For example, Sturrock et al. [108] found that relative proportions of migrating phenotypes that contributed to the spawning population differed between the wet and dry years in chinook salmon. In California's chinook salmon, the late migrating phenotype dominated, but other strategies played an important role for many years [76]. Kennedy and Crozier [147] observed that the marine survival of one sea winter Atlantic salmon was strongly influenced by the run timing, and during the observation period, later emigrating cohorts demonstrated increased survival. In lake-migrating sockeye salmon entering Lake Washington, juveniles migrating later in the season encountered higher zooplankton abundance and warmer water, but the optimal date for lake entry ranged across years by up to a month [149]. These examples show that the success of migratory phenotypes varies with environmental conditions. The warming of waters may highlight the importance of rare phenotypes in responding to climate change [76]. The loss of phenotypic diversity can, therefore, have an impact on population persistence in a warming climate.

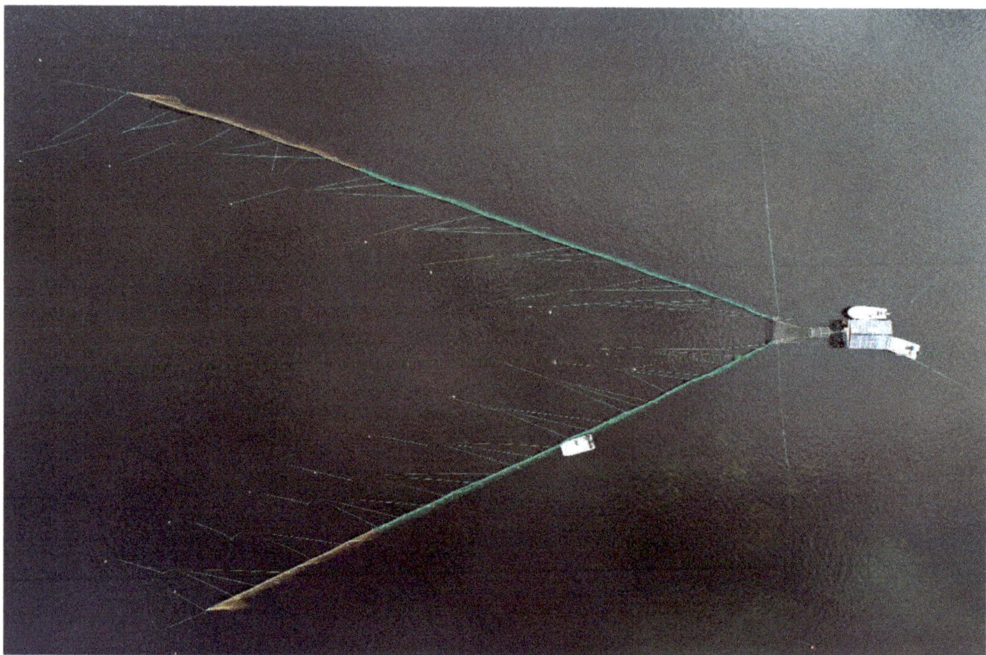

Figure 2. A smolt trap in the River Tornionjoki, Finland, to catch downstream migrating smolts and monitor their annual numbers and condition (Photo: Ville Vähä).

Anthropogenic use of fresh water, especially flow regulation with dam construction, has resulted in population declines and a loss of salmon life-history diversity [150,151]. Anthropogenic pressures, including climate change, affect the selection pressures in migratory salmonids, including on migrating smolt phenotypes. It would, therefore, be difficult to consider potential evolutionary responses to climate change without considering other human effects. For example, migration route selection at a hydropower plant intake has been found to be consistent with phenotypes, and those traits selecting turbines could potentially be eliminated from the population due to high mortality [152]. A long spill time is needed to protect the earliest and latest migrating phenotypes [131]. Low flows are expected to become more frequent, especially in the southern distribution area of salmonids, and during low flows, even small weirs can cause significant delays in smolt migration impacts [153]. River regulation practices are affected by climate-induced changes in temperature and precipitation, depending on the region, and they may also change selection pressures, affecting salmonid populations.

Survival during migration and patterns of mortality have the potential to yield important insights into population bottlenecks [154]. Smolts are vulnerable to predators during their downstream migration in the surface layer. In the southern River Minho (Spain/Portugal) and in the River Endrick (Scotland), the mortality of Atlantic salmon smolts by avian and piscine predators was high, demonstrating that the number of smolts lost in the river is likely to constrain population abundance in these rivers [154,155]. High inriver mortalities during downstream migration have also been found in Pacific salmon [156]. Climate change may create conditions that allow the successful spread of predators, including invasive species [157–159]. For example, increasing predator populations of cormorants

(*Phalacrocorax carbo sinensis* (Staunton, GL 1796)) and an invasive terrestrial predator in Europe, American mink (*Neovison vison* (Schreber, 1777), can cause elevated predation pressure on smolts [160–162]. Further, more detailed studies quantifying the impact of invasive species and climate change on smolt migration are needed for future management considerations.

6. Early Post-Smolt Survival

Most mortality between the smolt and adult stages is generally considered to occur during the first year of life at sea when survival, maturation, and migration trajectories are being defined [40,46,163,164]. Salmon's first year at sea, known as the post-smolt year, is characterized by variable mortality rates [165]. Mortality has often been considered to be highest during the first few months at sea [166,167]. Young salmonids are sensitive to variable climatic factors and food availability [168–170]. Reduced marine survival is widely accepted as an important contributor to the observed salmon population declines in recent decades [24,40,171,172]. Ocean climate variability during the first spring months of juvenile salmon migration to the sea seems to be central to the survival of North American populations, whereas summer climate variation appears to be important to adult recruitment variation for European populations [165]. In the Baltic Sea, marine survival estimates of salmon post-smolt were negatively correlated with temperature [173]. The anticipated warming due to global climate change will impose thermal conditions on salmon populations outside the historical context and will challenge the ability of many populations to persist [165].

The timing of salmon smolts' seaward migration and the size of smolts must be balanced with the marine conditions for the successful fulfillment of the life cycle [18,23,174]. Smolts' seaward migration should coincide with optimal thermal conditions at sea to maximize survival [2,40,175], but climate change has advanced the timing of salmon smolt migration and created a mismatch with optimal conditions for post-smolt growth and survival [18,147,176,177]. In the Gulf of St. Lawrence in the northwest Atlantic, the survival of sea-entering small smolts was found inferior to that of large smolts [178]. Smolt size can also influence the subsequent growth rate of Atlantic salmon at sea, with larger smolts showing slower growth [179]. Observations on brown trout in the River Imsa, Norway, suggest that an increased water temperature will induce seaward migration in the early spring, when sea growth and survival are poor [170].

Warmer temperatures in the North Atlantic have modified oceanic conditions, reducing the growth and survival of salmon by decreasing marine feeding opportunities [40,46,180,181]. Spring plankton blooms and, therefore, the peak of higher trophic resources available for salmon may be advanced in the season and may occur in different places [182–184], potentially creating a mismatch between salmon smolt migration and available resources [172,185]. A climate-driven shift in the zooplankton community composition towards more temperature-tolerant species with limited nutritional content may be associated with the decreased marine survival and growth of salmon smolts [169].

On a local scale, controlling climate change drivers is impossible. Proactive measures against population decline are, therefore, recommended [186]. These measures can include sustainable land use in the catchment and maintaining a diverse mosaic of habitats for salmonids [76,186]. Catchment scale conservation, including flow and connectivity restoration, is an important management priority for maintaining and improving juvenile salmonid, and thus smolt, production. Conserving headwater stream habitats maintains and increases the variability in habitats and the life history of salmonids to mitigate the effects of climate change. The freshwater environment is especially vulnerable to climate change effects because it is already exposed to numerous anthropogenic pressures, and water temperature and flow are highly climate dependent [187,188]. To secure the adaptive variability of a smolt, a safe downstream passage should be ensured at hydro dams, either with physical structures or sufficiently long spill time windows. Improved flow management is needed under climate change to avoid a further loss of phenotypic diversity in salmonids. It is important to integrate management throughout the life cycle, including both sea- and freshwater phases, to secure a positive outcome for the salmonid populations (Figure 3).

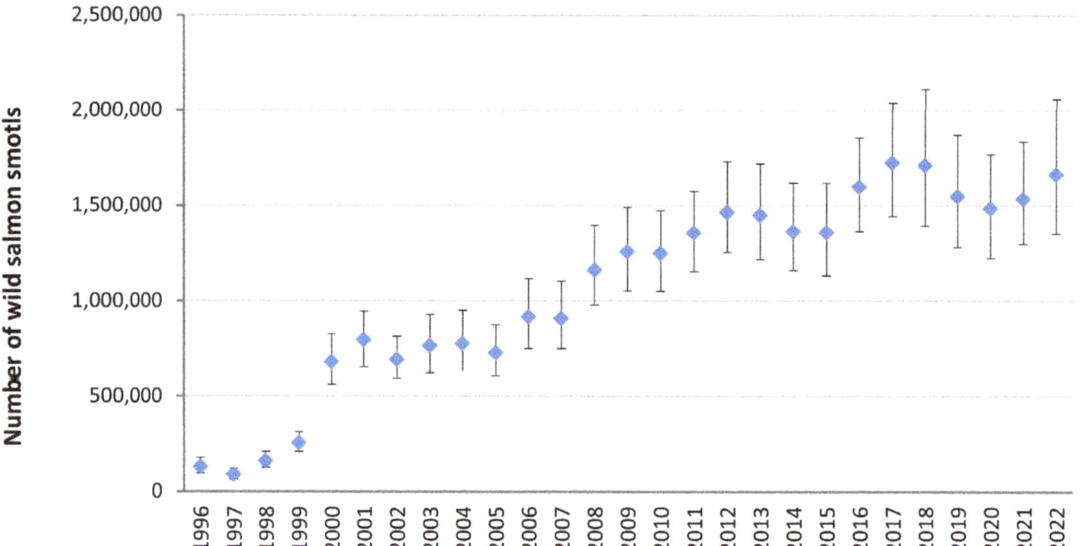

Figure 3. Toward a better future: After successful management actions in the Baltic Sea and in the river itself, the annual number of smolts migrating from the River Tornionjoki has increased substantially [189].

7. Conclusions

Atlantic salmon are already experiencing and responding to climate-change-induced warmer water temperatures at different scales. On a large scale, there are signs of salmon expanding their range northward, as expected due to a shift in the suitable thermal habitat, while southern populations are struggling more due to high temperatures and periodic droughts. While an increased temperature can have positive effects on the production of northern populations, increasing growth leading to earlier smolting also results in concerns about increased water quality problems via water brownification and eutrophication, particularly from human-impacted land areas with the effects of increased rain, especially during the winter. The risk of local extinction of salmon populations has increased, especially at the southern edge of salmon distribution. Low flow events, especially those typically associated with high water temperatures, are population bottlenecks. Mitigating the effects of climate change on a local scale to increase smolt production includes flow management and precautionary efforts to maintain and improve the ecological status of rivers. These measures are land-use planning and restoration on a catchment scale to diminish loading from the catchment and promote the conservation and restoration of instream habitats. Dense forests along riverbanks can decrease the water temperature.

Smolt characteristics depend in many ways on the factors acting in fresh water, and these characteristics affect post-smolt survival in the feeding area. Thermal heterogeneity in the river plays a significant role in survival and growth, and we should have better knowledge of the magnitude and location of cold-water refuges in streams. Mapping these areas with modern technology would help in conservation work. Maintaining a diverse mosaic of habitats and connectivity via conservation and restoration is crucial for mitigating climate change effects in rivers.

In response to increasing temperatures, an earlier migration timing of smolts is evident throughout the range of salmonids. This changes how and when species interact. It also restructures predator−prey interactions. To readjust to the changed, and still changing, conditions, it is important to maintain the widest possible smolt window to allow all existing phenotypes, whether early or late migrants, to prevail. Under climate change, different migrating phenological traits may be especially important to the future fitness of the species. This is especially important in regulated rivers, where the anthropogenic alteration of waterflow creates not only increased mortality but also artificial selection pressure on migrating smolts. For example, this would mean longer spill water times or keeping the downstream routes open throughout the migration period.

Predation creates a substantial impact on migrating smolts and, thus, on the entire population. Climate change enhances the spread of invasive species, including invasive predators, which can increase the total predation pressure on smolts. This emphasizes better control of invasive species, the prevention of their dispersal, and better control of their populations.

Finally, we agree with the previous literature, stating that collaboration and research among scientists and managers across life cycle stages and ecosystems are urgently needed to address the research gaps [27] and that the basic strategy to protect salmonids against the effects of climate change should be to ensure that the maximum number of wild smolts in the best condition leave rivers, e.g., [23,24].

Author Contributions: Conceptualization, T.V., T.S. and A.H.; methodology, T.V., T.S. and A.H.; resources, T.V., T.S. and A.H.; data curation, T.V., T.S. and A.H.; writing—original draft preparation, T.V., T.S. and A.H; writing—review and editing, T.V., T.S. and A.H.; visualization, T.V., T.S. and A.H.; supervision, T.V.; project administration, T.V.; funding acquisition, T.V. All authors have read and agreed to the published version of the manuscript.

Funding: This research was funded by the Natural Resources Institute Finland.

Data Availability Statement: No new data were created or analyzed in this study. Data sharing is not applicable to this article.

Conflicts of Interest: The authors declare no conflict of interest.

References

1. Dittman, A.H.; Quinn, T.P. Homing in Pacific Salmon: Mechanisms and Ecological Basis. *J. Exp. Biol.* **1996**, *199*, 83–91. [CrossRef]
2. McCormick, S.D.; Hansen, L.P.; Quinn, T.P.; Saunders, R.L. Movement, migration, and smolting of *Atlantic salmon* (*Salmo salar*). *Can. J. Fish. Aquat. Sci.* **1998**, *55*, 77–92. [CrossRef]
3. Quinn, T.P.; Myers, K.W. Anadromy and the marine migrations of Pacific salmon and trout: Rounsefell revisited. *Rev. Fish. Biol. Fish.* **2004**, *14*, 421–442. [CrossRef]
4. Lennox, R.J.; Chapman, J.M.; Souliere, C.M.; Tudorache, C.; Wikelski, M.; Metcalfe, J.D.; Cooke, S.J. Conservation physiology of animal migration. *Conserv. Physiopedia* **2016**, *4*, cov072. [CrossRef] [PubMed]
5. Thorstad, E.B.; Whoriskey, F.; Uglem, I.; Moore, A.; Rikardsen, A.H.; Finstad, B. A critical life stage of the *Atlantic salmon Salmo salar*: Behaviour and survival during the smolt and initial post-smolt migration. *J. Fish Biol.* **2012**, *81*, 500–542. [CrossRef]
6. Lucas, M.C.; Baras, E.; Thom, T.J.; Duncan, D.; Slavík, O. *Migration of Freshwater Fish*; Blackwell Scientific Publishing: Oxford, UK, 2001; p. 412.
7. Jonsson, B.; Jonsson, N. A review of the likely effects of climate change on anadromous *Atlantic salmon Salmo salar* and brown trout *Salmo trutta*, with particular reference to water temperature and flow. *J. Fish Biol.* **2009**, *75*, 2381–2447. [CrossRef]
8. Erkinaro, J.; Czorlich, Y.; Orell, P.; Kuusela, J.; Falkegård, M.; Länsman, M.; Pulkkinen, H.; Primmer, C.R.; Niemelä, E. Life history variation across four decades in a diverse population complex of *Atlantic salmon* in a large subarctic river. *Can. J. Fish. Aquat. Sci.* **2018**, *76*, 42–55. [CrossRef]
9. Veselov, A.E.; Kazakov, R.V.; Sysoyeva, M.I.; Bahmet, I.N. Ontogenesis of rheotactic and optomotor responses of juvenile *Atlantic salmon*. *Aquaculture* **1998**, *168*, 17–26. [CrossRef]
10. Björnsson, B.T.; Stefansson, S.O.; McCormick, S.D. Environmental endocrinology of salmon smoltification. *Gen. Comp. Endocrinol.* **2011**, *170*, 290–298. [CrossRef]
11. Piironen, J.; Kiiskinen, P.; Huuskonen, H.; Heikura-Ovaskainen, M.; Vornanen, M. Comparison of smoltification in *Atlantic salmon* (*Salmo salar*) from anadromous and landlocked populations under common garden conditions. *Ann. Zool. Fenn.* **2013**, *50*, 1–15. [CrossRef]
12. Nilsen, T.O.; Ebbeson, L.O.E.; Stefansson, S.O. Smolting in anadromous and landlocked strains of *Atlantic salmon* (*Salmo salar*). *Aquaculture* **2003**, *222*, 71–82. [CrossRef]

13. Crozier, L.G.; Burke, B.J.; Chasco, B.E.; Widener, D.L.; Zabel, R.W. Climate change threatens Chinook salmon throughout their life cycle. *Commun. Biol.* **2021**, *4*, 222. [CrossRef] [PubMed]

14. Olmos, M.; Massiot-Granier, F.; Prévost, E.; Chaput, G.; Bradbury, I.R.; Nevoux, M.; Rivot, E. Evidence for spatial coherence in time trends of marine life history traits of *Atlantic salmon* in the North Atlantic. *Fish Fish.* **2019**, *20*, 322–342. [CrossRef]

15. Lehnert, S.J.; Kess, T.; Bentzen, P.M.; Kent, P.; Lien, S.; Gilbey, J.; Clément, M.; Jeffery, N.W.; Waples, R.S.; Bradbury, I.R. Genomic signatures and correlates of widespread population declines in salmon. *Nat. Commun.* **2019**, *10*, 2996. [CrossRef] [PubMed]

16. Sobocinski, K.L.; Greene, C.M.; Anderson, J.H.; Kendall, N.W.; Schmidt, M.W.; Zimmerman, M.S.; Kemp, I.M.; Kim, S.; Ruff, C.P. A hypothesis-driven statistical approach for identifying ecosystem indicators of coho and Chinook salmon marine survival. *Ecol. Indic.* **2021**, *124*, 107403. [CrossRef]

17. Dadswell, M.; Spares, A.; Reader, J.; McLean, M.; McDermott, T.; Samways, K.; Lilly, J. The Decline and Impending Collapse of the *Atlantic salmon* (*Salmo salar*) Population in the North Atlantic Ocean: A Review of Possible Causes. *Rev. Fish. Sci. Aquac.* **2022**, *30*, 215–258. [CrossRef]

18. Otero, J.; L'Abée-Lund, J.H.; Castro-Santos, T.; Leonardsson, K.; Storvik, G.O.; Jonsson, B.; Dempson, B.; Russell, I.C.; Jensen, A.J.; Baglinière, J.-L.; et al. Basin-scale phenology and effects of climate variability on global timing of initial seaward migration of *Atlantic salmon* (*Salmo salar*). *Glob. Chang. Biol.* **2014**, *20*, 61–75. [CrossRef]

19. Beamish, R.J.; Mahnken, C. A critical size and period hypothesis to explain natural regulation of salmon abundance and the linkage to climate and climate change. *Prog. Oceanogr.* **2001**, *49*, 423–437. [CrossRef]

20. Huusko, A.; Hyvärinen, P. *Atlantic salmon* abundance and size track climate regimes in the Baltic Sea. *Boreal Environ. Res.* **2012**, *17*, 139–149.

21. Sear, D.; Langdon, P.; Leng, M.; Edwards, M.; Heaton, T.; Langdon, C.; Leyland, J. Climate and human exploitation have regulated *Atlantic salmon* populations in the River Spey, Scotland, over the last 2000 years. *Holocene* **2022**, *32*, 780–793. [CrossRef]

22. Spanjer, A.R.; Gendaszek, A.S.; Wulfkuhle, E.J.; Black, R.W.; Jaeger, K.L. Assessing climate change impacts on Pacific salmon and trout using bioenergetics and spatiotemporal explicit river temperature predictions under varying riparian conditions. *PLoS ONE* **2022**, *17*, e0266871. [CrossRef]

23. Russell, I.C.; Aprahamian, M.W.; Barry, J.; Davidson, I.C.; Fiske, P.; Ibbotson, A.T.; Kennedy, R.J.; Maclean, J.C.; Moore, A.; Otero, J.; et al. The influence of the freshwater environment and the biological characteristics of *Atlantic salmon* smolts on their subsequent marine survival. *ICES J. Mar. Sci.* **2012**, *69*, 1563–1573. [CrossRef]

24. Thorstad, E.B.; Bliss, D.; Breau, C.; Damon-Randall, K.; Sundt-Hansen, L.E.; Hatfield, E.M.C.; Horsburgh, G.; Hansen, H.; Maoiléidigh, N.Ó.; Sheehan, T.; et al. *Atlantic salmon* in a rapidly changing environment: Facing the challenges of reduced marine survival and climate change. *Aquatic Conserv. Mar. Freshw. Ecosyst.* **2021**, *3*, 2654–2665. [CrossRef]

25. Jonsson, N.; Jonsson, B.; Hansen, L.P. The relative role of density-dependent and density-independent survival in the life cycle of *Atlantic salmon Salmo salar*. *J. Anim. Ecol.* **1998**, *67*, 751–762. [CrossRef]

26. Jonsson, B.; Jonsson, M.; Jonsson, N. Optimal size at seaward migration in an anadromous salmonid. *Mar. Ecol. Prog. Ser.* **2016**, *559*, 193–200. [CrossRef]

27. Gillis, C.-A.; Ouellet, V.; Breau, C.; Frechette, D.; Bergeron, N. Assessing climate change impacts on North American freshwater habitat of wild *Atlantic salmon*: Urgent needs for collaborative research. *Can. Water Resour. J.* **2023**, *48*, 222–246. [CrossRef]

28. IPCC. Climate change 2022: Impacts, adaptation, and vulnerability. In *Contribution of Working Group II to the Sixth Assessment Report of the Intergovernmental Panel on Climate Change*; Pörtner, H.O., Roberts, D.C., Tignor, M., Poloczanska, E.S., Mintenbeck, K., Alegría, A., Craig, M., Langsdorf, S., Löschke, S., Möller, V., et al., Eds.; Cambridge University Press: Cambridge, UK; New York, NY, USA, 2022; p. 3056.

29. Thibeault, J.M.; Seth, A. Changing climate extremes in the Northeast United States: Observations and projections from CMIP5. *Clim. Chang.* **2014**, *127*, 273–287. [CrossRef]

30. Demaria, E.M.C.; Palmer, R.N.; Roundy, J.K. Regional Climate Change Projections of Streamflow in the Northeast and Midwest U.S. *J. Hydrol. Reg. Stud.* **2016**, *5*, 309–323. [CrossRef]

31. Hanssen-Bauer, I.; Førland, E.J. Long-term trends in precipitation and temperature in the Norwegian Arctic: Can they be explained by changes in the atmospheric circulation patterns? *Clim. Res.* **1998**, *10*, 143–153. [CrossRef]

32. Dore, M.H.I. Climate change and changes in global precipitation patterns: What do we know? *Environ. Int.* **2005**, *31*, 1167–1181. [CrossRef]

33. Guo, R.; Deser, C.; Terray, L.; Lehner, F. Human Influence on Winter Precipitation Trends (1921–2015) over North America and Eurasia Revealed by Dynamical Adjustment. *Geophys. Res. Lett.* **2019**, *46*, 3426–3434. [CrossRef]

34. Terrier, A.; Martin, P.; Girardin, M.P.; Périé, C.; Legendre, P.; Bergeron, Y. Potential changes in forest composition could reduce impacts of climate change on boreal wildfires. *Ecol. Appl.* **2013**, *23*, 21–35. [CrossRef] [PubMed]

35. Kazakov, R.V. History and *Atlantic salmon* fishery state in Russia. In *Atlanticheskyi Losos*; Kazakov, R.V., Ed.; Nauka: Sankt-Peterburg, Russia, 1998; pp. 335–380.

36. Nicola, G.G.; Elvira, B.; Jonsson, B.; Ayllón, D.; Almodóvar, A. Local and global climatic drivers of *Atlantic salmon* decline in southern Europe. *Fish. Res.* **2018**, *198*, 78–85. [CrossRef]

37. Todd, C.D.; Friedland, K.D.; MacLean, J.C.; Hazon, N.; Jensen, A.J. Getting into hot water? *Atlantic salmon* responses to climate change in freshwater and marine environments. In *Atlantic Salmon Ecology*; Aas, Ø., Klemetsen, A., Einum, S., Skurdal, J., Eds.; Blackwell Publishing Ltd.: Oxford, UK, 2011; pp. 409–443.

38. Hedger, R.D.; Sundt-Hansen, L.E.; Forseth, T.; Ugedal, O.; Diserud, O.H.; Kvambekk, Å.S.; Finstad, A.G. Predicting climate change effects on subarctic–Arctic populations of *Atlantic salmon* (*Salmo salar*). *Can. J. Fish. Aquat. Sci.* **2013**, *70*, 159–168. [CrossRef]

39. Hastings, R.A.; Rutterford, L.A.; Freer, J.J.; Collin, R.A.; Simpson, S.D.; Genner, M.J. Climate change drives poleward increases and equatorward declines in marine species. *Curr. Biol.* **2020**, *30*, 1572–1577. [CrossRef]

40. Gillson, J.P.; Bašić, T.; Davison, P.I.; Riley, W.D.; Talks, L.; Walker, A.M.; Russell, I.C. A review of marine stressors impacting *Atlantic salmon Salmo salar*, with an assessment of the major threats to English stocks. *Rev. Fish. Biol. Fish.* **2022**, *32*, 879–919. [CrossRef]

41. Jensen, A.J.; Karlsson, S.; Fiske, P.; Hansen, L.P.; Østborg, G.M.; Hindar, K. Origin and life history of *Atlantic salmon* (*Salmo salar*) near their northernmost oceanic limit. *Can. J. Fish. Aquat. Sci.* **2014**, *71*, 1740–1746. [CrossRef]

42. Bilous, M.; Dunmall, K. *Atlantic salmon* in the Canadian Arctic: Potential dispersal, establishment, and interaction with Arctic char. *Rev. Fish Biol. Fish.* **2020**, *30*, 463–483. [CrossRef]

43. Parrish, D.L.; Behnke, R.J.; Gephard, S.R.; McCormick, S.D.; Reeves, G.H. Why aren't there more *Atlantic salmon* (*Salmo salar*)? *Can. J. Fish. Aquat. Sci.* **1998**, *55*, 281–287. [CrossRef]

44. Almodóvar, A.; Ayllón, D.; Nicola, G.G.; Jonsson, B.; Elvira, B. Climate-driven biophysical changes in feeding and breeding environments explain the decline of southernmost European *Atlantic salmon* populations. *Can. J. Fish. Aquat. Sci.* **2021**, *76*, 1581–1595. [CrossRef]

45. Almodóvar, A.; Nicola, G.G.; Ayllón, D.; Leal, S.; Marchán, D.F.; Elvira, B. A Benchmark for *Atlantic salmon* Conservation: Genetic Diversity and Structure in a Southern European Glacial Refuge before the Climate Changed. *Fishes* **2023**, *8*, 321. [CrossRef]

46. Friedland, K.D.; MacLean, J.C.; Hansen, L.P.; Peyronnet, A.J.; Karlsson, L.; Reddin, D.G.; Maoiléidigh, Ó.N.; McCarthy, J.L. The recruitment of *Atlantic salmon* in Europe. *ICES J. Mar. Sci.* **2009**, *66*, 289–304. [CrossRef]

47. Mills, K.E.; Pershing, A.J.; Sheehan, T.F.; Mountain, D. Climate and ecosystem linkages explain widespread declines in North American *Atlantic salmon* populations. *Glob. Chang. Biol.* **2013**, *19*, 3046–3061. [CrossRef] [PubMed]

48. Berg, O.K. The formation of non-anadromous populations of *Atlantic salmon, Salmo salar* L., in Europe. *J. Fish Biol.* **1985**, *27*, 805–811. [CrossRef]

49. Ozerov, M.Y.; Veselov, A.J.; Lumme, J.; Primmer, C.R. Genetic structure of freshwater *Atlantic salmon* (*Salmo salar* L.) populations from the lakes Onega and Ladoga of northwest Russia and implications for conservation. *Conserv. Genet.* **2010**, *11*, 1711–1724. [CrossRef]

50. Lumme, J.; Ozerov, M.Y.; Veselov, A.E.; Primmer, C.R. The formation of landlocked populations of *Atlantic salmon*. In *Evolutionary Biology of the Atlantic Salmon*; Vladic, T., Pettersson, E., Eds.; CRC Press: Boca Raton, FL, USA, 2015; p. 297.

51. Kazakov, R.V. Distribution of *Atlantic salmon, Salmo salar* L., in freshwater bodies of Europe. *Aquac. Fish. Manag.* **1992**, *23*, 461–475. [CrossRef]

52. Leinonen, T.; Piironen, J.; Koljonen, M.-L.; Koskiniemi, J.; Kause, A. Restored river habitat provides a natural spawning area for a critically endangered landlocked *Atlantic salmon* population. *PLoS ONE* **2020**, *15*, e0232723. [CrossRef] [PubMed]

53. Segherloo, I.H.; Freyhof, J.; Berrebi, P.; Ferchaud, A.-L.; Geiger, M.; Laroche, J.; Levin, B.A.; Normandeau, E.; Bernatchez, L. A genomic perspective on an old question: Salmo trouts or Salmo trutta (Teleostei: Salmonidae)? *Mol. Phylogenet. Evol.* **2021**, *162*, 107204. [CrossRef]

54. Filipe, A.F.; Markovic, D.; Pletterbauer, F.; Tisseuil, C.; De Wever, A.; Schmutz, S.; Bonada, N.; Freyhof, J. Forecasting fish distribution along stream networks: Brown trout (*Salmo trutta*) in Europe. *Divers. Distrib.* **2013**, *19*, 1059–1071. [CrossRef]

55. Beaupré, J.; Boudreault, J.; Bergeron, N.E.; St-Hilaire, A. Inclusion of water temperature in a fuzzy logic *Atlantic salmon* (*Salmo salar*) parr habitat model. *J. Therm. Biol.* **2020**, *87*, 102471. [CrossRef]

56. Corey, E.; Linnansaari, T.; Cunjak, R.A. High temperature events shape the broadscale distribution of juvenile *Atlantic salmon* (*Salmo salar*). *Freshw. Biol.* **2023**, *63*, 534–545. [CrossRef]

57. Wilbur, N.M.; O'Sullivan, A.M.; MacQuarrie, K.T.B.; Linnansaari, T.; Curry, R.A. Characterizing physical habitat preferences and thermal refuge occupancy of brook trout (*Salvelinus fontinalis*) and *Atlantic salmon* (*Salmo salar*) at high river temperatures. *Riv. Res. Appl.* **2020**, *36*, 769–783. [CrossRef]

58. Schreck, C.B.; Tort, L. The Concept of Stress in Fish. *Fish Physiol.* **2016**, *35*, 1–34.

59. Zhang, X.; Li, H.-Y.; Deng, Z.D.; Leung, L.R.; Skalski, J.R.; Cooke, S.J. On the variable effects of climate change on Pacific salmon. *Ecol. Model.* **2019**, *397*, 95–106. [CrossRef]

60. Cunningham, C.J.; Westley, P.A.H.; Adkison, M.D. Signals of large scale climate drivers, hatchery enhancement, and marine factors in Yukon River Chinook salmon survival revealed with a Bayesian life history model. *Glob. Chang. Biol.* **2018**, *24*, 4399–4416. [CrossRef] [PubMed]

61. Hvidsten, N.A.; Diserud, O.H.; Jensen, A.J.; Jensås, J.G.; Johnsen, B.O.; Ugedal, O. Water discharge affects *Atlantic salmon Salmo salar* smolt production: A 27 year study in the River Orkla, Norway. *J. Fish Biol.* **2015**, *86*, 92–104. [CrossRef]

62. Gilbert, M.J.H.; Tierney, K.B. Warm northern river temperatures increase post-exercise fatigue in an Arctic migratory salmonid but not in a temperate relative. *Func. Ecol.* **2018**, *32*, 687–700. [CrossRef]

63. Kaushal, S.S.; Likens, G.E.; Jaworski, N.A.; Pace, M.L.; Sides, A.M.; Seekell, D.; Belt, K.T.; Secor, D.H.; Wingate, R.L. Rising stream and river temperatures in the United States. *Front. Ecol. Environ.* **2010**, *8*, 461–466. [CrossRef]

64. Isaak, D.J.; Wollrab, S.; Horan, D.; Chandler, G. Climate change effects on stream and river temperatures across the northwest U.S. from 1980–2009 and implications for salmonid fishes. *Clim. Chang.* **2012**, *113*, 499–524. [CrossRef]

65. Arevalo, E.; Maire, A.; Tétard, S.; Prévost, E.; Lange, F.; Marchand, F.; Josset, Q.; Drouineau, H. Does global change increase the risk of maladaptation of *Atlantic salmon* migration through joint modifications of river temperature and discharge? *Proc. R. Soc. B* **2021**, *288*, 20211882. [CrossRef]

66. Smalås, A.; Primicerio, R.; Kahilainen; Terentyev, P.; Kashulin, N.; Zubova, E.; Amundsen, P.-A. Increased importance of cool-water fish at high latitudes emerges from individual level responses to warming. *Authorea* **2023**, *13*, e10185. [CrossRef]

67. Solomon, D.J.; Lightfoot, G.W. *The Thermal Biology of Brown Trout and Atlantic Salmon*; Science Reports; Environment Agency: Bristol, UK, 2008; BS32 4UD; p. 42.

68. McGinnity, P.; Jennings, E.; deEyto, E.; Allott, N.; Samuelsson, P.; Rogan, G.; Whelan, K.; Cross, T. Impact of naturally spawning captive-bred *Atlantic salmon* on wild populations: Depressed recruitment and increased risk of climate-mediated extinction. *Proc. R. Soc. B* **2009**, *276*, 3601–3610. [CrossRef] [PubMed]

69. Piou, C.; Prevost, E. Contrasting effects of climate change in continental vs. oceanic environments on population persistence and microevolution of *Atlantic salmon*. *Glob. Chang. Biol.* **2013**, *19*, 711–723. [CrossRef]

70. Burton, T.; McKelvey, S.; Stewart, D.C.; Armstrong, J.D.; Metcalfe, N.B. Offspring investment in wild *Atlantic salmon* (*Salmo salar*): Relationships with smolt age and spawning condition. *Ecol. Freshw. Fish* **2013**, *22*, 317–332. [CrossRef]

71. Burt, J.M.; Hinch, S.G.; Patterson, D.A. The importance of parentage in assessing temperature effects on fish early life history: A review of the experimental literature. *Rev. Fish Biol. Fish.* **2011**, *21*, 377–406. [CrossRef]

72. Jonsson, B.; Jonsson, N. Egg incubation temperature affects the timing of the *Atlantic salmon Salmo salar* homing migration. *J. Fish Biol.* **2018**, *93*, 1016–1020. [CrossRef]

73. Thompson, J.N.; Beauchamp, D. Size-Selective Mortality of Steelhead during Freshwater and Marine Life Stages Related to Freshwater Growth in the Skagit River, Washington. *Trans. Am. Fish. Soc.* **2014**, *143*, 910–925. [CrossRef]

74. Jonsson, N.; Jonsson, B.; Hansen, L.P. Does climate during embryonic development influence parr growth and age of seaward migration in *Atlantic salmon* (*Salmo salar*)? *Can. J. Fish. Aquat. Sci.* **2005**, *62*, 2502–2508. [CrossRef]

75. Crozier, L.G.; Zabel, R.W.; Hockersmith, E.E.; Achord, S. Interacting effects of density and temperature on body size in multiple populations of Chinook salmon. *J. Anim. Ecol.* **2010**, *79*, 342–349. [CrossRef] [PubMed]

76. Cordoleani, F.; Phillis, C.C.; Sturrock, A.M.; FitzGerald, A.M.; Malkassian, A.; Whitman, G.E.; Weber, P.K.; Johnson, R.C. Threatened salmon rely on a rare life history strategy in a warming landscape. *Nat. Clim. Chang.* **2021**, *11*, 982–988. [CrossRef]

77. Fullerton, A.H.; Burke, B.J.; Lawler, J.J.; Torgersen, C.E.; Ebersole, J.L.; Leibowitz, S.G. Simulated juvenile salmon growth and phenology respond to altered thermal regimes and stream network shape. *Ecosphere* **2017**, *8*, e02052. [CrossRef] [PubMed]

78. Isaak, D.J.; Young, M.K. Cold-water habitats, climate refugia, and their utility for conserving salmonid fishes. *Can. J. Fish. Aquat. Sci.* **2023**, *80*, 1187–1206. [CrossRef]

79. Dugdale, S.J.; Bergeron, N.E.; St-Hilaire, A. Temporal variability of thermal refuges and water temperature patterns in *Atlantic salmon* rivers. *Remote Sens. Environ.* **2013**, *136*, 358–373. [CrossRef]

80. Battin, J.; Wiley, M.W.; Ruckelshaus, M.H.; Imaki, H. Projected impacts of climate change on salmon habitat restoration. *Proc. Natl. Acad. Sci. USA* **2007**, *104*, 6720–6725. [CrossRef]

81. Foley, K.M.; Rosenberger, S.A.; Mueter, F.J. Longitudinal Patterns of Juvenile Coho Salmon Distribution and Densities in Headwater Streams of the Little Susitna River, Alaska. *Trans. Am. Fish. Soc.* **2018**, *147*, 247–264. [CrossRef]

82. Jonsson, B.; Jonsson, N.; Brodtkorb, E.; Ingebrigtsen, P.-J. Life-history traits of Brown Trout vary with the size of small streams. *Func. Ecol.* **2001**, *15*, 310–317. [CrossRef]

83. Sutela, T.; Vehanen, T.; Jounela, P. Longitudinal patterns of fish assemblages in European boreal streams. *Hydrobiologia* **2020**, *847*, 3277–3290. [CrossRef]

84. Jespersen, H.; Rasmussen, G.; Pedersen, S. Severity of summer drought as predictor for smolt recruitment in migratory brown trout (*Salmo trutta*). *Ecol. Freshw. Fish* **2021**, *30*, 115–124. [CrossRef]

85. Mantua, N.; Tohver, I.; Hamlet, A. Climate change impacts on streamflow extremes and summertime stream temperature and their possible consequences for freshwater salmon habitat in Washington State. *Clim. Chang.* **2010**, *102*, 187–223. [CrossRef]

86. Kang, D.H.; Gao, H.; Shi, X.; ul Islam, S.; Déry, S.J. Impacts of a rapidly declining mountain snowpack on streamflow timing in Canada's Fraser River basin. *Sci. Rep.* **2016**, *6*, 19299. [CrossRef]

87. Caissie, D.C.; Breau, J.H.; Cameron, P. *Water Temperature Characteristics within the Miramichi and Restigouche Rivers*; Research Document 2012/165; Canadian Science Advisory Secretariat: Miramichi, NB, Canada, 2013.

88. Hayes, D.S.; Moreira, M.; Boavida, I.; Haslauer, M.; Unfer, G.; Zeiringer, B.; Greimel, F.; Auer, S.; Ferreira, T.; Schmutz, S. Life stage-specific hydropeaking flow rules. *Sustainability* **2019**, *11*, 1547. [CrossRef]

89. Huusko, A.; Greenberg, L.; Stickler, M.; Linnansaari, T.; Nykänen, M.; Vehanen, T.; Koljonen, S.; Louhi, P.; Alfredsen, K. Life in the Ice Lane: A Review of the Ecology of Salmonids during Winter. *River Res. Appl.* **2007**, *23*, 469–491. [CrossRef]

90. Mäki-Petäys, A.; Muotka, T.; Huusko, A.; Tikkanen, P.; Kreivi, P. Seasonal changes in habitat use and preference by juvenile brown trout, *Salmo trutta*, in a northern boreal river. *Can. J. Fish. Aquat. Sci.* **1997**, *54*, 520–530.

91. Magnuson, J.J.; Robertson, D.M.; Benson, B.J.; Wynne, R.H.; Livingstone, D.M.; Arai, T.; Assel, R.A.; Barry, R.G.; Card, V.; Kuusisto, K.; et al. Historical trends in lake and river ice cover in the Northern Hemisphere. *Science* **2000**, *289*, 1743–1746. [CrossRef] [PubMed]

92. Pörtner, H.O.; Peck, M.A. Climate change effects on fishes and fisheries: Towards a cause-and-effect understanding. *J. Fish Biol.* **2010**, *77*, 1745–1779. [CrossRef]

93. van Vliet, M.T.H.; Franssen, W.H.P.; Yearsley, J.R.; Ludwig, F.; Haddeland, I.; Lettenmaier, D.P.; Kabat, P. Global river discharge and water temperature under climate change. *Glob. Environ. Chang.* **2013**, *23*, 450–464. [CrossRef]

94. Härkönen, L.; Louhi, P.; Huusko, R.; Huusko, A. Wintertime growth in *Atlantic salmon* under changing climate: The importance of ice cover for individual growth dynamics. *Can. J. Fish. Aquat. Sci.* **2021**, *78*, 1479–1485. [CrossRef]

95. Graham, W.D.; Thorpe, J.E.; Metcalfe, N.B. Seasonal current holding performance of juvenile *Atlantic salmon* in relation to temperature and smolting. *Can. J. Fish. Aquat. Sci.* **1996**, *53*, 80–86. [CrossRef]

96. Enders, E.C.; Stickler, M.; Pennell, C.J.; Cote, D.; Alfredsen, K.; Scruton, D.A. Habitat use of *Atlantic salmon* parr (*Salmo salar* L.) during winter. In Proceedings of the 14th Workshop on the Hydraulics of Ice Covered Rivers, Quebec City, QC, Canada, 19–22 June 2007; CGU HS Committee on River Ice Processes and the Environment: Saskatoon, SK, Canada, 2007; pp. 20–22.

97. Cunjak, R.A.; Prowse, T.D.; Parrish, D.L. *Atlantic salmon* (*Salmo salar*) in winter: "the season of parr discontent"? *Can. J. Fish. Aquat. Sci.* **1999**, *55*, 161–180. [CrossRef]

98. Comte, L.; Olden, J.D.; Tedesco, P.A.; Ruhi, A.; Giam, X. Climate and land-use changes interact to drive long-term reorganization of riverine fish communities globally. *Proc. Natl. Acad. Sci. USA* **2021**, *118*, e2011639118. [CrossRef]

99. Tabari, H. Climate change impact on flood and extreme precipitation increases with water availability. *Sci. Rep.* **2020**, *10*, 13768. [CrossRef] [PubMed]

100. Harp, R.D.; Horton, D.E. Observed Changes in Daily Precipitation Intensity in the United States. *Geophys. Res. Lett.* **2022**, *49*, e2022GL099955. [CrossRef]

101. Cronin, L.; Regan, F.; Lucy, F.E. Detection of transient pollution events in an Irish river catchment in the context of increasing frequency and intensity of rainfall events due to climate change. *EGU Gen. Assem.* **2023**, *EGU23*, 15993.

102. Weyhenmeyer, G.; Prairie, Y.T.; Tranvik, L.J. Browning of Boreal Freshwaters Coupled to Carbon-Iron Interactions along the Aquatic Continuum. *PLoS ONE* **2014**, *9*, e88104. [CrossRef] [PubMed]

103. Kritzberg, E.S.; Hasselquist, E.M.; Škerlep, M.; Löfgren, S.; Olsson, O.; Stadmark, J.; Valinia, S.; Hansson, L.-A.; Laudon, H. Browning of freshwaters: Consequences to ecosystem services, underlying drivers, and potential mitigation measures. *Ambio* **2020**, *49*, 375–390. [CrossRef]

104. Sethi, S.A.; Carey, M.P.; Gerken, J.; Harris, B.P.; Wolf, N.; Cunningham, C.; Restrepo, F.; Ashline, J. Juvenile salmon habitat use drives variation in growth and highlights vulnerability to river fragmentation. *Ecosphere* **2022**, *13*, e4192. [CrossRef]

105. Vargas-Chacoff, L.; Regish, A.M.; Weinstock, A.; McCormick, S.D. Effects of elevated temperature on osmoregulation and stress responses in *Atlantic salmon Salmo salar* smolts in fresh water and seawater. *J. Fish Biol.* **2018**, *93*, 550–559. [CrossRef]

106. Thorpe, J.E. Maturation responses of salmonids to changing developmental opportunities. *Mar. Ecol. Prog. Ser.* **2007**, *335*, 285–288. [CrossRef]

107. Copeland, T.; Venditti, D.A. Contribution of three life history types to smolt production in a Chinook salmon (*Oncorhynchus tshawytscha*) population. *Can. J. Fish. Aquat. Sci.* **2009**, *66*, 1658–1665. [CrossRef]

108. Sturrock, A.M.; Wikert, J.D.; Heyne, T.; Mesick, C.; Hubbard, A.E.; Hinkelman, T.M.; Weber, P.K.; Whitman, G.E.; Glessner, J.G.; Johnson, R.C. Reconstructing the Migratory Behavior and Long-Term Survivorship of Juvenile Chinook Salmon under Contrasting Hydrologic Regimes. *PLoS ONE* **2015**, *10*, e0122380. [CrossRef]

109. Bassett, M.C.; Patterson, P.A.; Shrimpton, J.M. Temporal and spatial differences in smolting among *Oncorhynchus nerka* populations throughout fresh and seawater migration. *J. Fish Biol.* **2018**, *93*, 510–518. [CrossRef] [PubMed]

110. Elsner, R.A.; Shrimpton, J.M. Is the duration of the smolt window related to migration distance in coho salmon *Oncorhynchus kisutch*? *J. Fish Biol.* **2018**, *93*, 501–509. [CrossRef] [PubMed]

111. Bohlin, T.; Dellefors, C.; Faremo, U. Optimal Time and Size for Smolt Migration in Wild Sea Trout (*Salmo trutta*). *Can. J. Fish. Aquat. Sci.* **2011**, *50*, 224–232. [CrossRef]

112. Jutila, E.; Jokikokko, E.; Julkunen, M. Long-term changes in the smolt size and age of *Atlantic salmon, Salmo salar* L., in a northern Baltic river related to parr density, growth opportunity and postsmolt survival. *Ecol. Freshw. Fish* **2006**, *15*, 321–330. [CrossRef]

113. Gregory, S.D.; Ibbotson, A.T.; Riley, W.D.; Nevoux, M.; Lauridsen, R.B.; Russell, I.A.; Britton, J.R.; Gillingham, P.K.; Simmons, O.M.; Rivot, E. *Atlantic salmon* return rate increases with smolt length. *ICES J. Mar. Sci.* **2019**, *76*, 1702–1712. [CrossRef]

114. Debes, P.V.; Piavchenko, N.; Erkinaro, J.; Primmer, C.R. Genetic growth potential, rather than phenotypic size, predicts migration phenotype in *Atlantic salmon*. *Proc. R. Soc. B* **2020**, *287*, 20200867. [CrossRef] [PubMed]

115. Økland, F.; Jonsson, B.; Jensen, A.J.; Hansen, L.P. Is there a threshold size regulating seaward migration of brown trout and *Atlantic salmon*? *J. Fish Biol.* **1993**, *42*, 541–550. [CrossRef]

116. Vollset, K.W.; Lennox, R.J.; Lamberg, A.; Skaala, Ø.; Sandvik, A.D.; Sægrov, H.; Kvingedal, E.; Kristensen, T.; Jensen, A.J.; Haraldstad, T.; et al. Predicting the nationwide outmigration timing of *Atlantic salmon* (*Salmo salar*) smolts along 12 degrees of latitude in Norway. *Divers. Distrib.* **2021**, *27*, 1383–1392. [CrossRef]

117. Archer, L.C.; Hutton, S.A.; Harman, L.; Michael, N.; O'Grady, M.N.; Kerry, J.P.; Poole, W.R.; Gargan, P.; McGinnity, P.; Reed, T.E. The Interplay Between Extrinsic and Intrinsic Factors in Determining Migration Decisions in Brown Trout (*Salmo trutta*): An Experimental Study. *Front. Ecol. Evol.* **2019**, *7*, 222. [CrossRef]

118. Jonsson, B.; Jonsson, N.; Finstad, A.G. Effects of temperature and food quality on age at maturity of ectotherms: An experimental test of *Atlantic salmon*. *J. Anim. Ecol.* **2012**, *81*, 201–210. [CrossRef]

119. Jones, D.A.; Bergman, E.; Greenberg, L. Food availability in spring affects smolting in brown trout (*Salmo trutta*). *Can. J. Fish. Aquat. Sci.* **2015**, *72*, 1694–1699. [CrossRef]

120. Fängstam, H.; Berglund, I.; Sjöberg, M.; Lundqvist, H. Effects of size and early sexual maturity on downstream migration during smolting in Baltic salmon (*Salmo salar*). *J. Fish Biol.* **1993**, *43*, 517–529. [CrossRef]

121. Jeppesen, E.; Kronvang, B.; Meerhoff, M.; Søndergaard, M.; Hansen, K.M.; Andersen, H.E.; Lauridsen, T.L.; Liboriussen, L.; Beklioglu, M.; Ozen, A.; et al. Climate change effects on runoff, catchment phosphorus loading and lake ecological state, and potential adaptations. *J. Environ. Qual.* **2009**, *38*, 1930–1941. [CrossRef] [PubMed]

122. Pihlainen, P.; Zandersen, M.; Hyytiäinen, K.; Andersen, H.E.; Bartosova, A.; Gustafsson, B.; Jabloun, M.; McCrackin, M.; Meier, H.E.M.; Olesen, J.E.; et al. Impacts of changing society and climate on nutrient loading to the Baltic Sea. *Sci. Total Environ.* **2020**, *731*, 138935. [CrossRef] [PubMed]

123. Vehanen, T.; Sutela, T.; Aroviita, J.; Karjalainen, S.M.; Riihimäki, J.; Larsson, A.; Vuori, K.-M. Land use in acid sulphate soils degrades river water quality: Do the biological quality metrics respond? *Ecol. Indic.* **2022**, *141*, 109085. [CrossRef]

124. Nilsen, T.O.; Ebbesson, L.O.; Handeland, S.O.; Kroglund, F.; Finstad, B.; Angotzi, A.R.; Stefansson, S.O. Atlantic salmon (*Salmo salar* L.) smolts require more than two weeks to recover from acidic water and aluminium exposure. *Aquat. Toxicol.* **2013**, *142–143*, 33–44. [CrossRef]

125. Kastl, B.; Obedzinski, M.; Carlson, S.M.; Boucher, W.T.; Grantham, T.E. Migration in drought: Receding streams contract the seaward migration window of endangered salmon. *Ecosphere* **2022**, *13*, e4295. [CrossRef]

126. Bernard, B.; Mandiki, S.N.M.; Duchatel, V.; Rollin, X.; Kestemont, P. A temperature shift on the migratory route similarly impairs hypo-osmoregulatory capacities in two strains of *Atlantic salmon* (*Salmo salar* L.) smolts. *Fish Physiol. Biochem.* **2019**, *45*, 1245–1260. [CrossRef]

127. Hansen, L.P.; Jonsson, B. Salmon ranching experiments in the River Imsa: Effect of timing of *Atlantic salmon* (*Salmo salar*) smolt migration on survival to adults. *Aquaculture* **1989**, *82*, 367–373. [CrossRef]

128. Bjerck, H.B.; Urke, H.A.; Haugen, T.O.; Alfredsen, J.A.; Ulvund, J.B.; Kristensen, T. Synchrony and multimodality in the timing of *Atlantic salmon* smolt migration in two Norwegian fjords. *Sci. Rep.* **2021**, *11*, 6504. [CrossRef]

129. Achord, S.; Zabel, R.W.; Sandford, B.P. Migration Timing, Growth, and Estimated Parr-to-Smolt Survival Rates of Wild Snake River Spring–Summer Chinook Salmon from the Salmon River Basin, Idaho, to the Lower Snake River. *Trans. Am. Fish. Soc.* **2007**, *136*, 142–154. [CrossRef]

130. Sykes, G.E.; Johnson, C.J.; Shrimpton, J.M. Temperature and Flow Effects on Migration Timing of Chinook Salmon Smolts. *Trans. Am. Fish. Soc.* **2009**, *138*, 1252–1265. [CrossRef]

131. Frechette, D.M.; Hawkes, J.P.; Kocik, J.F. Managing for *Atlantic salmon* Smolt Run Timing Variability in a Changing Climate. *N. Am. J. Fish. Manag.* **2023**, *43*, 517–538. [CrossRef]

132. Antonsson, T.; Gudjonsson, S. Variability in Timing and Characteristics of *Atlantic salmon* Smolt in Icelandic Rivers. *Trans. Am. Fish. Soc.* **2002**, *131*, 643–655. [CrossRef]

133. Zydlewski, G.B.; Haro, A.; McCormick, S.D. Evidence for cumulative temperature as an initiating and terminating factor in downstream migratory behavior of *Atlantic salmon* (*Salmo salar*) smolts. *Can. J. Fish. Aquat. Sci.* **2005**, *62*, 68–78. [CrossRef]

134. Jonsson, B.; Ruud-Hansen, J. Water temperature as the primary influence on timing of seaward migrations of *Atlantic salmon* (*Salmo salar*) smolts. *Can.). Fish. Aquat. Sci.* **1985**, *42*, 593–595. [CrossRef]

135. Hvidsten, N.A.; Jensen, A.J.; Vivås, H.; Bakke, Ø. Downstream migration of *Atlantic salmon* smolts in relation to water flow, water temperature, moon phase and social interaction. *Nord. J. Freshw. Res.* **1995**, *70*, 38–48.

136. Persson, L.; Kagervall, A.; Leonardsson, K.; Royan, M.; Alanärä, A. The effect of physiological and environmental conditions on smolt migration in *Atlantic salmon Salmo salar*. *Ecol. Freshw. Fish* **2019**, *28*, 190–199. [CrossRef]

137. Harvey, A.C.; Glover, K.A.; Wennevik, V.; Skaala, Ø. *Atlantic salmon* and sea trout display synchronised smolt migration relative to linked environmental cues. *Sci. Rep.* **2020**, *10*, 3529. [CrossRef]

138. Zydlewski, G.B.; Stich, D.S.; McCormick, S.D. Photoperiod control of downstream movements of *Atlantic salmon Salmo salar* smolts. *J. Fish Biol.* **2014**, *85*, 1023–1041. [CrossRef]

139. Karppinen, P.; Hynninen, M.; Vehanen, T.; Vähä, J.-P. Variations in migration behaviour and mortality of *Atlantic salmon* smolts in four different hydroelectric facilities. *Fish. Manag. Ecol.* **2021**, *28*, 253–267. [CrossRef]

140. Aarestrup, K.; Nielsen, C.; Koed, A. Net ground speed of downstream migrating radio-tagged *Atlantic salmon* (*Salmo salar* L.) and brown trout (*Salmo trutta* L.) smolts in relation to environmental factors. *Hydrobiologia* **2002**, *483*, 95–102. [CrossRef]

141. Riley, W.D. Seasonal downstream movements of juvenile *Atlantic salmon*, *Salmo salar* L., with evidence of solitary migration of smolts. *Aquaculture* **2007**, *273*, 194–199. [CrossRef]

142. Olsén, K.H.; Petersson, E.; Ragnarsson, B.; Lundqvist, H.; Järvi, T. Downstream migration in *Atlantic salmon* (*Salmo salar*) smolt sibling groups. *Can. J. Fish. Aquat. Sci.* **2011**, *61*, 328–331. [CrossRef]

143. Kemp, P.S.; Williams, J.G. Illumination influences the ability of migrating juvenile salmonids to pass a submerged experimental weir. *Ecol. Freshw. Fish* **2009**, *18*, 297–304. [CrossRef]

144. Nielsen, C.; Holdensgaard, G.; Petersen, H.C.; Björnsson, B.T.; Madsen, S.S. Genetic differences in physiology, growth hormone levels and migratory behaviour of *Atlantic salmon* smolts. *J. Fish Biol.* **2001**, *59*, 28–44. [CrossRef]

145. Stewart, D.C.; Middlemas, S.J.; Youngson, A.F. Population structuring in *Atlantic salmon* (*Salmo salar*): Evidence of genetic influence on the timing of smolt migration in sub-catchment stocks. *Ecol. Freshw. Fish* **2006**, *15*, 552–558. [CrossRef]

146. Crozier, L.G.; Hendry, A.P.; Lawson, P.W.; Quinn, T.P.; Mantua, N.J.; Battin, J.; Shaw, R.G.; Huey, R.B. Potential responses to climate change in organisms with complex life histories: Evolution and plasticity in Pacific salmon. *Evol. Appl.* **2008**, *1*, 252–270. [CrossRef]
147. Kennedy, R.J.; Crozier, W.W. Evidence of changing migratory patterns of wild *Atlantic salmon Salmo salar* smolts in the River Bush, Northern Ireland, and possible associations with climate change. *J. Fish Biol.* **2010**, *76*, 1786–1805. [CrossRef]
148. Wilson, S.M.; Patterson, D.A.; Moore, J.M. Intra- and inter-population variation in sensitivity of migratory sockeye salmon smolts to phenological mismatch. *Mar. Ecol. Prog. Ser.* **2022**, *692*, 119–136. [CrossRef]
149. Hovel, R.A.; Fresh, K.L.; Schroder, S.L.; Litt, A.H.; Quinn, T.P. A wide window of migration phenology captures inter-annual variability of favourable conditions: Results of a whole-lake experiment with juvenile Pacific salmon. *Freshw. Biol.* **2019**, *64*, 46–55. [CrossRef]
150. Beechie, T.; Buhle, E.; Ruckelshaus, M.; Fullerton, A.; Holsinger, L. Hydrologic regime and the conservation of salmon life history diversity. *Biol. Conserv.* **2006**, *130*, 560–572. [CrossRef]
151. McClure, M.M.; Carlson, S.M.; Beechie, T.J.; Pess, G.R.; Jorgensen, J.C.; Sogard, S.M.; Sultan, S.E.; Holzer, D.M.; Travis, J.; Sanderson, B.L.; et al. Evolutionary consequences of habitat loss for Pacific anadromous salmonids: Salmonid habitat loss and evolution. *Evol. Appl.* **2008**, *1*, 300–318. [CrossRef] [PubMed]
152. Haraldstad, T.; Haugen, T.O.; Olsen, E.M.; Forseth, T.; Höglund, E. Hydropower-induced selection of behavioural traits in *Atlantic salmon (Salmo salar)*. *Sci. Rep.* **2021**, *11*, 16444. [CrossRef]
153. Gauld, N.R.; Campbell, R.N.; Lucas, M.C. Reduced flow impacts salmonid smolt emigration in a river with low-head weirs. *Sci. Total Environ.* **2013**, *458–460*, 435–443. [CrossRef] [PubMed]
154. Chavarie, L.; Honkanen, M.; Newton, M.; Lilly, J.M.; Greetham, H.R.; Adams, C.A. The benefits of merging passive and active tracking approaches: New insights into riverine migration by salmonid smolts. *Ecosphere* **2022**, *13*, e4045. [CrossRef]
155. Flávio, H.; Kennedy, R.; Ensing, D.; Jepsen, N.; Aarestrup, K. Marine mortality in the river? Atlantic salmon smolts under high predation pressure in the last kilometres of a river monitored for stock assessment. *Fish. Manag. Ecol.* **2020**, *27*, 92–101.
156. Welch, D.W.; Porter, A.D.; Rechisky, E.L. A synthesis of the coast-wide decline in survival of West Coast Chinook Salmon (*Oncorhynchus tshawytscha*, Salmonidae). *Fish Fish.* **2021**, *22*, 194–211. [CrossRef]
157. Dukes, J.S.; Mooney, H.A. Does global change increase the success of biological invaders? *Trends Ecol. Evol.* **1999**, *14*, 135–139. [CrossRef]
158. Rahel, F.J.; Olden, J.D. Assessing the Effects of Climate Change on Aquatic Invasive Species. *Conserv. Biol.* **2008**, *22*, 521–533. [CrossRef]
159. Mainka, S.A.; Howard, G.W. Climate change and invasive species: Double jeopardy. *Integr. Zool.* **2010**, *5*, 102–111. [CrossRef] [PubMed]
160. Koed, A.; Baktoft, H.; Bak, B.D. Causes of mortality of *Atlantic salmon (Salmo salar)* and brown trout (*Salmo trutta*) smolts in a restored river and its estuary. *River Res. Appl.* **2006**, *22*, 69–78. [CrossRef]
161. Jepsen, N.; Flávio, H.; Koed, A. The impact of Cormorant predation on *Atlantic salmon* and Sea trout smolt survival. *Fish. Manag. Ecol.* **2019**, *26*, 183–186. [CrossRef]
162. Vehanen, T.; Huusko, A.; Bergman, E.; Enefalk, Å.; Louhi, P.; Sutela, T. American mink (*Neovison vison*) preying on hatchery and wild brown trout (*Salmo trutta*) juveniles in semi-natural streams. *Freshw. Biol.* **2022**, *67*, 433–444. [CrossRef]
163. Hansen, L.P.; Quinn, T.P. The marine phase of the *Atlantic salmon (Salmo salar)* life cycle, with comparisons to Pacific salmon. *Can. J. Fish. Aquat. Sci.* **1998**, *55*, 104–118. [CrossRef]
164. Potter, E.C.E.; Crozier, W.W. A perspective on the marine survival of *Atlantic salmon*. In *The Ocean Life of Atlantic Salmon: Environmental and Biological Factors Influencing Survival*; Mills, D., Ed.; Fishing News Books: Oxford, UK, 2000; pp. 19–36.
165. Friedland, K.D.; Shank, B.V.; Todd, C.D.; McGinnity, P.; Nye, J.A. Differential response of continental stock complexes of *Atlantic salmon (Salmo salar)* to the Atlantic Multidecadal Oscillation. *J. Mar. Syst.* **2014**, *133*, 77–87. [CrossRef]
166. Salminen, M.; Kuikka, S.; Erkamo, E. Annual variability in survival of sea-ranched Baltic salmon, *Salmo salar* L.: Significance of smolt size and marine conditions. *Fish. Manag. Ecol.* **1995**, *2*, 171–184. [CrossRef]
167. Friedland, K.D.; Reddin, D.G.; McMenemy, J.R.; Drinkwater, K.F. Multidecadal trends in North American *Atlantic salmon (Salmo salar)* stocks and climate trends relevant to juvenile survival. *Can. J. Fish. Aquat. Sci.* **2003**, *60*, 563–583. [CrossRef]
168. Friedland, K.D.; Hansen, L.P.; Dunkley, D.A.; MacLean, J.C. Linkage between ocean climate, post-smolt growth, and survival of *Atlantic salmon (Salmo salar* L.) in the North Sea area. *ICES J. Mar. Sci.* **2000**, *57*, 419–429. [CrossRef]
169. Beaugrand, G.; Reid, P.C. Long-term changes in phytoplankton, zooplankton and salmon related to climate. *Glob. Chang. Biol.* **2003**, *9*, 801–817. [CrossRef]
170. Jonsson, B.; Jonsson, N. Migratory timing, marine survival and growth of anadromous brown trout in the River Imsa, Norway. *J. Fish Biol.* **2009**, *74*, 621–638. [CrossRef] [PubMed]
171. Chaput, G. Overview of the status of *Atlantic salmon (Salmo salar)* in the North Atlantic and trends in marine mortality. *ICES J. Mar. Sci.* **2012**, *69*, 1538–1548. [CrossRef]
172. Olmos, M.; Payne, M.R.; Nevoux, M.; Prévost, E.; Chaput, G.; Du Pontavice, H.; Guitton, J.; Sheehan, T.; Mills, K.; Rivot, E. Spatial synchrony in the response of a long range migratory species (*Salmo salar*) to climate change in the North Atlantic Ocean. *Glob. Chang. Biol.* **2020**, *26*, 1319–1337. [CrossRef]

173. Friedland, K.D.; Dannewitz, J.; Romakkaniemi, A.; Palm, S.; Pulkkinen, H.; Pakarinen, T.; Oeberst, R. Post-smolt survival of Baltic salmon in context to changing environmental conditions and predators. *ICES J. Mar. Sci.* **2017**, *74*, 1344–1355. [CrossRef]
174. Jonsson, B.; Jonsson, N.; Albretsen, J. Environmental change influences the life history of salmon *Salmo salar* in the North Atlantic Ocean. *J. Fish Biol.* **2016**, *88*, 618–637. [CrossRef]
175. Hvidsten, N.A.; Heggberget, T.G.; Jensen, A.J. Sea water temperatures at *Atlantic salmon* smolt entrance. *Nord. J. Freshw. Res.* **1998**, *74*, 79–86.
176. Simmons, O.M.; Gregory, S.D.; Gillingham, P.K.; Riley, W.D.; Scott, L.J.; Britton, J.R. Biological and environmental influences on the migration phenology of *Atlantic salmon Salmo salar* smolts in a chalk stream in southern England. *Freshw. Biol.* **2021**, *66*, 1581–1594. [CrossRef]
177. IPCC. Climate Change 2007: The Physical Science Basis. In *Contribution of Working Group I to the Fourth Assessment Report of the Intergovernmental Panel on Climate Change*; Solomon, S., Qin, D., Manning, M., Chen, Z., Marquis, M., Averyt, K.B., Tignor, M., Miller, H.L., Eds.; Cambridge University Press: Cambridge, UK; New York, NY, USA, 2007; p. 996.
178. Chaput, G.; Carr, J.; Daniels, J.; Tinker, S.; Jonsen, I.; Whoriskey, F. *Atlantic salmon* (*Salmo salar*) smolt and early post-smolt migration and survival inferred from multi-year and multi-stock acoustic telemetry studies in the Gulf of St. Lawrence, northwest Atlantic. *ICES J. Mar. Sci.* **2019**, *76*, 1107–1121. [CrossRef]
179. Jonsson, N.; Jonsson, B. Growth and sexual maturation in *Atlantic salmon Salmo salar* L. *J. Fish Biol.* **2007**, *71*, 245–252. [CrossRef]
180. Peyronnet, A.; Friedland, K.D.; Maoileidigh, N.O.; Manning, M.; Poole, W.R. Links between patterns of marine growth and survival of *Atlantic salmon Salmo salar*. *Fish Biol.* **2007**, *71*, 684–700. [CrossRef]
181. Todd, C.D.; Hughes, S.L.; Marshall, T.; MacLean, J.C.; Lonergan, M.E.; Biow, E.M. Detrimental effects of recent ocean surface warming on growth condition of *Atlantic salmon*. *Glob. Chang. Biol.* **2008**, *14*, 958–970. [CrossRef]
182. Edwards, M.; Beaugrand, G.; Hays, G.C.; Koslow, J.A.; Richardson, A.J. Multi-decadal oceanic ecological datasets and their application in marine policy and management. *Trends Ecol. Evol.* **2010**, *25*, 602–610. [CrossRef] [PubMed]
183. Malick, M.J.; Cox, S.P.; Mueter, F.J.; Peterman, R.M.; Bradford, M. Linking phytoplankton phenology to salmon productivity along a north–south gradient in the Northeast Pacific Ocean. *Can. J. Fish. Aquat. Sci.* **2015**, *72*, 697–708. [CrossRef]
184. Parmesan, C.; Yohe, G. A globally coherent fingerprint of climate change impacts across natural systems. *Nature* **2003**, *421*, 37–42. [CrossRef] [PubMed]
185. Cushing, D.H. Plankton Production and Year-class Strength in Fish Populations: An Update of the Match/Mismatch Hypothesis. *Adv. Mar. Biol.* **1990**, *26*, 249–293.
186. Nelson, K.C.; Palmer, M.A.; Pizzuto, J.E.; Moglen, G.E.; Angermeier, P.L.; Hilderbrand, R.H.; Dettinger, M.; Katharine Hayhoe, K. Forecasting the combined effects of urbanization and climate change on stream ecosystems: From impacts to management options. *J. Appl. Ecol.* **2009**, *46*, 154–163. [CrossRef]
187. Capon, S.J.; Stewart-Koster, B.; Bunn, S.E. Future of Freshwater Ecosystems in a 1.5 °C Warmer World. *Front. Environ. Sci.* **2021**, *9*, 784642. [CrossRef]
188. Woodward, G.; Perkins, D.M.; Brown, L.E. Climate change and freshwater ecosystems: Impacts across multiple levels of organization. *Philos. Trans. R. Soc. Lond. B Biol. Sci.* **2010**, *365*, 2093–2106. [CrossRef]
189. Palm, S.; Romakkaniemi, A.; Dannewitz, J.; Pakarinen, T.; Veneranta, L.; Huusko, R.; Isometsä, K.; Broman, A.; Miettinen, A. *Tornionjoen Lohi-, Meritaimen- ja Vaellussiikakannat—Yhteinen Ruotsalais-Suomalainen Biologinen Selvitys Sopivien Kalastussääntöjen Arvioimiseksi Vuodelle 2023*; Sveriges lantbruksuniversitet and Luonnonvarakeskus: Helsinki, Finland, 2023; p. 62.

 *fishes*

Article

Increased Temperature and Discharge Influence Overwinter Growth and Survival of Juvenile Salmonids in a Hydropeaking River: Simulating Effects of Climate Change Using Individual-Based Modelling

Johan Watz *, Joel Schill, Louis Addo, John J. Piccolo and Mahboobeh Hajiesmaeili

Department of Environmental and Life Sciences, Karlstad University, 651 88 Karlstad, Sweden
* Correspondence: johan.watz@kau.se; Tel.: +46-54-700-2267

Abstract: Climate change causes warming of rivers and may increase discharge, particularly during winter. Downstream of hydropower plants, fluctuating water temperature and flow create dynamic overwintering conditions for juvenile salmonids. We used inSTREAM 7.2-SD to simulate the effects of increased temperature (+2 °C) and discharge (+10%) on the overwinter growth and mortality of one-summer- and two-summer-old Atlantic salmon and brown trout in a river with a hydropeaking flow regime in a 2 × 2 design with replicated simulations. Water temperature had a major positive relationship with growth for both species and year classes, whereas increased flow alone had no major general effect on overwinter growth. For one-summer-old trout experiencing the high temperature regime, however, increased flow resulted in reduced growth. There were no major effects from temperature and flow on the survival rate of the two-summer-old fishes. On the other hand, there were significant interaction effects for the one-summer-olds, indicating that the effect of flow depended on temperature. For one-summer-old salmon, high flow resulted in increased survival in the low temperature regime, whereas it resulted in reduced survival in high temperature. In contrast, for one-summer-old trout, high flow resulted in reduced survival in the low temperature regime and increased survival in the high temperature. Different hydropower operation alternatives may interact with warming, affecting the relative competitive abilities of stream salmonids. Ecological models that predict the effects of different environmental conditions, such as temperature and flow regimes, may offer insight into such effects when in situ experiments are not feasible.

Keywords: flow; global warming; habitat; IBM; inSTREAM; salmon; trout

Key Contribution: Using the individual-based model inSTREAM 7.2-SD, we predicted how altered temperature and discharge regimes in a future climate would influence the growth and survival of brown trout and Atlantic salmon in the Gullspång River, Sweden, from September to April. Increased temperature and flow, and their interaction, affected the two species and the different age classes differently.

Citation: Watz, J.; Schill, J.; Addo, L.; Piccolo, J.J.; Hajiesmaeili, M. Increased Temperature and Discharge Influence Overwinter Growth and Survival of Juvenile Salmonids in a Hydropeaking River: Simulating Effects of Climate Change Using Individual-Based Modelling. *Fishes* **2023**, *8*, 323. https://doi.org/10.3390/fishes8060323

Academic Editor: Bror Jonsson

Received: 16 May 2023
Revised: 15 June 2023
Accepted: 16 June 2023
Published: 18 June 2023

1. Introduction

Climate change causes warming of rivers and streams [1], and during winter it may increase discharge, resulting in less snow and ice on streams at high latitudes. In rivers that do not freeze during winter, elevated water temperature and winter spates affect the riverine biota [1–3]. Fish that are winter active, such as salmonids, must cope with these changing winter conditions by adjusting their behaviour and physiology [4–6].

Many salmonid populations inhabit regulated rivers, and the operation of hydroelectric power plants, which affects downstream flow and temperature, plays a key role in ensuring that there is enough suitable rearing and spawning habitat for salmonid populations to survive [5,7]. In regulated rivers with hydropower generation that must meet sub-daily fluctuations in electricity demand, hydropeaking flow may contribute further to dynamic winter conditions by causing a fluctuation in water temperature and flow, which in turn affects ice dynamics [8] and fish habitat [4,9,10]. These changing environmental winter conditions affect the risk of stranding [11] and displacement [12] and may increase energy expenditure [4–6] in juvenile salmonids.

Typically, it is not generally feasible to test in situ the effects of different hydropower operation alternatives on salmonid population dynamics. In lieu of such field studies, ecological models that predict the effects of different environmental conditions, such as temperature and flow regimes, may offer insight. In particular, bioenergetic individual-based models (IBMs) have given biologically meaningful and mechanistically understandable explanations for observed ecological phenomena [13–15].

Here, we used inSTREAM 7.2-SD (an IBM of salmonids in a stream environment with sub-daily flow fluctuations) [16] to simulate the effects of increased temperature and discharge on the overwinter mortality and growth of one-summer- and two-summer-old Atlantic salmon and brown trout in the Gullspång River, Sweden. These two salmonid populations have high cultural, economic and conservation value [7]. This river is subject to sub-daily flow fluctuations because the hydropower plant that regulates the river operates with hydropeaking power generation [17]. Our study built on the work by Hajiesmaeili et al. [17], in which the effects of different hydropeaking and non-hydropeaking flows were compared. Here, we used the same study system to assess how winter conditions, including temperature change, may affect the two species. Specifically, we compared mortality and growth under current flow and temperature conditions and compared these to the growth and mortality in an environment with a 2 °C higher temperature and 10% higher discharge.

2. Material and Methods

2.1. Model Description and Study Site

The inSTREAM models are fed the following input: (1) a shapefile of cell geometry and habitat features imported from a geographical information system (GIS), (2) 2D bathymetry-based hydraulic modelling output to predict water velocity and depth in the cells at different discharges, (3) time-series information of turbidity, discharge and temperature (Figure 1), (4) parameters specific for the simulated reach and the fish species investigated and (5) the initial fish population at the start of the simulation (Table 1). The models provide different outputs at the individual level, such as growth and survival, as well as the selected cell position and the proportion of individuals displaying different behaviour (drift feeding, cruise feeding and hiding) at each time step. All choices are based on maximizing individual short-term fitness. These data can be summarized into population responses, such as population growth or decline, and spatial and temporal distribution patterns.

Using the NetLogo modelling software platform, the 7th version of inSTREAM is the most recent update that uses multiple time steps per day related to the light (dawn, day, dusk and night) and enables the user to incorporate additional flow change-dependent time steps (inSTREAM 7-SD [16,18]), thereby making this version suitable for simulations of rivers that have a hydropeaking flow regime [17]. At every time step, each individual fish selects its habitat cell and activity (drift feeding, search feeding or hiding) and experiences growth or weight loss based on its net energy intake. In addition, survival for each individual at each time step is determined in relation to mortality risks: high temperature, stranding (associated with an extremely shallow habitat), poor condition (starvation), and predation by terrestrial animals and other fish. Habitat selection (and consequently growth and survival) is modelled using a hierarchy of fish length. Selecting where to feed or hide is executed from the largest to the smallest fish, and individuals can only use food and velocity shelters that had not been used by larger fish. Growth is modelled as proportional to the net rate of energy intake, which was the difference between the energy from feeding and the metabolic costs.

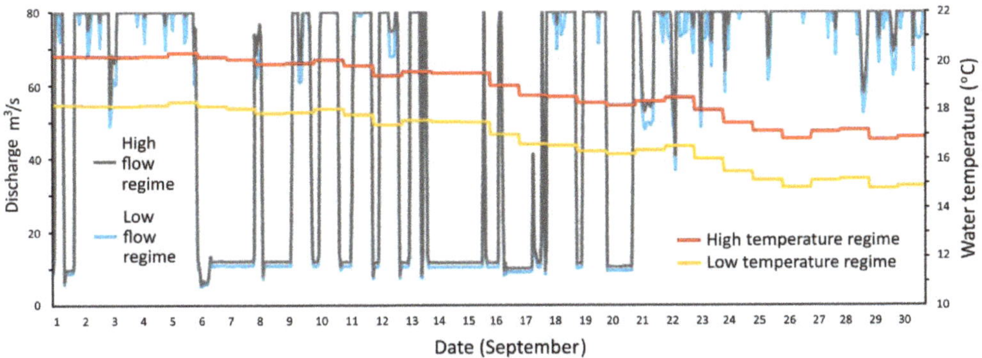

Figure 1. Effects of increased temperature and flow on growth and survival of Atlantic salmon and brown trout in a Swedish hydropeaking river was investigated by simulating population responses from 1 September to 30 April using the individual-based salmonid population model inSTREAM 7-SD. In the 2 × 2 design, each combination of low (baseline) and high temperature (+2 °C) and flow (+10%) regime was tested. As an example of how the treatments affect discharge and temperature, the figure shows the hydrograph and water temperature during the month of September.

Table 1. Initial size of one- and two-summers-old Atlantic salmon and brown trout population in a hydropeaking Swedish river used as input data for the individual-based salmonid population model inSTREAM 7-SD. Fish sizes were matched to electrofishing data.

Species	Starting Age	Number	Length (Min–Mode–Max; mm)
Atlantic salmon	one summer	810	6.2–9.0–14.3
	two summers	40	15.5–17.2–18.6
Brown trout	one summer	360	4.6–10.5–19.0
	two summers	20	19.7–22.5–23.5

Discharge affects water depth and velocity, which influences prey capture probability and energy expenditure, which in turn is affected by the availability of velocity shelters and fish size. The energy budget, in turn, affects growth rates and mortality risk. Various temperature effects are incorporated in the model, including metabolic rates and physical performance, which are represented through a bioenergetics approach. If the fish lose weight, they are vulnerable to poor condition (starvation) and predation because the decision of an individual fish on where and when to feed or hide depends on its own state. Predation risk from piscivores also increases with temperature due to increases in metabolic demands and feeding activity. Furthermore, temperature affects the maximum sustainable swimming speed, which influences the success of drift feeding. All the parameters in the model and detailed documentation of its formulation are described in the inSTREAM 7 user manual [16,18].

The 8 km-long Gullspång River, which connects Lakes Skagern and Vänern, serves a 5000 km² catchment area of mainly forested land and has a mean discharge at the mouth of 62 m³ s⁻¹. The river harbours migratory populations of land-locked, large-bodied Atlantic salmon and brown trout, and because of their high cultural, conservation and economic value [7,19], river restoration projects have been initiated to help these populations to recover. However, the efficiency of these restorations has not been thoroughly assessed, in particular in face of further climate change. Spawning and rearing habitats are limited to three rapids, and this study used the Lilla Åråsforsen rapids, as the study site (59.012 °N, 14.098 °E). Hydropeaking in this river is allowed from 20 August to 19 April, with a minimum base flow of 9 m³ s⁻¹. The maximum capacity of the hydropower plant is 230 m³ s⁻¹, but not all of this high discharge reaches Lilla Åråsforsen because a diversion weir upstream reduces the maximum flow. In our simulations, we assumed that all water over 80 m^3 s^{-1} would be directed to the diversion channel and not reach Lilla Åråsforsen.

The model was calibrated using electrofishing data on growth for the different year classes and adjusting drift and benthic food availability in addition to aquatic and terrestrial predation risk because these parameters have been shown to affect model output the most [17,18,20]. A detailed description of the hydrodynamic modelling (using MIKE 21; DHI Sweden), Lilla Åråsforsen model application description, model calibration and a map of the area can be found in Hajiesmaeili et al. [17]. The model covered 24,000 m^2 (5500 cells) and was populated by 810 one-summer-old and 40 two-summer-old salmon and 360 one-summer-old and 20 two-summer-old trout with sizes according to those electrofished in the same reach [17].

2.2. Flow and Temperature Scenarios

We used two temperature and two discharge time series in a 2 $\times$ 2 full factorial design. For the temperature time series, which represented the current temperature, we used modelled data based on air temperature for 2013–2014 [21]. The temperature model was validated by empirical water temperature data from 2019 to 2021. We used the temperature time series from 2013–2014 in our simulations because these years were not unusually warm or cold. To create the time series that represented an increased temperature regime, we used the first temperature time series and added 2 °C to all data points (Figure 1). In arctic rivers with ice dynamics, climate change will likely have less effect on winter water temperatures because these rivers will stay frozen [1]. However, in the Gullspång River, an increase can be expected because it is rarely ice covered and typically has winter water temperatures > 0 °C (SMHI's Vattenwebb [22]), and it will likely be affected by a milder and wetter winter. The discharge time series that represented the current flow regime was based on data having a high temporal resolution (1 h^{-1}) provided by the hydropower operator, and originating from years that were not unusually dry or wet (2013–2014). For the time series 1 September to 30 April, which represented a future flow regime in a climate with wetter winters, we used the first discharge time series and added a 10% discharge to all data points (Figure 1). We could not reliably estimate when conditions similar to those in our scenarios may occur. Nevertheless, as a comparison, modelling results from the Norwegian River Mandalselva (at approximately the same latitude as the Gullspång River) predict a substantially larger increase than 2 °C and 10% more discharge within 100 years [23].

2.3. Data Analysis

For each of the four combinations of temperature and flow regimes in our 2 $\times$ 2 design, we carried out five replicated simulation runs using different random seeds each time for a total of 20 runs. We calculated the overwinter growth and survival for each species and year class based on the mean mass and number of individuals at the start of the simulation (1 September) and the end (30 April) (Table 1). Specifically, we calculated the mean instantaneous growth rate (g) as

$$g = (\ln(M_{end}) - \ln(M_{start}))/\Delta t$$

where M_{end} and M_{start} are the mean body masses at the end and the start of the simulation, and Δt is the duration of the simulation. Specific growth rate (SGR, % per day) was calculated per Crane et al. [24]:

$$SGR = 100 \times (e^g - 1)$$

Survival rates were calculated as the proportion of live fish at the end of the simulations. We arcsine square root- transformed the proportions to achieve normal distribution. Levene's test for equality of variances showed that variances were similar among the groups ($p > 0.05$). We argue that inferential statistical methods based on null hypothesis testing may be relevant when the model used for the simulations is complex; the prediction of the treatment effects is not trivial; and the results are presented as both statistical significance and effects size [25]. To analyse the effect of increased temperature and flow (and their interaction term) on the overwinter growth and survival of juvenile salmon and trout in the Gullspång River, we used two-way ANOVAs and analyzed the data using SPSS Statistics 28 (IBM, Armonk, NY, USA).

3. Results

3.1. Specific Growth Rates

Both the salmon and trout grew during the simulated period 1 September to 30 April the following year. The mean lengths of the salmon at the start of the simulations were 9.8 and 17.1 cm for the one-summer-olds and two-summer-olds, respectively, and 14.1 and 27.9 cm, respectively, at the end. The corresponding values for trout were 11.4 and 22.0 cm at the start and 17.3 and 29.7 cm at the end. The mean body mass growth (SGR) across all treatment combinations was 0.52% day^{-1} for the one-summer-olds and 0.60% day^{-1} for two-summers-olds. One-summer-old trout had higher growth (0.62% day^{-1}), whereas two-summer-old trout had lower growth (0.39% day^{-1}).

Increased flow had no major effect on overwinter growth (Table 2; Figure 2). Only for one-summer-old trout experiencing a high temperature regime did the increased flow result in reduced SGR (Figure 2), as indicated by a significant interaction term in the ANOVA (Table 2). For one-summer-old trout in high temperature, mean SGR decreased from 0.68% day^{-1} for the low flow regime to 0.65% day^{-1} for high flow. Water temperature had a major positive relationship with growth for both species and year classes (Figure 1; Table 2).

Table 2. Results from two-way ANOVAs, presenting the effects of temperature and flow regimes on specific growth and survival rates of one-summer-old and two-summer-old Atlantic salmon and brown trout. Growth and survival rates were extracted from simulations (n = 5) of salmonid populations in a Swedish hydropeaking river in low (baseline) and high (+2 °C increase from baseline) temperature regimes and low (baseline) high (+10% from baseline) flow regimes in a 2 × 2 design. The individual-based salmonid population model inSTREAM SD-7 was used for the simulations. The *p* values in boldface indicate significant effects (α = 0.05).

Variable	Population	Source of Variation	*F*	df	*p*	η_p^2
Specific	Atlantic salmon, one summer	Temperature	112.65	1, 16	**<0.001**	0.876
growth		Flow	<0.01	1, 16	1.000	<0.001
rate		Temperature × Flow	0.67	1, 16	0.426	0.040
	Brown trout, one summer	Temperature	136.97	1, 16	**<0.001**	0.895
		Flow	2.80	1, 16	0.114	0.149
		Temperature × Flow	4.98	1, 16	**0.040**	0.237
	Atlantic salmon, two summers	Temperature	13.13	1, 16	**0.002**	0.451
		Flow	0.04	1, 16	0.851	0.002
		Temperature × Flow	1.78	1, 16	0.201	0.100
	Brown trout, two summers	Temperature	8.08	1, 16	**0.012**	0.336
		Flow	1.29	1, 16	0.272	0.075
		Temperature × Flow	<0.01	1, 16	1.000	<0.001
Survival	Atlantic salmon, one summer	Temperature	158.24	1, 16	**<0.001**	0.908
rate		Flow	0.20	1, 16	0.661	0.012
		Temperature × Flow	11.84	1, 16	**0.030**	0.425
	Brown trout, one summer	Temperature	7.26	1, 16	0.160	0.312
		Flow	0.10	1, 16	0.752	0.006
		Temperature × Flow	5.29	1, 16	**0.035**	0.248

Table 2. *Cont.*

Variable	Population	Source of Variation	F	df	p	η_p^2
	Atlantic salmon, two summers	Temperature	0.90	1, 16	0.358	0.053
		Flow	0.07	1, 16	0.800	0.004
		Temperature × Flow	0.36	1, 16	0.555	0.022
	Brown trout, two summers	Temperature	0.18	1, 16	0.679	0.011
		Flow	0.18	1, 16	0.679	0.011
		Temperature × Flow	1.60	1, 16	0.224	0.091

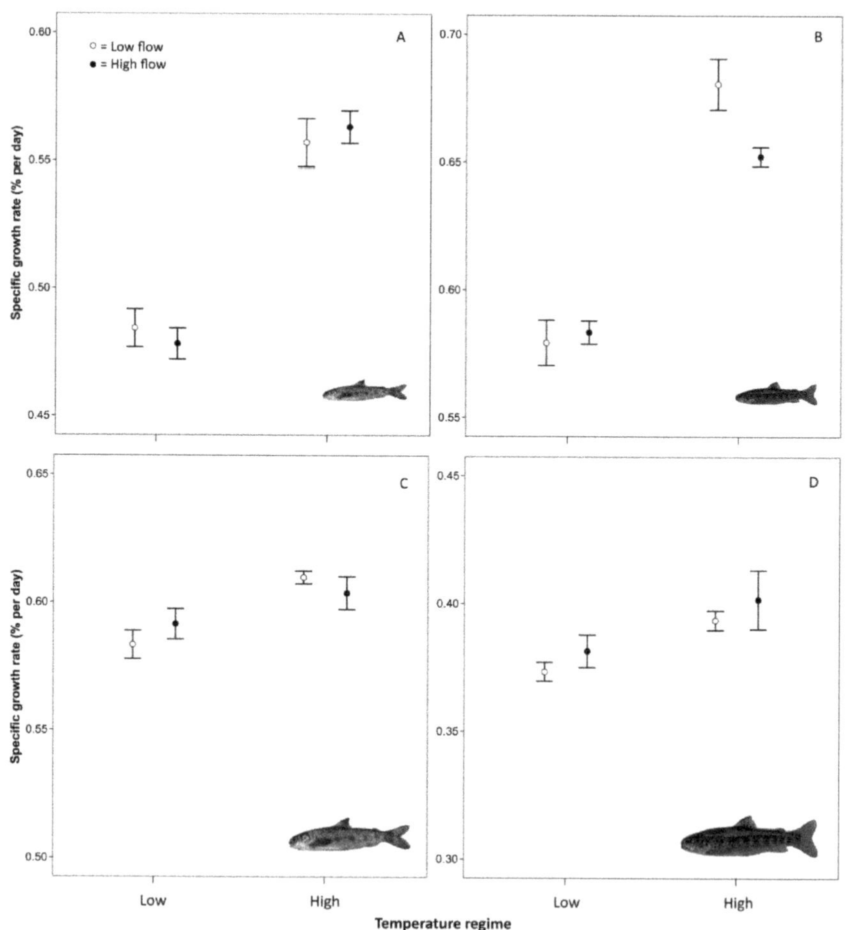

Figure 2. Mean specific growth rate (1 September–30 April) of (**A**) one-summer-old Atlantic salmon and (**B**) brown trout and (**C**) two-summer-old Atlantic salmon and (**D**) brown trout estimated from simulations (n = 5) of a Swedish hydropeaking river in low (baseline) and high (+2 °C increase from baseline) temperature regimes. Open circles represent a low (baseline), and filled circles a high (+10% from baseline) flow regime. The individual-based salmonid population model inSTREAM SD-7 was used for the simulations. Error bars indicate ± 1 SE.

3.2. Survival

One-summer-olds had lower mean survival rates (salmon: 40%; trout: 50%) than did the two-summers-olds (salmon: 84%; trout: 92%). Temperature and flow regimes did not affect the survival rates of the two-summer-olds (Figure 3; Table 2). For the analysis of survival rates for one-summer-old fish of both species, there was significant interaction between the temperature and flow regime (Figure 3; Table 2), indicating that the effect of the flow depended on the temperature. For one-summer-old salmon, high flow resulted in increased survival in low temperatures, whereas it resulted in reduced survival in high temperatures. For one-summer-old trout, the pattern was the reverse; high flow resulted in reduced survival in the low-temperature and increased survival in the high-temperature regimes (Figure 3).

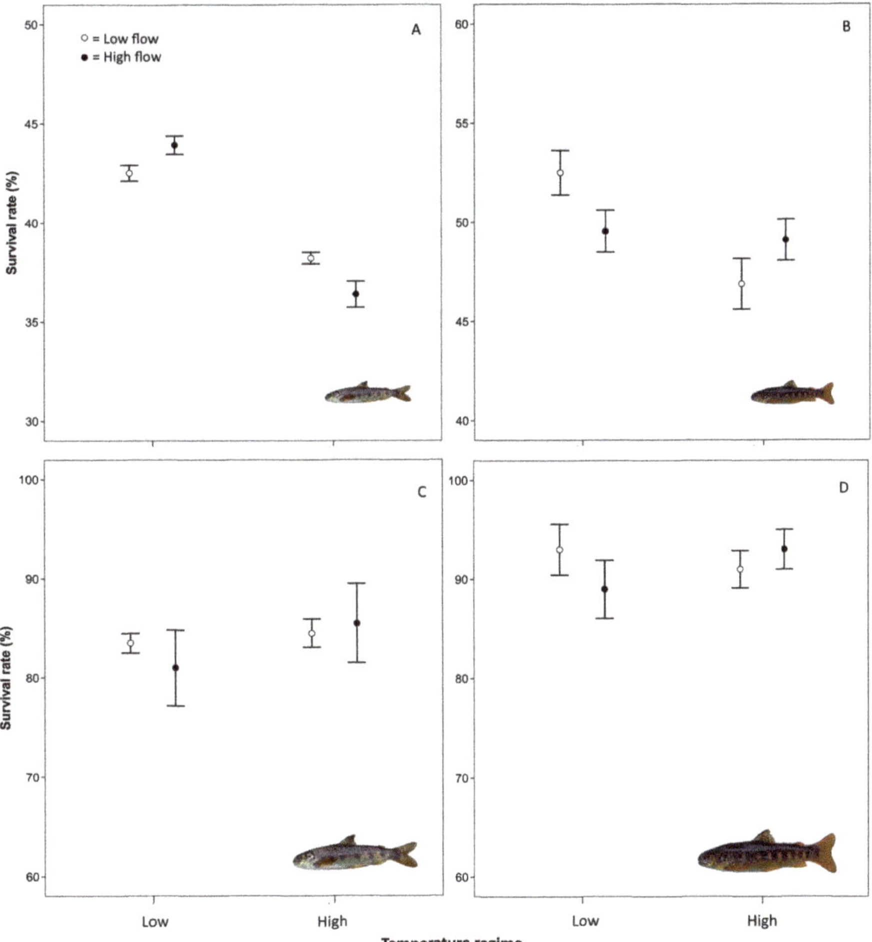

Figure 3. Mean survival rate (1 September–30 April) of (**A**) one-summer-old Atlantic salmon and (**B**) brown trout and (**C**) two-summer-old Atlantic salmon and (**D**) brown trout estimated from simulations (n = 5) of a Swedish hydropeaking river in low (baseline) and high (+2 °C increase from baseline) temperature regimes. Open circles represent a low (baseline) and filled circles a high (+ 10% from baseline) flow regime. The individual-based salmonid population model inSTREAM SD-7 was used for the simulations. Error bars indicate ± 1 SE.

4. Discussion

We used the salmonid population IBM inSTREAM 7.2-SD to simulate the effects of an increase in temperature and flow on overwintering juvenile Atlantic salmon and brown trout in a Swedish hydropeaking river. The results indicated the potential effects of climate change on two threatened salmonid populations in the Gullspång River and highlighted that the youngest size class (one-summer old) was the life stage most vulnerable to warming.

Increased winter temperature due to climate warming resulted in faster growth. Salmonid growth typically ceases at 1–3 °C [26], and with an increase of 2 °C in the Gullspång River, the period with little-to-no potential growth (water temperatures ≤ 3 °C) was considerably shorter by approximately 70% during the winter of 2013–2014). On the other hand, high winter temperatures resulting in increased metabolic rates required access to food resources and foraging opportunities to avoid starvation, and this feeding activity may have increased predation risk [4,6]. Although increased winter temperatures resulted in higher prey capture success for drift-feeding salmonids [27], foraging positions with a low predation risk may be lacking. Therefore, our result that the one-summer-old Atlantic salmon suffered from increased mortality in the high temperature regime was expected because small fish are typically more vulnerable to both predation and starvation compared to large fish [4,28].

Flow had only minor effects on growth and survival; however, in interaction with temperature it did affect one-summer-old Atlantic salmon survival. The combination of high flow and low temperature resulted in the highest survival rates, whereas high flow with high temperature resulted in the lowest. This may be worrisome because both temperature and flow will likely continue to increase. Here, we added 10% to the baseline scenario (empirical flow data from 2013–2014) as a constant addition. However, the potentially added discharge in northern rivers during winter will unlikely be released evenly over the year. Further simulations to investigate how an increased yearly discharge may be released over the year, i.e., different scenarios of hydropower generation schemes, may be worthwhile to find measures to minimize Atlantic salmon parr mortality in the Gullspång River. For one-summer-old brown trout, the pattern was the reverse: the highest survival was achieved under the low-flow and low-temperature regimes, whereas low flow and high temperature resulted in the lowest survival rates but the highest growth. Therefore, it is possible that by adjusting the flow under different warming scenarios, the relative competitive abilities of salmonid species will change [29,30].

In a previous study employing the same inSTREAM model, Hajiesmaeili et al. [17] tested the effects of hydropeaking scenarios with different baseflows. They found that increased flow generally had a negative effect on the survival of both species over the course of a whole year due to the increased aquatic predation facilitated by increased water depth. However, increasing the minimum base flow within hydropeaking flow scenarios had positive effects on the predicted growth of both species [17]. In the present study, focusing on the winter season, we demonstrated that these flow effects can be mediated by temperature at least for the one-summer-old fish.

The modelled species-specific differences in response to flow and temperature was likely driven by the size differences between species, a pattern reflecting observed life history variation [17]. In the Gullspång River, brown trout spawn in the fall earlier than Atlantic salmon, and their eggs hatch and emerge earlier in the spring. This difference in hatching date results in brown trout being larger than Atlantic salmon at the alevin, fry and parr life stages. Key inSTREAM factors, such as drift-feeding performance, predation risk, metabolic rate and habitat selection depend on the body size in the model [20]. Therefore, specific differences in these factors may relate to the different intra- (one- vs. two-summer-old) and interspecies (Atlantic salmon vs. brown trout) effects of flow and temperature regimes. The contrasting interaction effects of temperature and flow on one-summer-old Atlantic salmon and brown trout may be partly caused by brown trout outcompeting Atlantic salmon for the best feeding positions.

The inSTREAM modelling framework has the potential to develop hypothesis-driven research that may help to answer questions relating to complex ecological processes such as how sympatric species may respond to interacting factors like climate change and hydropeaking. This study, for example, highlighted the need for developing species-specific ecophysiological parameters for salmon and trout in the Gullspång River. Although conducting such ecophysiological studies would be a nontrivial task [31], further developed model capabilities may yield even more detailed and realistic results for these two populations. The alternative, replicated field studies under varying streamflow and temperature regimes, would clearly be impractical and require considerable resources. Thus, ecological modelling approaches will likely remain a key element of any research efforts that aim to assess the complex realities of river management, especially under future climate change scenarios.

5. Conclusions

In the Gullspång River, potential winter conditions with increased water temperatures and altered flows will likely influence the relative competitive abilities of Atlantic salmon and brown trout. In a warmer future, juvenile brown trout may dominate the most favourable feeding positions because of its competitive advantage of being larger at any given time and being a relatively more aggressive species [32] than the Atlantic salmon. On the other hand, the thermal preference of brown trout is lower than that of Atlantic salmon [26], but this physiological difference was not considered in the model we used due to a lack of standardized tests across species, size, temperature and water velocity [33]. Further modelling studies to investigate the potential non-linear effects and tipping points (using more than two levels of each treatment) should prove to be useful, together with species-specific physiological parameters, to assess the combined effects of different climate-change scenarios. We demonstrated potential trends in growth and survival of sympatric salmonid populations in hydropeaking rivers [34] caused by a future environment with warmer and wetter winters in temperate areas, which may influence the relative competitive abilities of juvenile salmonid species [30,32].

Author Contributions: Conceptualization, J.W.; methodology, software, validation J.W., L.A., J.J.P. and M.H.; formal analysis, J.W. and J.S.; investigation, J.S.; writing—original draft preparation, J.W.; writing—review and editing, all authors.; visualization, J.W.; supervision, J.W. and J.J.P.; project administration, J.W. and J.J.P.; funding acquisition, J.J.P. All authors have read and agreed to the published version of the manuscript.

Funding: The study was funded by the Swedish research council Formas (grant number 2019-00543) and the Knowledge Foundation (KK-Stiftelsen; grant number 20170129).

Institutional Review Board Statement: Not applicable.

Informed Consent Statement: Not applicable.

Data Availability Statement: The data that support the findings of this study are available from the corresponding author upon reasonable request.

Acknowledgments: We thank Sam Shry for the fish photos used in the figures. The IBM used in the study, inSTREAM 7-SD, is freely available and can be downloaded from The Humboldt Mathematics Department (https://ecomodel.humboldt.edu/individual-based-ecological-modeling-humboldt-state-university; accessed on 15 June 2023).

Conflicts of Interest: The authors declare no conflict of interest.

References

1. van Vliet, M.T.; Franssen, W.H.; Yearsley, J.R.; Ludwig, F.; Haddeland, I.; Lettenmaier, D.P.; Kabat, P. Global river discharge and water temperature under climate change. *Glob. Environ. Chang.* **2013**, *23*, 450–464. [CrossRef]
2. Isaak, D.J.; Wollrab, S.; Horan, D.; Chandler, G. Climate change effects on stream and river temperatures across the northwest US from 1980–2009 and implications for salmonid fishes. *Clim. Chang.* **2012**, *113*, 499–524. [CrossRef]

3. O'Briain, R. Climate change and European rivers: An eco-hydromorphological perspective. *Ecohydrology* **2019**, *12*, e2099. [CrossRef]

4. Huusko, A.R.I.; Greenberg, L.; Stickler, M.; Linnansaari, T.; Nykänen, M.; Vehanen, T.; Koljonen, S.; Louhi, P.; Alfredsen, K. Life in the ice lane: The winter ecology of stream salmonids. *River Res. Appl.* **2007**, *23*, 469–491. [CrossRef]

5. Jonsson, B.; Jonsson, N. A review of the likely effects of climate change on anadromous Atlantic salmon Salmo salar and brown trout Salmo trutta, with particular reference to water temperature and flow. *J. Fish Biol.* **2009**, *75*, 2381–2447. [CrossRef]

6. Heggenes, J.; Alfredsen, K.; Bustos, A.A.; Huusko, A.; Stickler, M. Be cool: A review of hydro-physical changes and fish responses in winter in hydropower-regulated northern streams. *Environ. Biol. Fishes* **2018**, *101*, 1–21. [CrossRef]

7. Watz, J.; Aldvén, D.; Andreasson, P.; Aziz, K.; Blixt, M.; Calles, O.; Bjørnås, K.L.; Olsson, I.; Österling, M.; Stålhammar, S.; et al. Atlantic salmon in regulated rivers: Understanding river management through the ecosystem services lens. *Fish Fish.* **2022**, *23*, 478–491. [CrossRef]

8. She, Y.; Hicks, F.; Andrishak, R. The role of hydro-peaking in freeze-up consolidation events on regulated rivers. *Cold Reg. Sci. Technol.* **2012**, *73*, 41–49. [CrossRef]

9. Linnansaari, T.; Alfredsen, K.; Stickler, M.; Arnekleiv, J.V.; Harby, A.; Cunjak, R.A. Does ice matter? Site fidelity and movements by Atlantic salmon (*Salmo salar* L.) parr during winter in a substrate enhanced river reach. *River Res. Appl.* **2009**, *25*, 773–787. [CrossRef]

10. Watz, J.; Bergman, E.; Piccolo, J.J.; Greenberg, L. Ice cover affects the growth of a stream-dwelling fish. *Oecologia* **2016**, *181*, 299–311. [CrossRef]

11. Saltveit, S.J.; Halleraker, J.H.; Arnekleiv, J.V.; Harby, A. Field experiments on stranding in juvenile Atlantic salmon (*Salmo salar*) and brown trout (*Salmo trutta*) during rapid flow decreases caused by hydropeaking. *Regul. Rivers Res. Manag.* **2001**, *17*, 609–622. [CrossRef]

12. Saltveit, S.J.; Bremnes, T.; Lindå, O.R. Effect of sudden increase in discharge in a large river on newly emerged Atlantic salmon (*Salmo salar*) and brown trout (*Salmo trutta*) fry. *Ecol. Freshw. Fish* **1995**, *4*, 168–174. [CrossRef]

13. Jager, H.I.; DeAngelis, D.L. The confluences of ideas leading to, and the flow of ideas emerging from, individual-based modeling of riverine fishes. *Ecol. Model.* **2018**, *384*, 341–352. [CrossRef]

14. Railsback, S.F.; Harvey, B.C.; Ayllón, D. Contingent trade-off decisions with feedbacks in cyclical environments: Testing alternative theories. *Behav. Ecol.* **2020**, *31*, 1192–1206. [CrossRef]

15. Railsback, S.F.; Harvey, B.C.; Ayllón, D. Importance of the daily light cycle in population–habitat relations: A simulation study. *Trans. Am. Fish. Soc.* **2021**, *150*, 130–143. [CrossRef]

16. Railsback, S.F.; Harvey, B.C.; Ayllón, D. *InSTREAM 7.3 User Manual: Model Description, Software Guide, and Application Guide*; USDA Forest Service, Pacific Southwest Research Station: Albany, CA, USA, 2022.

17. Hajiesmaeili, M.; Addo, L.; Watz, J.; Railsback, S.F.; Piccolo, J.J. Individual-based modelling of hydropeaking effects on brown trout and Atlantic salmon in a regulated river. *River Res. Appl.* **2023**, *39*, 522–537. [CrossRef]

18. Railsback, S.F.; Ayllón, D.; Harvey, B.C. InSTREAM 7: Instream flow assessment and management model for stream trout. *River Res. Appl.* **2021**, *37*, 1294–1302. [CrossRef]

19. Piccolo, J.J.; Norrgård, J.R.; Greenberg, L.A.; Schmitz, M.; Bergman, E. Conservation of endemic landlocked salmonids in regulated rivers: A case-study from Lake Vänern, Sweden. *Fish Fish.* **2012**, *13*, 418–433. [CrossRef]

20. Railsback, S.F.; Harvey, B.C.; Jackson, S.K.; Lamberson, R.H. *InSTREAM: The Individual-Based Stream Trout Research and Environmental Assessment Model*; (General Technical Report. PSW-GTR-218); USDA Forest Service, Pacific Southwest Research Station: Albany, CA, USA, 2009.

21. Bjørnås, K.L.; Railsback, S.F.; Calles, O.; Piccolo, J.J. Modeling Atlantic salmon (*Salmo salar*) and brown trout (*S. trutta*) population responses and interactions under increased minimum flow in a regulated river. *Ecol. Eng.* **2021**, *162*, 106182. [CrossRef]

22. Strömbäck, L.; Hjerdt, N.; Eriksson, L.B.; Lewau, P. Vattenwebb: A transparent service to support decision makers in achieving improved water status. In *10th IFIP WG 5.11 International Symposium, Proceedings of the ISESS 2013, Neusiedl Am See, Austria, 9–11 October 2013*; Hřebíček, J., Schimak, G., Kubásek, M., Rizzoli, A.E., Eds.; IFIP International Federation for Information Processing: Laxenburg, Austria, 2013.

23. Sundt-Hansen, L.E.; Hedger, R.D.; Ugedal, O.; Diserud, O.H.; Finstad, A.G.; Sauterleute, J.F.; Tøfte, L.; Alfredsen, K.; Forseth, T. Modelling climate change effects on Atlantic salmon: Implications for mitigation in regulated rivers. *Sci. Total Environ.* **2018**, *631*, 1005–1017. [CrossRef] [PubMed]

24. Crane, D.P.; Ogle, D.H.; Shoup, D.E. Use and misuse of a common growth metric: Guidance for appropriately calculating and reporting specific growth rate. *Rev. Aquac.* **2020**, *12*, 1542–1547. [CrossRef]

25. Heard, S.B. Of Course You Can Do Significance Testing on Simulation Data! Scientist Sees Squirrel. 2019. Available online: https://scientistseessquirrel.wordpress.com/2019/05/07/of-course-you-can-do-significance-testing-on-simulation-data/ (accessed on 11 May 2023).

26. Elliott, J.; Elliott, J.A. Temperature requirements of Atlantic salmon Salmo salar, brown trout Salmo trutta and Arctic charr Salvelinus alpinus: Predicting the effects of climate change. *J. Fish Biol.* **2010**, *77*, 1793–1817. [CrossRef]

27. Watz, J.; Piccolo, J.; Bergman, E.; Greenberg, L. Day and night drift-feeding by juvenile salmonids at low water temperatures. *Environ. Biol. Fishes* **2014**, *97*, 505–513. [CrossRef]

28. Hurst, T.P. Causes and consequences of winter mortality in fishes. *J. Fish Biol.* **2007**, *71*, 315–345. [CrossRef]

29. Heggenes, J.; Gunnar Dokk, J. Contrasting temperatures, waterflows, and light: Seasonal habitat selection by young Atlantic salmon and brown trout in a boreonemoral river. *Regul. Rivers Res. Manag.* **2001**, *17*, 623–635. [CrossRef]
30. Watz, J.; Otsuki, Y.; Nagatsuka, K.; Hasegawa, K.; Koizumi, I. Temperature-dependent competition between juvenile salmonids in small streams. *Freshw. Biol.* **2019**, *64*, 1534–1541. [CrossRef]
31. Zillig, K.W.; Lusardi, R.A.; Cocherell, D.E.; Fangue, N.A. Interpopulation variation in thermal physiology among seasonal runs of Chinook salmon. *Can. J. Fish. Aquat. Sci.* **2023**, *80*, 1–13. [CrossRef]
32. Hesthagen, T.; Larsen, B.M.; Bolstad, G.; Fiske, P.; Jonsson, B. Mitigation of acidified salmon rivers–effects of liming on young brown trout *Salmo trutta*. *J. Fish Biol.* **2017**, *91*, 1350–1364. [CrossRef]
33. Piccolo, J.J.; Hughes, N.F.; Bryant, M.D. Water velocity influences prey detection and capture by drift-feeding juvenile coho salmon (*Oncorhynchus kisutch*) and steelhead (*Oncorhynchus mykiss irideus*). *Can. J. Fish. Aquat. Sci.* **2008**, *65*, 266–275. [CrossRef]
34. Addo, L.; Hajiesmaeili, M.; Piccolo, J.J.; Watz, J. Growth and mortality of sympatric Atlantic salmon and brown trout fry in fluctuating and stable flows. *Ecol. Freshw. Fish* **2023**, *32*, 282–290. [CrossRef]

Article

Long-Term Trends in Freshwater and Marine Growth Patterns in Three Sub-Arctic Atlantic Salmon Populations

Nico Alioravainen *, Panu Orell and Jaakko Erkinaro

Natural Resources Institute Finland (Luke), P.O. Box 413, FI-90014 Oulu, Finland; panu.orell@luke.fi (P.O.); jaakko.erkinaro@luke.fi (J.E.)
* Correspondence: nico.alioravainen@luke.fi; Tel.: +358-29-5322700

Abstract: The rapid warming of the Northern hemisphere has especially challenged the evolvability of anadromous fish species, such as Atlantic salmon (*Salmo salar*), which must cope with drastically different environments depending on their life-history stage. We studied the long-term trends in, and the effects of environmental factors and life-history traits on, Atlantic salmon growth rates in both freshwater and in the ocean using c. 35,000 scale samples collected across 48 years from spawners returning to three tributaries of the subarctic River Teno in the northernmost parts of Finland and Norway (70° N). The freshwater growth has decreased in all three populations and spending more than three juvenile years in freshwater before the sea migration comes at the expense of growth. On the other hand, returning mature salmon (one-sea-winter, 1SW) showed increased growth at the sea with increasing marine temperatures, which results in larger sizes at return in 1SW spawners. We did not observe such trends in growth rates in larger, two-sea-winter salmon. Here, we report the contrasting responses in Atlantic salmon growth rates to a warming climate depending on the life-history stage.

Keywords: climate change; growth rate; marine ecosystems; river ecosystems; scales

Key Contribution: Long-term trends in Atlantic salmon growth rates in both freshwater and marine environments were studied using scale samples collected across half a century in three sub-arctic populations. The decrease in freshwater growth was associated with rising temperatures, whereas sea growth increased during the same period.

Citation: Alioravainen, N.; Orell, P.; Erkinaro, J. Long-Term Trends in Freshwater and Marine Growth Patterns in Three Sub-Arctic Atlantic Salmon Populations. *Fishes* **2023**, *8*, 441. https://doi.org/10.3390/fishes8090441

Academic Editor: Bror Jonsson

Received: 21 June 2023
Revised: 28 August 2023
Accepted: 29 August 2023
Published: 30 August 2023

1. Introduction

The global climate change is affecting all natural habitats and ecosystems worldwide, and both observed and predicted effects are the most pronounced at high latitudes: the Arctic has been shown to be warming at least twice as fast as the rest of the globe on average [1]. Climate change can reduce the viability of species and the associated biodiversity loss will affect ecosystem functions and services [2,3]. Such changes may result in ecological regime shifts where functions and structures of ecosystems may change and then persist at a new equilibrium [4–6]. The adaptation capabilities of organisms under the changing climate are the key to avoid reductions in survival or reproductive rates. Adaptative responses to climate change may include physiological or behavioural changes, through phenotypic plasticity, microevolution, or their combination, that reduce the extent of mismatch between the species' phenotype and the environment [7,8].

Considering that the changing climate strongly affects various environments, anadromous fish is a special group of animal species that must cope with a variety of different habitats and highly variable environments during their life cycle, across long geographic distances and a wide salinity gradient from fresh water to full saline sea water, e.g., [9]. Such directed migrations between habitats are typically triggered by seasonal changes in the environment. The life cycle of anadromous fish typically include reproduction, egg incubation and juvenile rearing in freshwater (streams, rivers, lakes), downstream migration through varying salinities of river outlets and brackish water estuaries, ocean migration in full sea water for effective feeding, growth, and maturation, and return migration back to freshwater for reproduction through the same environments, e.g., [10–12]. It has recently been suggested that warming climatic conditions across northern latitudes in Iceland and northern Norway have already strongly influenced the populations of anadromous salmonid fish species [13].

Atlantic salmon (*Salmo salar*, L. 1759) is an iconic, well-studied species among migratory fish that utilize and experience all the different aquatic habitats mentioned above during its life cycle. Starting from the freshwater phases of the species, changes in river temperatures are of special concern as they affect the juvenile food base and growth rate, phenology of life cycle, timing of migrations, reproductive success of spawners, and ultimately survival at various life stages. In their recent review on climate change effects on the freshwater phase of Atlantic salmon, Gillis et al. [14] concluded that climate change affects and will continue to affect instream habitats across all seasons and render challenging conditions, especially through changes in water temperature and discharge, for all Atlantic salmon life stages in freshwater. Juvenile Atlantic salmon emigrate from freshwater after having achieved a certain size and physiological stage (referred to as smolts at this life stage) in one or more years, and subsequently adapt to first estuarine and eventually full sea water and vastly increase their growth rate in the ocean. The timing of downstream migration differs among areas and rivers [15–17], but it has been shown that the initiation of the smolt seaward migration takes place earlier and earlier in association with the increasing river and ocean temperatures and has occurred 2.5 days earlier per decade since the early 1960s across the basins of the North Atlantic Ocean and the Baltic Sea [16]. Changes in phenology of habitat shifts of juvenile salmon may lead to a mismatch with the critical environmental conditions (temperature, food availability) in freshwater and at sea [15,18].

The fast growth of Atlantic salmon at sea and concurrent decisions to either return to fresh water for reproduction after one year or stay for one or more additional years in the ocean to increase more in size is dictating the key life-history patterns [10,19]. Although a strong effect of genes has been shown to largely control the age at maturity (=years spent at sea), environmental conditions play a role as well [19,20]. Recent analysis of long-term time-series data have shown temporal trends in life-history characteristics, both in years spent in fresh water and seawater, and in iteroparity, i.e., patterns in occurrence of repeated spawning [21–24]. In addition, the genetic background affecting the life-history decisions have changed over the decades in some populations in response to environmental conditions, especially the marine food web [25].

In this study, a 48-year time series data sets from three Atlantic salmon populations within a large subarctic catchment were used to investigate the long-term patterns in both freshwater and marine growth rates. Given the recent climate warming of the sub-Arctic area, we searched for possible environmental changes linked with the growth patterns of Atlantic salmon, in both freshwater and marine environments, using both abiotic and biotic environmental variables.

2. Materials and Methods

2.1. Study Area

The River Teno system (Norwegian: Tana, Sami: Deatnu) is located in northern Europe (68–70° N, 25–27° E). It forms the border between northern Finland and Norway, draining into the Tanafjord at the Barents Sea (Figure 1). More than 1100 km of different stretches of the system is accessible to anadromous Atlantic salmon, including the main stem, the large headwater branches, and numerous smaller tributaries (Figure 1). Genetic studies have revealed a highly structured population complex consisting of nearly 30 demographically independent, genetically distinct, and temporally stable population segments in tributaries and different parts of the main stem [26]. In addition, life-history variation of the Teno salmon populations is among the widest, if not the widest, within a single river system [23]. Until very recently, the River Teno has been one of the few remaining large river systems that still supports various forms of recreational and net fisheries in the river that has yielded annual freshwater catches between 80 and 250 t, or 20,000–60,000 individual adult Atlantic salmon [23,27]. However, in recent years, the population status has strongly declined [28]. Typically, the majority (50–60%) of the salmon catch has been taken in recreational rod fishery, the next most important gear being weir and stationary gill nets, and the smallest proportion being taken with drift nets [23]. The net fisheries are practiced by locals, mostly native Sámi, and are based on special fishing rights connected to land use, ownership, or inherited rights. In addition to the extensive fishery in the main stem, salmon fishing is also operational in most of the tributaries. In most tributaries, little or no net fishing is operated or allowed, although in some larger tributaries, net fishing forms a significant share of the catch, like in the rivers studied in the present study, the Pulmankijoki, Utsjoki, and Inarijoki (Figure 1). This study is focusing on these three tributaries where long-term scale sample collections are available, which are genetically distinct [26], and represent the lower, mid, and upper parts of the catchment (Figure 1).

Figure 1. The River Teno catchment in northern Finland and Norway. The studied Atlantic salmon populations were from the tributaries Pulmankijoki, Utsjoki, and Inarijoki.

Salmon stocks of the River Teno system are managed and fisheries regulated by bilateral agreements between Finland and Norway, with the aim of conserving the wild stocks but also supporting sustainable fisheries. The border river stretches (Teno main stem, Inarijoki; Figure 1) are managed bilaterally, but tributary regulations are under national legislation in both countries. Stocking of reared fish or eggs is strictly forbidden in the entire river system.

2.2. Scale Sampling and Analyses

Scale samples were collected from 35,065 Atlantic salmon harvested over a 48-year period (1972–2020) in the Teno river system by recreational tourist anglers and local fishers using various net and rod fishing methods cf. [23]. The samples were collected by a network of trained fisherman, equipped with standard measuring boards and scales, and was established by the Finnish Game and Fisheries Research Institute (currently: Natural Resources Institute Finland (LUKE)) in the 1970s, and later complemented on the Norwegian side by the County Government of Finnmark and the Tana River Fish

Management. The network has been maintained and developed with the aim of covering all fishing methods, the entire fishing season, and different parts of the Teno river system cf. [23,27]. Based on our previous studies, the sampling from June to August virtually covers the entire run timing of these salmon populations [29,30]. The fishermen measured the length and mass of the fish in their retained catches, and recorded the sex of the fish, date of capture, and location and fishing gear used. Scales were dried and archived in paper envelopes at room temperature at the Teno River Fisheries Research station of LUKE in Utsjoki, Finland.

The river and sea age (1SW: one-sea-winter salmon; 2SW: two-sea-winter salmon, etc.) and possible previous spawning history were determined via assessment of scale patterns by trained experts following the internationally agreed guidelines for Atlantic salmon scale reading [31].

We focused on three main populations with the most coherent time series of scales and catch statistics cf. [23,26]. In general, the Teno salmon show large variability in life-history strategies, but here, we focused on the most abundant ones in these tributary populations, i.e., fish that have gone through smolting at the age of three, four, or five years, and stayed at sea on their feeding migration for one or two years before return to their natal rivers (hereafter 1SW, n = 16,032; and 2SW, n = 1912), resulting in six different life-history combinations.

For the growth analyses, a smaller subset of samples (n = 17,944) was analysed. The growth of the freshwater stage was defined as the total length (mm) from the nucleus to the edge of the first sea-summer growth zone (Figure 2). Respectively, the increments during sea migration were measured from the edge of the first sea-summer growth zone to the edge of the first or second sea-winter growth zone (Figure 2). For the marine phase, each alternate increments of fast- and slow-growing zones of the scales were measured (in mm). The total radius of scales correlated with the total length of the fish (Pearson's r = 0.60, t = 32.958, d.f. = 1893, p < 0.001). As the back-calculation of juvenile lengths from adult scales may result in life-history specific under- or over-estimations [32], and Jonsson et al. [33] showed that the potential mismatch between back-calculated juvenile length and actual length can be within ±2 cm, we therefore abandoned such calculations.

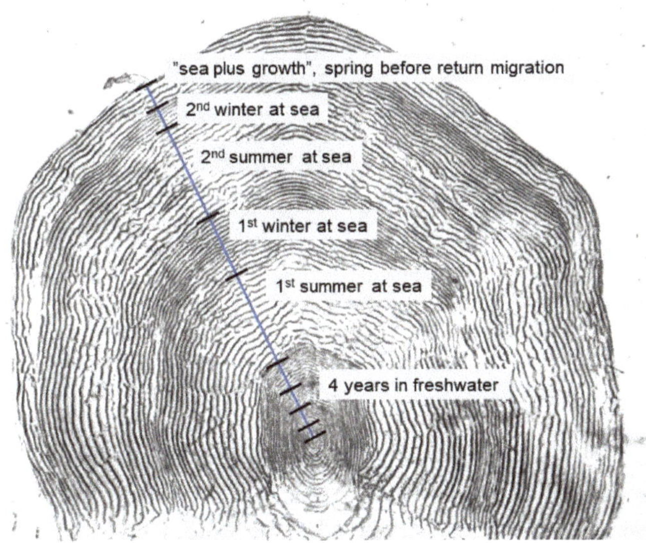

Figure 2. A sample scale from a 2SW Atlantic salmon showing growth zones in four freshwater years and two years at sea (summer and winter zones separately).

2.3. Environmental Data and Curation

The longest available temperature time series for the Utsjoki region was collected at Kevo Meteorological Station (69.75° N, 27.00° E, Finnish Meteorological Institute). The data consist of monthly mean air temperatures from the beginning of the 1960s, which we used as a proxy for a climatic impact on fish growth at the freshwater stage. The temperature data indicate how the annual 10-year rolling mean temperature has risen in the area, e.g., from -1.95 °C in 1975 to -0.26 °C in 2022.

To assess the effects of a changing climate on salmon growth at the sea stage, we used sea temperature data that are available via the ICES Data Portal [34]. We used sea temperature data collected in the Barents Sea area that is likely included in the feeding area for 1SW and 2SW Teno salmon. According to the report on ocean climate [34], the Barents Sea mean temperature has remained above the long-term average since the early 2000s, and in 2020, the mean temperatures of surface, deeper, and bottom waters were warmer than the 1981–2010 average by 1.4, 0.5, and 0.7 °C, respectively. There are two available water temperature data sets from the Barents Sea: one from the Eastern Kola Section from the surface layer (0–200 m depth; 71.5° N, 33.5° E; collected by Polar Institute of Marine Research, Russia [34]). Via the Pinro website (http://www.pinro.vniro.ru, accessed on 4 March 2022), the monthly mean temperature data are available from 1951 to 2017, which were used for the model where monthly mean temperatures were used as covariates (Supplementary Table S3). Another temperature data set was collected from the Western Barents Sea/Atlantic Inflow at the Fugløya–Bear Island Section from a depth of 50–200 m (73° N, 20° E; collected by Institute of Marine Research, Norway [34]).

We also tested the association between sea growth of 1SW and 2SW salmon with the capelin (*Mallotus villosus*, Cuvier, 1829) stock variation. The high capelin biomass has been previously linked with a higher probability for late maturation in Atlantic salmon [25]. We used the acoustic trawl survey-based Barents Sea capelin stock assessment data that are available via the ICES Data Portal [35] and used both log-transformed total and immature capelin biomass and rolling average of two consecutive years when fitting the growth models for 1SW and 2SW fish.

2.4. Statistical Analysis

For data wrangling and to visualize the time series of fish growth, we used the Tidyverse-package family. To analyse the freshwater growth, we especially wanted to investigate whether a certain age at the river has a substantial effect on the total growth. We used linear mixed-effects models (LMM hereafter; lme4-package), where we first fitted birth year, smolt year, and the growth increment year as random factors. We preferred to apply the mixed effects models for growth analyses, which allows us to replace the fixed year effects with the random year effects, thus avoiding the year effects being auto-correlated with environmental effects [36]. For the fixed effects, we followed a model selection procedure, where we compared the goodness of fit of various fixed effect structures based on AIC and deviance (Supplementary Table S1). Based on the selection procedure, we ran the final model to obtain the estimates and their 95% confidence intervals and statistical probabilities. The sea growth analyses were otherwise similar, but only birth year and smolt year were used as random factors. Then, we followed a similar procedure for fixed effects (Supplementary Table S1). Sexes were merged in the analyses, as they did not indicate divergent trends in growth patterns, except in Inarijoki 2SW salmon. All the analyses were performed in the RStudio environment using R version 4.2.1.

3. Results

3.1. Freshwater Growth

Salmon parr growth in freshwater has decreased within the past c. 50 years, and the size difference among three-, four-, and five-year-old smolts seems to have disappeared (Figure 3). The best fitting LMM indicated the importance of air temperature in mid-summer and autumn impacts on juvenile salmon growth (Figure 4). Warmer air temperatures, especially the higher mean temperatures in July or September, seemed to indicate slower growth for juvenile salmon (Figure 4). The age of fish and the annual increase in size were strongly associated showing that the first freshwater growing season has the highest impact on the total growth during the juvenile years in the river (Figure 4). The fixed effect estimates from LMM indicate that the back-calculated size of fish at the smolt stage differed significantly between populations, and the smolt size was smaller in 2SW compared to 1SW salmon (Figure 4).

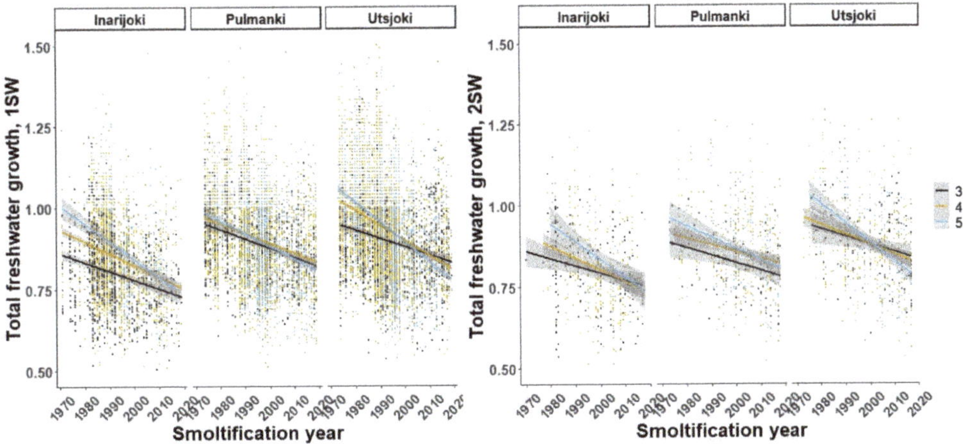

Figure 3. Total freshwater scale growth (mm) of 1SW (n = 16,032), and 2SW (n = 1912) salmon, including smolt age groups of 3, 4, or 5 from three different populations. The trend lines represent the linear regression with 95% confident intervals.

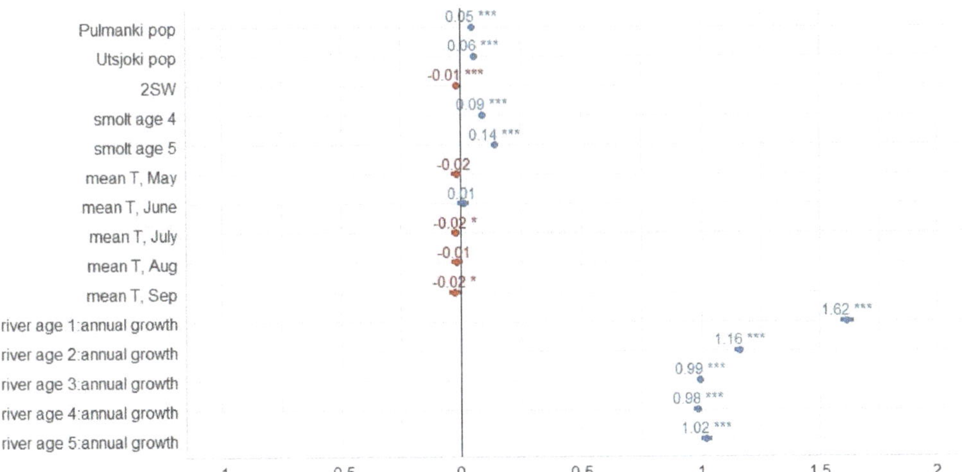

Figure 4. The fixed effect estimates of LMM of the total growth in freshwater, FWG, based on scale measures. The estimates are in relation to the 0 point of the x-axis that represents the intercept (Inarijoki pop., 1SW, 3 y-o-smolt). Mean T refers to monthly mean air temperatures. Asterisks indicate the statistical significance at the level of $\alpha < 0.05$ ("*") or $\alpha < 0.001$ ("***"), Negative estimate values are shown in red, positive in blue.

3.2. Marine Growth

When looking at the sea growth (SG) of the 1SW fish, the faster the smolt has been growing in the freshwater (i.e., smolt age 3), the larger the positive effect on sea growth the freshwater growth (FWG) had (Figure 5). The old smolting age on the other hand showed a negative impact on the sea growth (Figure 5 and Supplementary Table S2). The SG of 1SW salmon was significantly different between the three populations with the Inarijoki population showing the highest growth rate (Figure 5). The annual mean sea water temperature in the Barents Sea showed a positive association with fish growth (Figure 5). The models including the total or immature capelin biomass did not increase the goodness of fit of the model (Supplementary Table S1). When looking at the monthly mean temperatures separately, it seems that the mean temperature in September especially had a positive association with sea growth in 1SW fish (Supplementary Table S3).

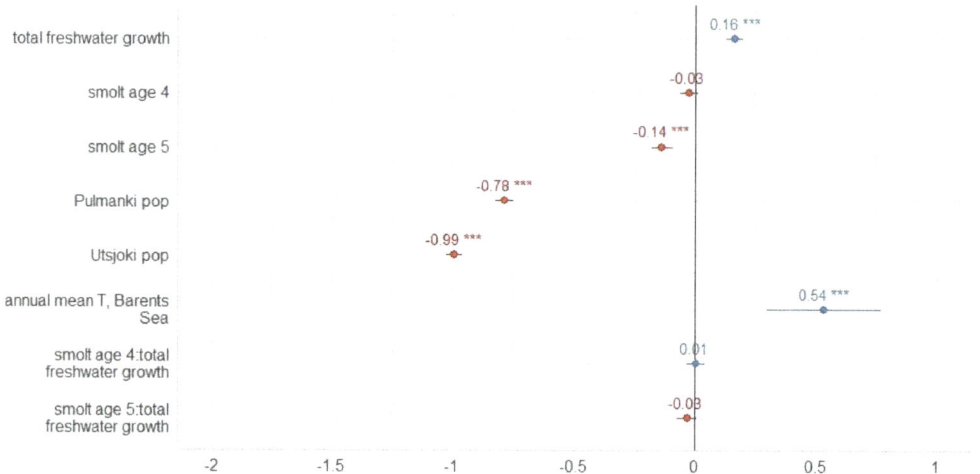

Figure 5. The fixed effect estimates of LMM of the sea growth of 1SW salmon based on scale measures. The estimates are in relation to the 0 point of the *X*-axis that represents the intercept (Inarijoki pop., 3 y-o-smolt). Asterisks indicate the statistical significance at the level of $\alpha < 0.001$ ("***"), Negative estimate values are shown in red, positive in blue.

Similarly, with 1SW fish, the capelin stock status nor the sea water temperature were strongly associated with SG of 2SW salmon (Supplementary Table S4). Congruently with the SG of 1SW salmon, the smolt size showed a positive association with the marine growth of 2SW salmon, and Utsjoki and Pulmanki fish were growing slower than Inarijoki salmon (Supplementary Table S4). Smolt age and FWG showed an interesting antagonistic interaction, where the older and the larger the smolts were, the slower the growth at the sea was (Supplementary Table S4).

There was an increasing trend in the total length (cm) and mass (kg) of the returning spawners since the 1970s in both 1SW and 2SW in all three populations except Inari 2SW, where size development was slightly negative (Figure 6). The total length of the returning spawners indicates concurrently that the later the fish smolt, the smaller they are at return to the tributaries, and the differences among smolt ages are potentially increasing in the Inari and Utsjoki populations (Figure 7).

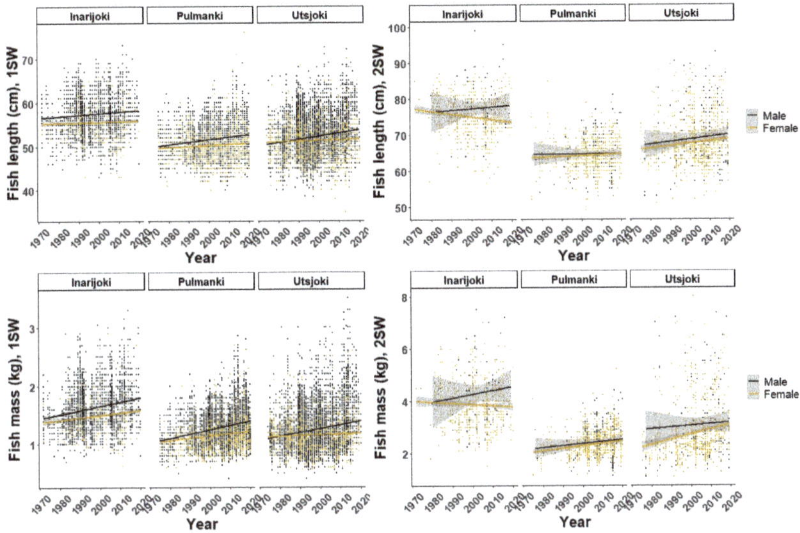

Figure 6. Total length (upper panel) and mass (lower panel) of male and female 1SW and 2SW salmon (cm) at capture from three different populations (*n* = 35,065). The trend lines represent the linear regression with 95% confident intervals.

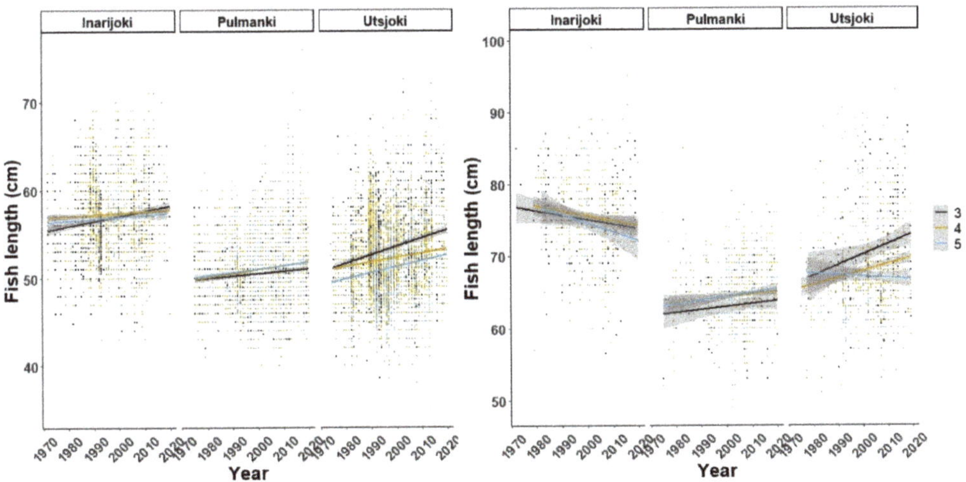

Figure 7. Total length of 1SW and 2SW salmon (cm) at capture from three different populations representing smolt ages of 3, 4, or 5 (*n* = 35,065). The trend lines represent the linear regression with 95% confident intervals.

4. Discussion

The long-term trends in Atlantic salmon growth, based on more than 35,000 catch samples and near 18,000 scales with detailed growth measurements, collected over the past 48 years, clearly indicate the life-history stage and temperature-dependent changes in growth rates. The freshwater growth has markedly decreased, whereas the sea growth of 1SW has undoubtedly increased resulting in larger sized 1SW spawners. The growth reduction was more pronounced in smolt ages four and five than in age three smolts, indicating that spending more years in the river comes with a growth cost, possibly due to the impact of rising temperatures. The diverged growth patterns in different life-history stages likely reflect biotic and abiotic responses to increasing temperature that differ substantially between riverine and marine ecosystems.

The Increasing water temperatures during the active feeding and growing season of juvenile Atlantic salmon between mid-summer and early autumn, e.g., [37], may affect their metabolism and restrict growth at the freshwater stage. Although the temperatures in the River Teno area may not yet reach critical levels for juvenile Atlantic salmon, cf. [38], the rapid change in sub-arctic environments may cause more and more frequent surpassing of the optimal range of water temperatures for juvenile salmon growth (14–20 °C) [39], and induce varying ecological responses (see, e.g., [14,40] for reviews).

As the riverine environments may react rapidly to the temperature changes, especially in the areas of shallow and slow-running flow, the parr must either adapt to higher temperatures or escape to the refugia that buffer against the rising water temperatures even during the peak temperatures or longer waves, e.g., [14]. In such areas, food may come a limiting resource for growth via density-dependent intra-specific competition, and it has been observed that salmon parr may cease feeding altogether in cool water refugia [41], which potentially impede juvenile growth especially during summer peak temperatures, as was the case in July in our study. On the other hand, decreased feeding activity in juvenile salmon in autumn [37] may result in decreased growth (Figure 4), potentially driven by increased temperatures in September and increased basic metabolism of fish [42].

Our time series data indicate that multiple years spent in freshwater in recent years will come with a cost in the form of reduced growth compared to the situation a few decades ago. Moreover, multi-river-year smolts also grow slower in the sea and may return for spawning at a smaller size than younger, three-year-old smolts, e.g., [43]. Nevertheless, this seems to be rather population specific (c.f. Pulmanki, Figure 7). The fast growth in freshwater and early smolting seem to result in fast growth at the sea, supporting the life-history theory [44] linking fast growth with early maturation (1SW). Alternatively, if older smolts emigrate earlier, and younger smolts later in the migration window [15], there is a potential for environmental mismatch at the sea leading to growth reduction during the first year at the sea. This might cause a smaller maturation size as well. As the first year at the river shows the largest effect on total growth at the river stage, it is plausible that the unfavourable temperature conditions in the first year might have cumulative effects on the later development of juvenile fish, restricting growth in later years. This can potentially have vast cascading effects on intra-specific resource competition for habitat use and food. Competition within conspecifics may increase if formerly different-sized age-classes have utilised divergent niches, but reduced growth in old parr may drive them to compete for resources with the younger parr.

A large body of research suggests a reduction in the productivity of several marine ecosystems, including those in the Northern Hemisphere, resulting in changes in community composition and population abundances across all trophic levels [45–47]. In contrast to freshwater, the ocean environment might be less sensitive for peak air temperatures and react more slowly to the warming climate. However, the water temperature in the Barents Sea has increased in recent decades, e.g., [34,48], and such a change might have induced the observed increase in growth of 1SW fish. In contrast, the growth of 2SW fish did not covary with sea temperature, and no significant long-term trend in growth was detected within our time series. It is possible that 2SW fish have less growth potential with increased temperature since higher temperatures will increase the metabolism of larger fish more than in smaller fish, and thereby 2SW fish must allocate their energy differently compared to 1SW fish. Recent changes in the North Atlantic environment have already been shown to negatively affect growth [47] and alter genetic architecture [25] especially in larger salmon (2SW and multi-sea-winter, MSW). Tréhin et al. [49] showed that after the decline in growth during the first year at sea, growth remained stable during the later periods at sea among returning salmon, which suggests a similar lack of growth response in 2SW salmon as in our results.

Although a decline in size (total length and/or mass) of Atlantic salmon (mostly 1SW) at return to freshwater has recently been reported across several areas and rivers in the northeast Atlantic area [49–51], all three salmon populations in this study showed a long-term increase in body length for both 1SW and 2SW (with one exception) fish. Changes in marine environmental conditions, typically resulting in changes in the growth and size of salmon, appear to affect in different directions in different areas. The increase in sea water temperature in the Barents and Norwegian seas [47,48] and other possible changes in the marine environment appear to improve, at least for the time being, the marine growth of salmon in the northern populations. A slight increasing trend in growth for northern Norwegian salmon populations has been documented, whereas a decline was detected in salmon populations further south in Norway [47]. A recent study indicates, however, that the Arctic oceanic environment may change and become warmer even faster than earlier predicted, and such a scenario may further affect the growth of Atlantic salmon in an unknown direction [52].

Other interesting differences in Atlantic salmon growth patterns between different parts of the North Atlantic Ocean emerged when comparing our results to growth analyses carried out elsewhere. Both Vollset et al. and Tréhin et al. [47,49] showed recent, abrupt reductions in Atlantic salmon growth during the first year at sea in fish from a French and several southern Norwegian populations, respectively. In addition, a general decrease in marine growth rate in a Scottish salmon population has recently been documented [51]. Such rapid changes in growth indicate a large-scale shift in conditions for Atlantic salmon: a marked decrease in the extent of Arctic water in the Norwegian Sea, a subsequent warming of spring water temperature before Atlantic salmon entering the sea, and an approximately 50% reduction in zooplankton across large geographic areas of the northeast Atlantic Ocean [47]. They further concluded that these phenomena suggest an ecosystem-scale regime shift in the northeast Atlantic Ocean.

In addition, during the recent years, the proportion of early maturing 1SW salmon decreased in the Norwegian populations [47] (see also [21]). Similarly, Jonsson et al. [50] reported a decline in the proportion of 1SW salmon since the 1980s in a small Norwegian salmon population. Tréhin et al. [49] concluded that changes in age at maturity in a French salmon population support the hypothesis of a sex-specific probabilistic reaction norm: the individual probability to return after one year at sea is increasing with changes in marine growth rate. However, the proportion of both 2SW salmon and previous spawners in both the Utsjoki and Inarijoki populations have increased since the 1970s, whereas the proportion of 1SW fish has slightly decreased and showed more variable patterns in recent years compared to the past [23]. The observed patterns in growth and sea age composition in the tributary populations of the River Teno suggest more complex linkages between the variables, and alternative factors should be explored as candidate drivers of these dynamics.

If temperature acts as a growth-limiting factor, the growth of post-smolt fish may show a positive correlation with Sea Surface Temperature (SST). Previous studies have established a link between post-smolt growth and the increasing SST [53], which appears to be population specific [54]. The evidence from northeast Atlantic populations shows a negative correlation between sea growth and SST [47,54]. However, in parallel to our findings, northwest populations indicate a positive association between post-smolt growth and sea temperatures in sub-arctic and arctic marine environments [53,54]. This growth pattern can arise from either a physiological response to an optimal temperature or from the influence of certain ecosystem factors on growth. While previous evidence has established a direct link between growth and post-smolt survival [55], there are conflicting findings that suggest post-smolt growth might not always be the primary determinant of post-smolt survival [54,56]. Warming conditions during certain stages of the salmon life cycle have been associated with reduced adult recruitment. However, warming during other stages has been observed to promote salmon growth and is assumed to amplify the reproductive output of spawning fish [53]. The situation in which elevated SST leads to enhanced post-smolt growth but triggers a decline in spawner recruitment demands further investigation, particularly regarding the potential latent effects of environmental factors, and genotype–environment interactions. Furthermore, a comprehensive understanding of the sea migration routes of sub-arctic salmon populations is essential for understanding climate effects on their dynamics.

Forage fishes defined here as small- or intermediate-sized marine pelagic species that are important or the primary food source for predators, play important roles in marine ecosystems by providing energy transfer between trophic levels [57]. In the Barents Sea and other northeast parts of the Atlantic Ocean, one of the key species is capelin, which has suffered from several population collapses over the past decades largely because of overfishing [58]. Moreover, the drastic variation in capelin population in the Barents Sea has been associated with the changes in life-history composition of the River Teno salmon and the genetic architecture behind it [25]. Changes in the food base were also associated with a reduction in marine growth of salmon in southern Norwegian populations, accompanied by a simultaneous decrease in the growth of Atlantic mackerel (*Scomber scombrus*, L. 1758) in the corresponding sea areas [47]. In our study, however, no link between capelin abundance in the Barents Sea and marine growth patterns of salmon was found. It should be remembered that the salmon sampled are the survivors who made their return to the natal rivers, and the growth patterns and other biological characteristics of these fish may not entirely reflect those that did not survive and may have suffered from changes in the environment, such as collapses of capelin stock in this case. Czorlich et al. [25] showed that the capelin collapses especially affected the large MSW salmon, and it can be speculated that the smaller 1-2SW salmon analysed in the present study may have successfully utilize other prey items, e.g., [59], and their growth rates benefited from the increasing sea water temperatures despite the periods of low capelin stock. In the sea, growing salmon shift from foraging on invertebrates to a piscivorous diet [60]. Therefore, the effect of temperature on growth may be complicated by the availability of suitable prey, which ultimately determines growth opportunities [24].

This study investigated growth patterns of Atlantic salmon from three genetically distinct populations [26], although they represent tributaries of the same large catchment. Temporal patterns in both freshwater and marine growth were largely consistent between populations, although there were differences in absolute levels. The largest 1SW and 2SW salmon were those of the Inarijoki population whereas the 1SW fish were similar in size in Utsjoki and Pulmanki, and 2SW salmon were slightly larger in the Utsjoki population. Differences in marine growth are difficult to interpret since little or nothing is known about certain aspects, e.g., the exact marine feeding areas of the salmon in these tributary populations. There are likely genetic components and local adaptations dictating marine feeding migration and growth, and a genetic component in the ontogenetic diet during the ocean feeding phase in Atlantic salmon [61]. Moreover, Jonsson et al. [20] showed that 1SW salmon tend to be larger in size in rivers with higher proportions of MSW salmon in the population, a pattern consistent with the size differences documented in the present study.

5. Conclusions

In conclusion, the rapidly progressing global change, especially in the northern areas, is affecting the anadromous fish which utilize both freshwater and ocean waters during their life cycle. Our results indicate contrasting responses in Atlantic salmon growth rates to warming climate: freshwater growth decreased whereas marine growth has largely increased. The predicted continuing and accelerating increase in water temperature and other changes in the aquatic ecosystems in the north call for carefully designed, long-term monitoring of migratory fish populations at various phases of their life cycle. There is strong evidence pointing towards the Arctic ecosystems warming more rapidly than those further south [62], which is posing particular needs for careful monitoring of changes in environmental conditions and anadromous fish populations in the northern areas, cf. [13,63].

Supplementary Materials: The following supporting information can be downloaded at: https://www.mdpi.com/article/10.3390/fishes8090441/s1, Table S1. Model selection procedure for FGW and SG models presenting AIC, LogLikelihood and Deviance values. The selected models are presented in bold. FWG = Freshwater growth, RG = River growth, SG = Sea growth, SST = Sea surface temperature. Rainfall and air temperatures were standardized N (0, 1). Log-transformation was used for capelin biomass (BM) and rolling averages over two years of capelin biomass (BM2y) and immature capelin biomass (Imm2y). The final models are in bold. Table S2. Summary of Linear mixed-effects models for Freshwater growth of 1SW and 2SW fish, and sea growth of 1SW fish. Table S3. Summary for LMM of 1SW fish sea growth with monthly means as covariates. SST refers sea surface temperature (°C). Table S4. Summary of Linear mixed-effects model for sea growth of 2SW fish.

Author Contributions: Conceptualization, J.E. and P.O.; Investigation, J.E. and P.O.; Data curation, P.O. and J.E.; Formal analysis, N.A.; Writing—original draft preparation, N.A. and J.E.; Writing—review and editing, N.A., J.E. and P.O. All authors have read and agreed to the published version of the manuscript.

Funding: This research received no external funding.

Institutional Review Board Statement: Only the fisheries data were used in this study meaning that local fishermen have delivered the samples taken from captured and culled fish to the Natural Resources Institute Finland's collections. Living fish were not used in this study.

Informed Consent Statement: Not applicable.

Data Availability Statement: The scripts for the statistical analyses will be available online via osf.io (https://doi.org/10.17605/OSF.IO/GJ76N). The original data will be available on the request in accordance with the data policy of Luke.

Acknowledgments: We thank the large number of fishermen fishing the tributaries of the River Teno system who collected the large set of salmon scale samples and phenotypic data over a nearly 50-year period, and our skilful personnel who have analysed the scale material, especially, Jari Haantie, Jorma Ollila, and Matti Kylmäaho. In addition, Jorma Kuusela is thanked for organizing the complex salmon scale database.

Conflicts of Interest: The authors declare no conflict of interest.

References

1. Rantanen, M.; Karpechko, A.Y.; Lipponen, A.; Nordling, K.; Hyvärinen, O.; Ruosteenoja, K.; Vihma, T.; Laaksonen, A. The Arctic has warmed nearly four times faster than the globe since 1979. *Commun. Earth Environ.* **2022**, *3*, 168. [CrossRef]
2. Thomas, C.; Cameron, A.; Green, R.; Bakkenes, M.; Beaumont, L.J.; Collingham, Y.C.; Erasmus, B.F.N.; de Siqueira, M.F.; Grainger, A.; Hannah, L.; et al. Extinction risk from climate change. *Nature* **2004**, *427*, 145–148. [CrossRef] [PubMed]
3. Pacifici, M.; Visconti, P.; Butchart, S.H.M.; Watson, J.E.M.; Cassola, F.M.; Rondinini, C. Species' traits influenced their response to recent climate change. *Nat. Clim. Change* **2017**, *7*, 205–208. [CrossRef]
4. Beaugrand, G.; Kirby, R.R. How Do Marine Pelagic Species Respond to Climate Change? Theories and Observations. *Annu. Rev. Mar. Sci.* **2018**, *10*, 169–197. [CrossRef] [PubMed]
5. Biggs, R.; Carpenter, S.R.; Brock, W.A. Turning back from the brink: Detecting an impending regime shift in time to avert it. *Proc. Natl. Acad. Sci. USA* **2009**, *106*, 826–831. [CrossRef]
6. Beaugrand, G.; Edwards, M.; Brander, K.; Luczak, C.; Ibanez, F. Causes and projections of abrupt climate-driven ecosystem shifts in the North Atlantic. *Ecol. Lett.* **2008**, *11*, 1157–1168. [CrossRef]
7. Foden, W.B.; Young, B.E.; Akçakaya, H.R.; Garcia, R.A.; Hoffmann, A.A.; Stein, B.A.; Thomas, C.D.; Wheatley, C.J.; Bickford, D.; Carr, J.A.; et al. Climate change vulnerability assessment of species. *WIREs Clim. Change* **2019**, *10*, e551. [CrossRef]
8. Radchuk, V.; Reed, T.; Teplitsky, C.; van de Pol, M.; Charmantier, A.; Hassall, C.; Adamík, P.; Adriaensen, F.; Ahola, M.P.; Arcese, P.; et al. Adaptive responses of animals to climate change are most likely insufficient. *Nat. Commun.* **2019**, *10*, 3109. [CrossRef]
9. Todd, C.D.; Friedland, K.D.; MacLean, J.C.; Hazon, N.; Jensen, A.J. Getting into hot water? Atlantic salmon responses to climate change in freshwater and marine environments. In *Atlantic Salmon Ecology*; Blackwell Publishing: Hoboken, NJ, USA, 2011; pp. 409–443.
10. Klemetsen, A.; Amundsen, P.-A.; Dempson, J.B.; Jonsson, B.; Jonsson, N.; O'Connell, M.F.; Mortensen, E. Atlantic salmon *Salmo salar* L., brown trout *Salmo trutta* L. and Arctic charr *Salvelinus alpinus* (L.): A review of aspects of their life histories. *Ecol. Freshw. Fish* **2003**, *12*, 1–59. [CrossRef]

11. Jonsson, B.; Jonsson, N. Ecology of Atlantic Salmon and Brown Trout. In *Habitat as a Template for Life Histories*; Fish & Fisheries Series; Springer: Berlin/Heidelberg, Germany, 2011; p. 708.

12. Quinn, T.P. *The Behavior and Ecology of Pacific Salmon and Trout*; University of British Columbia Press: Vancouver, BC, Canada, 2018.

13. Svenning, M.-A.; Falkegård, M.; Dempson, J.B.; Power, M.; Bårdsen, B.-J.; Gudbergsson, G.; Fauchald, P. Temporal changes in the relative abundance of anadromous Arctic charr, brown trout, and Atlantic salmon in northern Europe: Do they reflect changing climates? *Freshw. Biol.* **2022**, *67*, 64–77. [CrossRef]

14. Gillis, C.-A.; Ouellet, V.; Breau, C.; Frechette, D.; Bergeron, N. Assessing climate change impacts on North American freshwater habitat of wild Atlantic salmon-urgent needs for collaborative research. *Can. Water Resour. J.* **2023**, *48*, 222–246. [CrossRef]

15. Thorstad, E.B.; Whoriskey, F.; Uglem, I.; Moore, A.; Rikardsen, A.H.; Finstad, B. A critical life stage of the Atlantic salmon *Salmo salar*: Behaviour and survival during the smolt and initial post-smolt migration. *J. Fish Biol.* **2012**, *81*, 500–542. [CrossRef]

16. Otero, J.; L'Abée-Lund, J.H.; Castro-Santos, T.; Leonardsson, K.; Storvik, G.O.; Jonsson, B.; Dempson, B.; Russell, I.C.; Jensen, A.J.; Baglinière, J.-L.; et al. Basin-scale phenology and effects of climate variability on the global timing of initial seaward migration of Atlantic salmon (*Salmo salar*). *Glob. Change Biol.* **2014**, *20*, 61–75. [CrossRef] [PubMed]

17. Teichert, N.; Benitez, J.-P.; Dierckx, A.; Tétard, S.; de Oliveira, E.; Trancart, T.; Feunteun, E.; Ovidio, M. Development of an accurate model to predict the phenology of Atlantic salmon smolt spring migration. *Aquat. Conserv. Mar. Freshw. Ecosyst.* **2020**, *30*, 1552–1565. [CrossRef]

18. Vehanen, T.; Sutela, T.; Huusko, A. Potential impact of climate change on salmonid smolt ecology. *Fishes* **2023**, *8*, 332. [CrossRef]

19. Mobley, K.B.; Aykanat, T.; Czorlich, Y.; House, A.; Kurko, J.; Miettinen, A.; Moustakas-Verho, J.; Salgado, A.; Sinclair-Waters, M.; Verta, J.-P.; et al. Maturation in Atlantic salmon (*Salmo salar*, Salmonidae): A synthesis of ecological, genetic, and molecular processes. *Rev. Fish Biol. Fish.* **2021**, *31*, 523–571. [CrossRef]

20. Jonsson, N.; Hansen, L.P.; Jonsson, B. Variation in age, size and repeat spawning of adult Atlantic salmon in relation to river discharge. *J. Anim. Ecol.* **1991**, *60*, 937–947. [CrossRef]

21. Otero, J.; Jensen, A.J.; L'Abée-Lund, J.H.; Stenseth, N.C.; Storvik, G.O.; Vøllestad, L.A. Contemporary ocean warming and freshwater conditions are related to later sea age at maturity in Atlantic salmon spawning in Norwegian rivers. *Ecol. Evol.* **2012**, *2*, 2192–2203. [CrossRef] [PubMed]

22. Reid, J.E.; Chaput, G. Spawning history influence on fecundity, egg size, and egg survival of Atlantic salmon (*Salmo salar*) from the Miramichi River, New Brunswick, Canada. *ICES J. Mar. Sci.* **2012**, *69*, 1678–1685. [CrossRef]

23. Erkinaro, J.; Czorlich, Y.; Orell, P.; Kuusela, J.; Länsman, M.; Falkegård, M.; Pulkkinen, H.; Primmer, C.; Niemelä, E. Life history variation across four decades in a diverse population complex of Atlantic salmon in a large subarctic river. *Can. J. Fish. Aquat. Sci.* **2019**, *76*, 42–55. [CrossRef]

24. Jonsson, B.; Jonsson, N.; Finstad, A.G. Effects of temperature and food quality on age at maturity of ectotherms: An experimental test of Atlantic salmon. *J. Anim. Ecol.* **2013**, *82*, 201–210. [CrossRef] [PubMed]

25. Czorlich, J.; Aykanat, T.; Erkinaro, J.; Orell, P.; Primmer, C.R. Rapid evolution in salmon life-history induced by direct and indirect effects of fishing. *Science* **2022**, *376*, 420–423. [CrossRef]

26. Vähä, J.P.; Erkinaro, J.; Falkegård, M.; Orell, P.; Niemelä, E. Genetic stock identification of Atlantic salmon and its evaluation in a large population complex. *Can. J. Fish. Aquat. Sci.* **2017**, *74*, 327–338. [CrossRef]

27. Niemelä, E.; Erkinaro, J.; Julkunen, M.; Hassinen, E.; Länsman, M.; Brørs, S. Temporal variation in abundance, return rate and life histories of previously spawned Atlantic salmon in a large subarctic river. *J. Fish Biol.* **2006**, *68*, 1222–1240. [CrossRef]

28. Tana Monitoring and Research Group. Status of the River Tana Salmon Populations 2022. Report of the Working Group on Salmon Monitoring and Research in the Tana River System. Available online: http://urn.fi/URN:NBN:fi-fe202301306407 (accessed on 15 May 2023).

29. Niemelä, E.; Orell, P.; Erkinaro, J.; Dempson, J.B.; Brørs, S.; Svenning, M.A.; Hassinen, E. Previously spawned Atlantic salmon ascend a large subarctic river earlier than their maiden counterparts. *J. Fish Biol.* **2006**, *69*, 1151–1163. [CrossRef]

30. Vähä, J.-P.; Erkinaro, J.; Niemelä, E.; Primmer, C.R.; Saloniemi, I.; Johansen, M.; Svenningm, M.; Brørs, S. Temporally stable population-specific differences in run timing of one-sea-winter Atlantic salmon returning to a large river system. *Evol. Appl.* **2010**, *4*, 39–53. [CrossRef] [PubMed]

31. ICES. *Report of the Workshop on Age Determination of Salmon (WKADS)*; ICES Document CM 2011/ACOM:44; ICES: Galway, Ireland, 2011.

32. Hanson, N.N.; Smith, G.W.; Middlemas, S.J.; Todd, C.D. Precision and accuracy of Dahl-Lea back-calculated smolt lengths from adult scales of Atlantic salmon *Salmo salar*. *J. Fish Biol.* **2019**, *94*, 183–186. [CrossRef] [PubMed]

33. Jonsson, B.; Jonsson, M.; Jonsson, N. Optimal size at seaward migration in an anadromous salmonid. *Mar. Ecol. Prog. Ser.* **2016**, *559*, 193–200. [CrossRef]

34. Gonzalez-Pola, C.; Larsen, K.M.H.; Fratantoni, P.; Beszczynska-Möller, A. ICES Report on ocean climate 2020. *ICES Coop. Res. Rep. CRR* **2022**, *356*, 121. [CrossRef]

35. ICES. Capelin (*Mallotus villosus*) in Subareas 1 and 2 (Northeast Arctic), Excluding Division 2.a West of 5° W (Barents Sea Capelin). In *ICES Advice: Recurrent Advice*; ICES: Copenhagen, Denmark, 2021. [CrossRef]

36. Weisberg, S.; Spangler, G.; Richmond, L.S. Mixed effects models for fish growth. *Can. J. Fish. Aquat. Sci.* **2010**, *67*, 269–277. [CrossRef]
37. Erkinaro, H.; Erkinaro, J. Feeding of Atlantic salmon, *Salmo salar* L., parr in the subarctic River Teno and three tributaries, northernmost Finland. *Ecol. Freshw. Fish* **1998**, *7*, 13–24. [CrossRef]
38. Corey, E.; Linnansaari, T.; Cunjak, R.A. High-temperature events shape the broadscale distribution of juvenile Atlantic salmon (*Salmo salar*). *Freshw. Biol.* **2023**, *68*, 534–545. [CrossRef]
39. Jonsson, B.; Forseth, T.; Jensen, A.J.; Næsje, T.F. Thermal performance of juvenile Atlantic salmon, *Salmo salar* L. *Funct. Ecol.* **2001**, *15*, 701–711. [CrossRef]
40. Jonsson, B.; Jonsson, N. A review of the likely effects of climate change on anadromous Atlantic salmon Salmo salar and brown trout *Salmo trutta*, with particular reference to water temperature and flow. *J. Fish Biol.* **2009**, *75*, 2381–2447. [CrossRef]
41. Breau, C.; Cunjak, R.A.; Bremset, G. Age-specific aggregation of wild juvenile Atlantic salmon *Salmo salar* at cool water sites during high-temperature events. *J. Fish Biol.* **2007**, *71*, 1179–1191. [CrossRef]
42. Fry, F.E.J. The Effect of Environmental Factors on the Physiology of Fish. *Fish Physiol.* **1971**, *6*, 1–98. [CrossRef]
43. Erkinaro, J.; Dempson, J.B.; Julkunen, M.; Niemelä, E. Importance of ontogenetic habitat shifts to juvenile output and life history of Atlantic salmon in a large subarctic river: An approach based on analysis of scale characteristics. *J. Fish Biol.* **1997**, *51*, 1174–1185. [CrossRef] [PubMed]
44. Roff, D.A. *Evolution of Life Histories*; Chapman and Hall: New York, NY, USA, 1993.
45. Mills, K.E.; Pershing, A.J.; Sheehan, T.F.; Mountain, D. Climate and ecosystem linkages explain widespread declines in North American Atlantic salmon populations. *Glob. Change Biol.* **2013**, *19*, 3046–3061. [CrossRef]
46. Capuzzo, E.; Lynam, C.P.; Barry, J.; Stephens, D.; Forster, R.M.; Greenwood, N.; McQuatters-Gollop, A.; Silva, T.; van Leeuwen, S.M.; Engelhard, G.H. A decline in primary production in the North Sea over 25 years, associated with reductions in zooplankton abundance and fish stock recruitment. *Glob. Change Biol.* **2018**, *24*, e352–e364. [CrossRef] [PubMed]
47. Vollset, K.W.; Urdal, K.; Utne, K.; Thorstad, E.B.; Fiske, P. Ecological regime shift in the Northeast Atlantic Ocean revealed from the unprecedented reduction in marine growth of Atlantic salmon. *Sci. Adv.* **2022**, *8*, eabk2542. [CrossRef]
48. Pasanen, L.; Laukkanen-Nevala, P.; Launonen, I.; Prusov, S.; Holmström, L.; Niemelä, E.; Erkinaro, J. Extraction of sea temperature in the Barents Sea by a scale-space multiresolution method–prospects for Atlantic salmon. *J. Appl. Stat.* **2017**, *44*, 2317–2336. [CrossRef]
49. Tréhin, C.; Rivot, E.; Lamireau, L.; Meslier, L.; Besnard, A.-L.; Gregory, S.D.; Nevoux, M. Growth during the first summer at sea modulates sex-specific maturation schedule in Atlantic salmon. *Can. J. Fish. Aquat. Sci.* **2021**, *78*, 659–669. [CrossRef]
50. Jonsson, B.; Jonsson, N.; Albretsen, J. Environmental change influences the life history of salmon *Salmo salar* in the North Atlantic Ocean. *J. Fish Biol.* **2016**, *88*, 618–637. [CrossRef] [PubMed]
51. Todd, C.D.; Hanson, N.N.; Boehme, L.; Revie, C.W.; Marques, A.R. Variation in post-smolt growth pattern of wild one sea-winter salmon (*Salmo salar* L.), and its linkage to surface warming in the eastern North Atlantic Ocean. *J. Fish Biol.* **2020**, *98*, 6–16. [CrossRef] [PubMed]
52. Kim, Y.-H.; Min, S.-K.; Gillett, N.P.; Notz, D.; Malinina, E. Observationally-constrained projections of an ice-free Arctic even under a low emission scenario. *Nat. Commun.* **2023**, *14*, 3139. [CrossRef]
53. Friedland, K.D.; Todd, C.D. Changes in Northwest Atlantic Arctic and Subarctic conditions and the growth response of Atlantic salmon. *Polar Biol.* **2012**, *35*, 593–609. [CrossRef]
54. Friedland, K.D.; Chaput, G.; MacLean, J.C. The emerging role of climate in post-smolt growth of Atlantic salmon. *ICES J. Mar. Sci.* **2005**, *62*, 1338–1349. [CrossRef]
55. Peyronnet, A.; Friedland, K.D.Ó.; Maoiléidigh, N.; Manning, M.; Poole, W.R. Links between patterns of marine growth and survival of Atlantic salmon (*Salmo salar*, L.). *J. Fish Biol.* **2007**, *71*, 684–700. [CrossRef]
56. Friedland, K.D.; Moore, D.; Hogan, F. Retrospective growth analysis of Atlantic salmon (*Salmo salar*) from the Miramichi River, Canada. *Can. J. Fish. Aquat. Sci.* **2009**, *66*, 1294–1308. [CrossRef]
57. Pikitch, E.K.; Rountos, K.J.; Essington, T.E.; Santora, C.; Pauly, D.; Watson, R.; Sumaila, U.R.; Boersma, P.D.; Boyd, I.L.; Conover, D.O.; et al. The global contribution of forage fish to marine fisheries and ecosystems. *Fish Fish.* **2014**, *15*, 43–64. [CrossRef]
58. Hjerman, D.Ø.; Ottersen, G.; Stenseth, N.C. Competition among fishermen and fish causes the collapse of Barents Sea capelin. *Proc. Natl. Acad. Sci. USA* **2004**, *101*, 11679–11684. [CrossRef] [PubMed]
59. Utne, R.U.; Diaz Pauli, B.; Haugland, M.; Jacobsen, J.A.; Maoileidigh, N.; Melle, W.; Broms, C.T.; Nøttestad, L.; Holm, M.; Thomas, K.; et al. Poor feeding opportunities and reduced condition factor for salmon post-smolts in the Northeast Atlantic Ocean. *ICES J. Mar. Sci.* **2021**, *78*, 2844–2857. [CrossRef]
60. Jacobsen, J.A.; Hansen, L.P. Feeding habits of wild and escaped farmed Atlantic salmon, *Salmo salar* L., in the Northeast Atlantic. *ICES J. Mar. Sci.* **2001**, *58*, 916–933. [CrossRef]
61. Aykanat, T.; Rasmussen, M.; Ozerov, M.; Niemelä, E.; Paulin, L.; Vähä, J.-P.; Hindar, K.; Wennevik, V.; Pedersen, T.; Svenning, M.-A.; et al. Life-history genomic regions explain differences in Atlantic salmon marine diet specialization. *J. Anim. Ecol.* **2020**, *89*, 2677–2691. [CrossRef] [PubMed]

62. Prowse, T.D.; Wrona, F.J.; Reist, J.D.; Gibson, J.J.; Hobbie, J.E.; Levesque, L.M.J.; Vincent, W.F. Climate change effects on hydroecology of Arctic freshwater ecosystems. *Ambio* **2006**, *35*, 347–358. [CrossRef]
63. Heino, J.; Culp, J.M.; Erkinaro, J.; Goedkoop, W.; Lento, J.; Ruhland, K.M.; Smol, J.P. Abruptly and irreversibly changing Arctic freshwaters urgently require standardized monitoring. *J. Appl. Ecol.* **2020**, *57*, 1192–1198. [CrossRef]

 *fishes*

Review

The Role of Cold-Water Thermal Refuges for Stream Salmonids in a Changing Climate—Experiences from Atlantic Canada

Tommi Linnansaari [1,2,*], Antóin M. O'Sullivan [1], Cindy Breau [3], Emily M. Corey [2], Elise N. Collet [1], R. Allen Curry [1,2] and Richard A. Cunjak [1,2]

[1]   Canadian Rivers Institute, Faculty of Forestry and Environmental Management, University of New Brunswick, P.O. Box 4400, Fredericton, NB E3B 5A3, Canada; aosulliv@unb.ca (A.M.O.); elise.collet@unb.ca (E.N.C.); racurry@unb.ca (R.A.C.); cunjak@unb.ca (R.A.C.)

[2]   Department of Biology, University of New Brunswick, P.O. Box 4400, Fredericton, NB E3B 5A3, Canada; emily.corey@unb.ca

[3]   Fisheries and Oceans Canada, 343 Université Avenue, Moncton, NB E1C 9B6, Canada; cindy.breau@dfo-mpo.gc.ca

*   Correspondence: tommi.linnansaari@unb.ca

Citation: Linnansaari, T.; O'Sullivan, A.M.; Breau, C.; Corey, E.M.; Collet, E.N.; Curry, R.A.; Cunjak, R.A. The Role of Cold-Water Thermal Refuges for Stream Salmonids in a Changing Climate—Experiences from Atlantic Canada. *Fishes* **2023**, *8*, 471. https://doi.org/10.3390/fishes8090471

Academic Editor: Bror Jonsson

Received: 24 August 2023
Revised: 19 September 2023
Accepted: 19 September 2023
Published: 21 September 2023

Abstract: Thermal refuges are becoming increasingly influential for dictating the population status and spatial distribution of cold-water stenotherm salmonids in the mid- to southern extent of their range. The global climate is predicted to continue to warm, and therefore, the overall thermal suitability of freshwater habitats for stream salmonids is predicted to decline in concert. However, stream and river thermal heterogeneity will offer considerable resiliency for these populations. Thermal refuges are formed by many physical processes; common natural refuges include cold tributary plumes, groundwater springs, alcoves, and hyporheic upwellings. However, many anthropogenically formed refuges (such as stratified reservoirs or cold-water tailrace outflows) also exist in hydropower-regulated rivers. The significance of these refuges to stream salmonids depends on their size and temperature differential, but also other habitat characteristics such as their depth, flow velocity, Froude number, and many biotic factors within the refuges. Modern technologies such as drone-mounted thermal infrared cameras and other remote sensing techniques allow for the efficient identification of such refuges, and inexpensive options include the identification of refuges during ice cover using orthophotographs. Behavioural thermoregulation, i.e., salmonids aggregating in cold-water refuges, can be either facultative or obligate and the timing of these events is governed by life stage, species, and population-specific physiologically regulated cumulative thresholds that are inherently related to the recent thermal history, or hysteresis, of each individual. Salmonids appear to have an excellent spatial cognition for locating and relocating cold-water refuges, and their spatial distribution is largely affected by the availability of the cold-water refuges during the warm-water period in many thermally stressed rivers. Gregarious behaviour is the norm for salmonid fishes within the thermal refuges; however, the size/microhabitat hierarchy appears to dictate the within-refuge distribution at the micro-scale. There continues to be a great impetus for protecting—and in carefully determined cases creating—cold-water refuges in the future. A thorough understanding of what a "goldilocks" refuge is for various salmonids and their different life stages will be imperative as cold-water restoration is gaining popularity. Finally, disentangling the roles of the climate-induced and landscape activity-induced warming potential of fluvial freshwater will be important to ensure continued environmentally responsible landscape activities in future waterscapes.

Keywords: climate change; salmonids; thermal refuge; behavioural thermoregulation; remote sensing

Key Contribution: This paper synthesizes over two decades of work in Atlantic Canada regarding cold-water refuge use by stream salmonids with comparisons to similar research on stream salmonids elsewhere. The article introduces the concept of facultative and obligate behavioural thermoregulation, and therefore, expands the previous concept by allowing for a better understanding of the reasons and timing of the cold-water refuge-seeking behaviour.

---

1. Introduction

Climate change has become the environmental phenomenon epitomizing the early 21st century and it has directed much of the research questions and hypotheses examined by contemporary environmental scientists. Climate change is, and likely more so in the future, causing a variety of changes in climate patterns, the frequency of extreme observations, and the severity of climate-correlated events [1]. Thousands of research articles, especially those since the 1990s, have been written about how predicted changes in the future climate will affect aquatic resources, including cold-water stenothermic fishes and particularly stream salmonids (e.g., *Salmo*, *Oncorhynchus*, *Salvelinus* spp.; e.g., the current special issue [2], and reviews such as [3] for *Salmo* spp. [4] for winter perspectives [5]). Generally speaking, warming air temperature is predicted to lead into concomitant increases in water temperature in both freshwater [6,7] and in oceanic systems [8]. It is the warming of water that is predicted to have the largest cascading effects on the ecology of stream salmonids in summer [3,9]. In winter, at least in seasonal climates, the water temperature may still generally remain at or close to 0 °C, but the duration of this relatively stable period may become shorter with potential consequences for winter-adapted salmonids [10,11]. Climate change is also predicted to cause changes in other weather parameters, such as changes in precipitation patterns with concomitant effects on stream discharge and the frequency and severity of ice break-up [1,12]. Many environmental parameters affecting freshwater resources act in concert in response to climate change, and some of these effects may be cumulative (e.g., less summer precipitation → lower discharge → higher cumulative effect on water temperature) [13], whereas some may be antagonistic (e.g., higher water temperature → lower ice thickness → less severe ice break-ups) [4].

One of the main concerns regarding climate warming is its potential effect on cold-water stenothermic fishes, such as the Salmonidae—a socio-economically important family of anadromous, potamodromous, and resident fishes in many jurisdictions [3]. With warming water, a declining overall habitat suitability for a number of salmonids has been documented [5,14–17]. However, it has been long recognized that despite the general trend of warming freshwater temperatures, rivers and streams possess an inherent thermal habitat heterogeneity wherein the generally uniform temperature of a given river reach may be "punctuated" with cold-water inputs of various types [18–22] and that these cold-water anomalies have the potential to act as refuges for salmonids during otherwise adverse temperature conditions in summer [13,23–25]. With a warming climate, the ecological significance of these refuges has increased for many salmonid populations in the southern-to-middle extent of their ranges, and related research has similarly increased in the last two decades for both the physical and biological aspects governing thermal refuges and their utility for salmonids [26,27].

In this article, our aim was not to conduct a quantitative meta-analysis or a systematic review of cold-water habitat use or the thermal capabilities of stream salmonids. Other recent reviews in related topics exist, for example, the work by [2,26–28]. Rather, we synthesized some of the insights the authors observed and concluded in the rivers of Atlantic Canada over the last two decades while conducting research into cold-water refuges and stream salmonid behavioural thermoregulation (see definition below). Further, we related this synthesis to the research of others for corollaries, parallels, or at times contrasting or supplementary perspectives. Our work mostly focused on Atlantic salmon *Salmo salar* L., 1758, and the examples that follow are primarily focused on this species, although the discussed concepts may be applicable more widely to other stream salmonids. We further narrowed down the focus by only considering thermal refuges from the perspective of the behaviour related to cold-water stenothermic salmonids in response to warm-water conditions, i.e., summer, while recognizing that the thermal heterogeneity, and often the very same thermal anomalies, may play a crucial, yet drastically different role in the winter biology of salmonids [4,29,30].

We start with a short description of the physical aspects of typical thermal refuges in Atlantic Canada and their detection and mapping using contemporary tools at different scales.

We then review the concept of behavioural thermoregulation for common stream salmonids and synthesize the movement ecology and behavioural aspects related to using cold-water refuges by salmonids. We conclude by providing insights related to the management of cold-water refuges with respect to the conservation of salmonid populations and offer some future directions for research and the conservation/restoration of cold-water refuges.

2. Thermal Refuges—Physical Environment

As implied by the name, thermal refuges are defined as cold-water patches across the riverscape that provide thermal relief for cold-water fishes during extreme heat events [18,31]. Such refuges may exist naturally, or may be a product of anthropogenic alterations, for example in (hydropower) regulated rivers [22].

In natural rivers, the predominant thermal refuge type described in the literature (but not the most abundant category by number [32]) is the tributary-derived thermal refuge [13,24,33]. These refuges occur at the confluence of a cold-water tributary and the warmer main river (Figure 1a,b). The temperature differential between the cold-water tributaries and the main river can be caused by multiple factors [7], but commonly includes elevational differences, groundwater influx, or sufficient shading from the riparian vegetation. The exact size of the cold-water plume, and thus, the thermal refuge is dictated by the amount of cold-water inflow relative to the main river discharge, the temperature differential between the two mixing waterbodies, and a complex set of hydrodynamic physical properties [21,34]. Generally speaking, however, the tributary-derived cold-water plumes are often the most commonly used and sizeable thermal refuges for salmonids and may exceed hundreds of square metres (Figure 1b). These refuges, in fact, may have the potential to even qualify as refugia (as opposed to just refuges; as per the terminology defined in [26]), i.e., a thermal refuge so significant that it may act as a last 'stronghold' wherein a salmonid population in a cold tributary can persist generation after generation while the remainder of a river may become uninhabitable for the species.

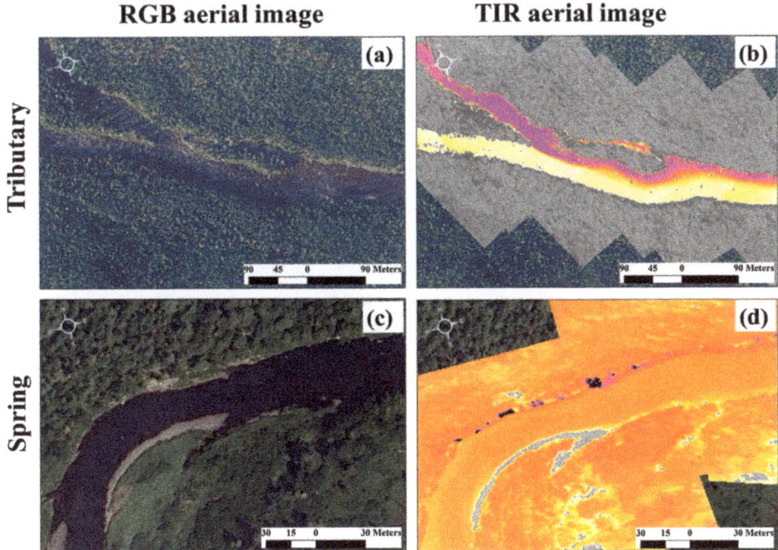

Figure 1. *Cont.*

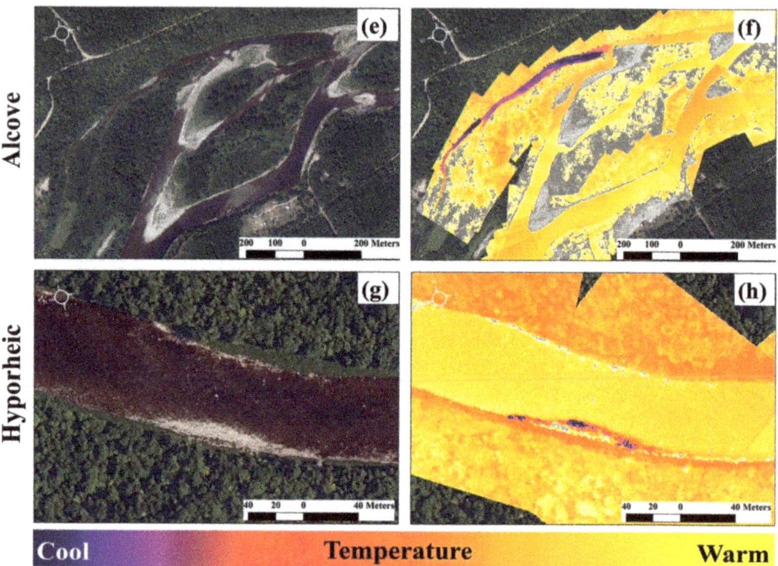

Figure 1. Illustrations of four thermal refuge classes, where the red–green–blue (RGB) aerial image is paired with a corresponding thermal infrared (TIR) image. A tributary confluence thermal refuge is shown in (**a**,**b**), where the influence of the tributary is > 200 m in the main river. Panels (**c**,**d**) display a groundwater spring refuge, where the influence of the spring has a markedly small footprint. An alcove thermal refuge is presented in (**e**,**f**). These areas can be groundwater dominant and have low dissolved oxygen values and can also be disconnected from the main river during low flow. The last thermal refuge class is the hyporheic discharge refuge (**g**,**h**). These present as very fine-scale discharges and are likely to be the least energetically favorable thermal refuges for salmonids due to the hydrological processes that govern their presence [35]. The example images are from the Miramichi River, NB, Canada.

There are several other important thermal refuge types that have been observed to hold salmonids during thermally stressful conditions. These include (1) groundwater springs (Figure 1c,d [25]); (2) groundwater alcoves (Figure 1e,f [34,36]), and (3) hyporheic discharge points (Figure 1g,h [37]), as well as other types of cold-water patches and/or thermal refuges as described by [19,21,26,32].

Groundwater springs can be excellent thermal refuges for salmonids [25]; however, these refuge types represent fine-scale (small) cold-water patches. Although high densities of fishes can be observed in springs, they generally have a limited capacity to hold large quantities of salmonids due to their limited area and depth [18,19,37] (Figure 1d).

Other thermal refuges can also hold salmonids during thermally stressful conditions; however, these refuge types can be limited by other physical characteristics. For instance, alcove thermal refuges (Figure 1e,f) may have favorable low velocities and macrophytes that offer protection from predation. However, a low dissolved oxygen content can limit their usefulness [34,38]. Further, these refuges can become disconnected from the main river during low discharge (Figure 1f)—a critical criterion differentiating a cold-water patch from a thermal refuge. The least efficient thermal refuge type, and most under-reported in the literature, is a hyporheic discharge thermal refuge (Figure 1g,h). This is, perhaps, unsurprising as hyporheic discharge points are associated with changes in hydraulic gradients [39] and typically occur along gravel bars and at the downstream end of a riffle [35]. These areas are likely to be shallow (Figure 1h) if they present along a gravel bar or will be associated with downwelling currents from the riffle, which can induce higher velocities that would increase swimming costs [40].

In addition to thermal refuges in natural rivers, anthropogenically created thermal refuges may be found in (hydropower) regulated rivers. First, hydropower dams generate impoundments such as reservoirs or "headponds" that—depending on their size, depth, and flow regulation regime of the affiliated dam—may essentially act as lentic rather than lotic waterbodies and can thermally stratify due to their relatively slow flows and large depths (relative to the same body of water in an unimpounded condition). In rivers that frequently exceed the thermal tolerances of stream salmonids, the thermal stratification in impoundments can act as a thermal refuge, e.g., a 45 m deep, 37 km long reservoir in the Wolastoq ǀ Saint John River in NB, Canada [41]. In regulated river systems, thermal refuges (or alternatively 'cold-water pollution', sensu [42]) may also be created downstream of impoundments via the release of cold hypolimnetic water from deep intakes into the tailrace or other adjacent areas [43]. In the Wolastoq ǀ Saint John River system, for example, cold hypolimnetic water is mixed with groundwater that is further pumped into a biodiversity facility associated with the Mactaquac Dam. The outflow of this facility serves as an important thermal refuge for adult Atlantic salmon [44]. Finally, thermal refuges may be created in regulated systems where inter-system water transfers are used to route the flows between tributaries of different sizes in an effort to funnel flows to central hydropower generation units. Such inter-system transfers are common in Norway, for example [11,45]. It is important to note that while anthropogenically created thermal refuges may offer survival benefits to salmonids in regulated rivers with regard to protection from excessive water temperature, the repercussions of manipulated river temperatures may be far reaching, and a system-level assessment of the effects should be undertaken in cases where anthropogenic thermal manipulations occur [42].

3. Detection and Mapping of Thermal Refuges

The methods for detecting and mapping thermal refuges vary at spatial and temporal scales (Figure 2). The most common method for detecting and further understanding the function and utility of thermal refuges has been via the deployment of thermographs, or temperature loggers [23,36,46] (Figure 2a). These data loggers are relatively inexpensive and provide researchers with insights into how a thermal refuge's temperature regime varies through time using high temporal resolution. Temperature loggers have been deployed across broad spatial scales to better understand how different spatial locations across entire catchments may, or may not, provide thermal refuge for salmonids [27,47,48]. Typically, the data from a large network of temperature loggers is assembled and various modelling tools are then used to provide meaningful inference about the spatial patterns of the water temperature [49,50]. Spatial statistical network (SSN) models have proven useful in many cases because of the spatial autocorrelation of the water temperature data in dendritic networks of streams and rivers [50,51]. However, catchments with complex geologies may severely affect the simple predictable continuum of spatial autocorrelation patterns, and the current generation of SSN models (where spatial autocorrelation is assumed to be positive) may fail to perform well in such complex hydro-topo-geological settings [52]. In such hydro-geological complex situations, the water temperature logger data are better inferred using other modelling tools, such as machine learning Random Forest algorithms [52]. Temperature loggers are also naturally used at finer spatial scales—e.g., a tributary confluence—to capture the spatio-temporal variability of thermal refuges [53,54]. The trade-off with data loggers, however, is that they only represent a single point across a thermally heterogenous riverscape [5].

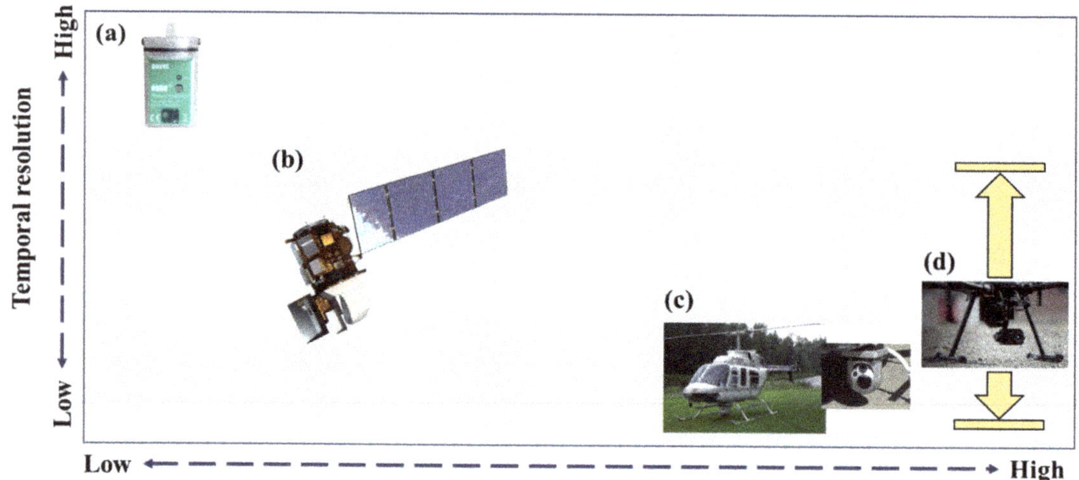

Figure 2. A schematic depicting both the temporal and spatial scale that different temperature sensors can capture. Different spatio-temporal resolution combinations of water temperature are achieved by using temperature loggers (**a**), satellites (**b**) and manned- (**c**) and unmanned (**d**) airborne thermal infrared imagers. The use of these sensors is discussed in depth in the main text.

In recent decades, remote sensing has become a critical tool in the river ecologist's/hydrologist's toolbox. Satellites, such a Landsat 8 with two thermal bands, can provide spatially continuous data across broad swaths of rivers [55]. However, this comes at the cost of temporal resolution as the revisit rate for Landsat 8 is ~16 days [56] (Figure 2b). The major limitation of using thermal satellite images is their resolution. For instance, the Landsat 8 thermal bands have a 100 m ground sampling distance (GSD). While useful for large rivers, these data are not adequate for mapping the fine-scale heterogeneity that characterizes thermal refuges, especially in smaller streams that are often obscured by riparian vegetation. In a seminal paper, Torgersen et al. [31] illustrated how airborne thermal infrared imagers (or TIR) could be used to map entire riverscapes at a sub-metre resolution to identify thermal refuges (Figure 2c). Even though Torgersen et al. [31] were not the first to use TIR for mapping river temperatures [57], they formalized the process that has become common in mapping thermal refuges and other cold-water patches. Beginning in the 2010s, a sharp rise in the use of TIR for mapping potential thermal refuges commenced. Dugdale et al., [32,58] collected airborne TIR to classify potential thermal refuges across an entire catchment (the Restigouche River, NB, Canada). Similarly, airborne TIR has been used to elucidate the landscape controls on potential thermal refuges [52,59,60], whereas others have coupled TIR data with temperature loggers to develop broader understandings of thermalscape heterogeneity [22,61]. Further, others still have used airborne TIR to identify the location of thermal refuges to guide behavioural ecology studies [25,62]. Nonetheless, airborne TIR imagery remains expensive and provides data for only a snapshot in time.

Recently, relatively cheap TIR imagers (mostly uncooled microbolometers) have been paired with unmanned aerial vehicles, or drones, to map the reach-scale thermal heterogeneity of rivers [63,64] (Figure 2d). The efficacy of drone-based thermal mapping lies in its ability for users to collect high spatial resolution data (<0.2 m) at a high temporal resolution. For example, O'Sullivan et al. [65] coupled a TIR imager and a drone to gain repeated insights into how adult salmon hydraulic habitat selection varied as a function of temperature over a three-week period in the summer. Morgan and O'Sullivan [66] found that the holding positions of two salmonid species in a thermal plume differed by age class and species, and this was explained by the fine-scale thermal variability mapped via drone-based TIR. On a cautionary side, drone-based TIR imagers can suffer from large thermal drift [63]. However, the recent work by O'Sullivan and Kurylyk [67] found that the addition of tin foil to the outside of the TIR imager and flying the drone during clear sky conditions can largely alleviate the temperature bias related to thermal drift.

In northern climates where rivers freeze during winter, true colour (red, green, blue—RGB) satellite imagery can also be used to identify (summer) thermal refuges. Relatively deep groundwater will not freeze in the winter. Using this simple fact, O'Sullivan et al. [68] showed how sub-metre resolution (0.5 m) true colour satellite images collected when ice cover exists (ICE) can be used to identify potential (summer) thermal refuges with very high success (Figure 3). This method is illustrated by the identification of both tributary and spring-derived thermal refuges by comparing true colour summertime aerial images (Figure 3a,d), their corresponding airborne TIR images (Figure 3b,e), and winter satellite RGB images (Figure 3c,f). Since the ICE method utilizes freely available Google Earth imagery, it provides a much more inexpensive method for identifying potential thermal refuges in comparison to the expensive TIR method, making the identification of potential thermal refuges accessible to almost everyone, provided that iced-over conditions exist in winter.

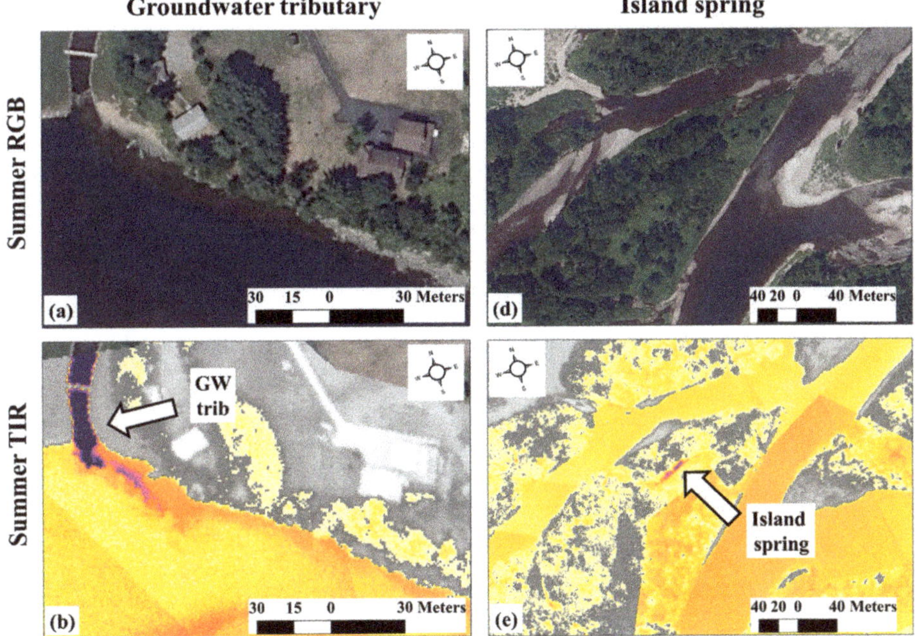

Figure 3. *Cont.*

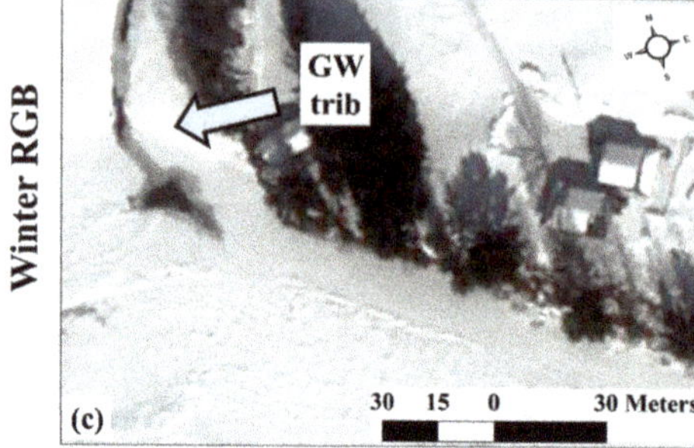

Figure 3. In frozen environments, discrete groundwater (GW) discharges (dependent upon the depth the groundwater is sourced from [68]) can remain free-flowing during iced-over conditions. This is detailed for a groundwater tributary in panels (**a–c**) and for an island spring in panels (**d–f**), where the thermal images (**b,e**) reflect the summer conditions, and the high-resolution satellite images (Worldview 3) reflect the winter conditions (**c,f**).

4. Behavioural Thermoregulation; Elaborating the Concept

In the widest sense, behavioural thermoregulation takes place continuously in all free ranging poikilotherms, including fishes. Fish will generally align themselves in habitats that are conductive to their optimized metabolism, and ideally, fish will live in areas where their aerobic capacity is optimized (i.e., "optimal temperature for growth"; T_{opt}). Another article in this special issue synthesized these optimal ranges for many stream salmonids [2]. All the movement of fish into such a suitable thermal habitat is, in a sense, behavioural thermoregulation: that is, fish use cognitive processes to select habitats within suitable or even optimal temperature ranges. However, in the thermal stress literature, including the main purpose of this current article (and to which the term henceforth refers), behavioural thermoregulation is a concept used to highlight the behaviour where stream salmonids aggregate (often in high numbers) in a patch of generally cooler water than the ambient water temperature during conditions when the ambient water temperature exceeds some physiologically disadvantageous criteria. The purpose of behavioural thermoregulation in response to heat stress appears to be two tiered and is reflected by the timing when different life stages start increasingly using cold-water refuges.

The two tiers of behavioural thermoregulation are herein defined as facultative and obligate behavioural thermoregulation. Conceptually, facultative thermoregulation is proposed to become increasingly common (yet, remains "optional") under conditions when the water temperature exceeds the species/life stage/population-specific *pejus* water temperature (T_{pejus}). This demarcates a threshold when the water temperature exceeds the upper limit of an optimal temperature range and when the thermal conditions start to get increasingly worse for the individual fish (Figure 4). Facultative thermoregulation may continue to be observed slightly beyond the critical temperature (T_{Crit}; Figure 4); a temperature threshold at which anaerobiosis starts to dominate the energy supply as the aerobic scope approaches zero. Within this range of water temperatures, the thermal conditions in the ambient river are increasingly beyond those that are physiologically suitable. Therefore, salmonids achieve a better thermal habitat by seeking out cold-water refuges in comparison to remaining in the ambient water temperature (but where they could still survive).

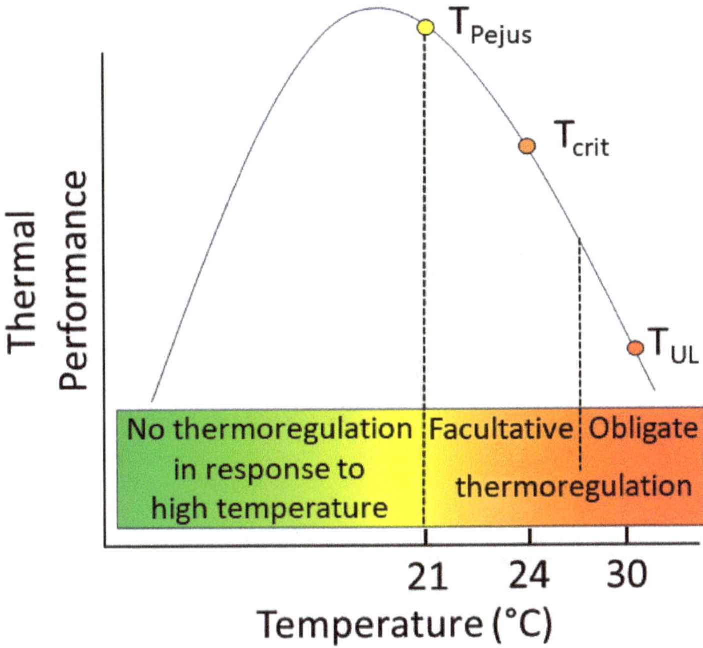

Figure 4. Conceptual model of behavioural thermoregulation in response to high water temperature events differentiating non-aggregating behaviour and facultative vs. obligate behavioural thermoregulation. Facultative thermoregulation is proposed to (voluntarily) begin to occur under conditions when the water temperature exceeds the species/life stage/population-specific *pejus* water temperature (T_{pejus}) and may be observed slightly beyond the critical temperature (T_{Crit}). Obligate behavioural thermoregulation takes over at some temperature > T_{Crit} but in advance of the ultimate lethal temperature (T_{UL}) and continues at temperatures exceeding the initial trigger temperature because a fish can no longer survive without behavioural thermoregulation. The example approximates the published literature values for juvenile Atlantic salmon to illustrate the conceptual difference between the two modes of behavioural thermoregulation [69,70].

Contrastingly, obligate behavioural thermoregulation occurs at some time (but not necessarily at a crisp "threshold") when the ambient water temperature exceeds T_{Crit}. As the water temperature continues to rise, and anaerobic metabolism produces accumulating concentrations of by-products (e.g., lactate) in fishes' bodies, causing stream salmonids to enter a mode of obligate behavioural thermoregulation; a phenomenon that is necessary for the animal to avoid imminent death (Figure 4). As the metabolic by-products start to accumulate at an accelerating rate when the water temperatures exceed T_{Crit}, the timing of the initiation of obligatory thermoregulation occurs at some temperature around T_{Crit} but, supported by observations in field, well in advance of the ultimate lethal temperature (T_{UL}), i.e., the temperature that fish cannot tolerate for more than 10 min [70]. Obligate behavioural thermoregulation continues under all water temperatures exceeding the initial trigger temperature (Figure 4).

It is important to recognize that not all the life stages of a salmonid species necessarily respond to high temperature events by both facultative and obligate thermoregulation. This is highly related to life-stage-specific behavioural patterns. For example, juvenile Atlantic salmon establish territories that are defended against conspecifics [71]. Due to their territoriality, juvenile Atlantic salmon may remain in their territories even when conditions exceed the T_{Crit} values established in the laboratory [69], and generally aggregate (at least initially but see the concept of thermal hysteresis below; [72]) when temperatures exceed T_{Crit} by 2–4 °C. This means that the juvenile life stage due to their inherent territorial nature seems to forgo facultative behavioural thermoregulation, and only exhibits obligate thermoregulation under conditions where remaining in territories becomes impossible due to the imminent probability of physiologically induced death (Figure 5). By contrast, adult Atlantic salmon return from the ocean and ascend their natal rivers typically in three distinct migratory phases [73]. Especially during the active upstream migratory phase, it appears common to encounter adult salmon in cold-water refuges soon after T_{pejus} is exceeded. However, and importantly, under these conditions adult Atlantic salmon can also be commonly encountered in ambient water temperatures, especially in reaches where cold-water refuges are not immediately close by. Such circumstances highlight facultative behavioural thermoregulation. Adult salmon do not have to remain in the cold-water refuge; however, if a refuge is available it may offer a thermal habitat that is physiologically more favorable than the ambient river water, provided that other physical habitat needs are met (e.g., depth, velocity, dissolved oxygen). For example, in the Northwest Miramichi River, NB, we observed adult Atlantic salmon beginning to use thermal refuges at 19 °C. However, other adult salmon continued to remain in ambient water temperatures up to 26 °C during the early part of their run (i.e., June [74]). This would represent a case of facultative behavioural thermoregulation, although it is also possible that the fish had to cope with the temperature stress due to the lack of a nearby cold-water refuge. However, in the Southwest Miramichi River, NB, we witnessed very strong obligatory behavioural thermoregulation in adult Atlantic salmon already at a water temperature slightly exceeding 19 °C (Figure 5 [65]), albeit this was observed in August by adults that had likely been in the river for a number of weeks already, and therefore, may have had pre-existing "thermal fatigue", i.e., previously accumulated physiological by-products in their bodies [25,72]. We also observed a similar transition from facultative to obligate behavioural thermoregulation for brook trout *Salvelinus fontinalis* (Mitchill, 1814) in the tributaries of the Miramichi River (Figure 5) [25]. A relatively straightforward way to differentiate whether behavioural thermoregulation is facultative or obligate under a given situation can be inferred from the number of individual fishes in cold-water refuges vs. those remaining in ambient water temperatures. During facultative use, the individuals were often readily observed in both the thermal refuge and in ambient river habitats, whether their territories (e.g., juvenile salmonids [23,25]) or in non-refuge pool/deep run mesohabitats for adults (e.g., adult Atlantic salmon [74]). During obligate behavioural thermoregulation, effectively all the individuals of a given life stage aggregated in thermal refuges—typically en masse—or were in the process of doing so (Figure 5) and temperature-induced mortalities were also observed during these events (Figure 6). However, care has to be taken when using such field observations to determine the severity of behavioural thermoregulation because the relative abundance of fish in the area naturally dictates how many fish will be observed in a thermal refuge. That is, even one fish in a thermal refuge may represent obligate behavioural thermoregulation "aggregation" if it happens to be the lone fish present in the river reach. Such a consideration would have appeared ridiculous historically, but sadly with dwindling population densities in areas such as the Inner and Outer Bay of Fundy and even in many rivers in Atlantic Canada, small aggregations of very few Atlantic salmon should not be dismissed as non-important events.

Figure 5. Examples of obligate behavioural thermoregulation in the Miramichi River, NB, Canada. (**a,b**) Adult Atlantic salmon in size-segregated aggregation in an inflow of 14.3 °C and when the adjacent main river temperature was 19.1 °C. (**c**) An aggregation of 1+ and 2+ Atlantic salmon parr in an inflow of 21.0 °C; main river 31.5 °C. (**d**) An aggregation of adult brook trout in an inflow of 20.4 °C; main river 26.0 °C.

Figure 6. Example of adult Atlantic salmon mortality in direct response to a high temperature stress event in the Little Southwest Miramichi River, NB, Canada when the water temperature exceeded 30 °C, illustrating the grave consequences to salmonid populations if access to cold-water refuges is compromised.

When do stream salmonids have to aggregate in thermal refuges, i.e., what factors affect the timing of obligate behavioural thermoregulation? At the proximate level, the need for stream salmonids for obligate thermoregulation has an intricate physiological basis and the basic thermal biology of salmonids is described thoroughly elsewhere. Cardiac function/collapse or possibly hypoxia in the brain play a pivotal role as a limiting factor for upper temperature tolerance [75–77]. Since the need for behavioural thermoregulation is rooted in the physiological demands of the fish, there is variability in the timing when this phenomenon occurs. First, different stream salmonids have their species-specific thresholds for thermal habitat suitability (see this special issue [2]), and not surprisingly, the water temperatures at which various authors have observed behavioural thermoregulation either in situ or in experimental settings reflect these inter-species differences (Table 1). Some of the variability in the aggregation temperatures observed in the previous work stemmed from the fact that the data did not always represent a so-called "threshold" when behavioural thermoregulation was triggered, but simply described the temperatures at which the behaviour was observed to occur (Table 1). Additional variability may result from the authors who possibly describe both facultative and obligate behavioural thermoregulation. Similarly, it has been well established that the need for behavioural thermoregulation is linked to body size, translating into different timings for the phenomenon between the life stages within each species (Table 1). In effect, the larger, and therefore, typically older life stages respond earlier (i.e., in lower water temperatures) than smaller, younger life stages (Table 1). Breau et al. [69] discussed this size-dependent difference in timing for behavioural thermoregulation in the context of a larger surface-to-volume ratio in the younger life stages, and the consequent improved ability to uptake oxygen across the skin surface. A larger surface-to-volume ratio also creates a faster heat transfer, and therefore, a lower internal body temperature; one of the main predictions of Allen's rule originally drawn for endotherms [78], but with a high relevance to ectotherms in thermally heterogenous environments, such as avoiding overheating in hot microhabitats [79]. In addition to species- and size-specific differences, there is also evidence that the timing when obligate behavioural thermoregulation is initiated may differ between populations within species, as would be expected due to the population-specific physiological differences in thermal tolerance [80–82]. For example, Corey et al. [13] examined the timing of obligate behavioural thermoregulation in prominent Atlantic salmon rivers in Eastern Canada that differed in their extent of warming and concluded that the aggregation response was not universal across rivers. They observed that the onset temperature for juvenile Atlantic salmon aggregations in thermal refuges occurred 1 °C sooner in the relatively cooler river (Little Southwest Miramichi; LSW) in comparison to the warmer Ouelle River, indicating a population-specific phenotypic plasticity in thermal abilities [13]. The preliminary data similarly suggested that in comparisons of a warm (LSW) to a relatively "cool" river (Restigouche, NB, Canada), the same pattern held wherein the thermoregulation was initiated earlier in the cooler river [83], which validated the assumptions made by Breau [82]. To predict that such differences exist between the populations of different thermal backgrounds, and especially across latitudinal clines, seems logical because these variations may be related to differences in the metabolic rates across a latitudinal gradient, as often described for many ectotherms (see e.g., [84] and references therein). Some laboratory experiments conducted on Atlantic salmon suggested that, with respect to their thermal capabilities, the differences between the populations did not appear to be evidence of local adaptation but rather were a result of high cardiac plasticity wherein the acclimation temperature (i.e., differences in the ambient water temperature between rivers) allowed for cardiac functions to compensate for the warmer temperatures [85]. However, laboratory experiments have also shown that there was a high phenotypic variation between individuals, families [86], and populations [80] with respect to the thermal tolerance where the variation had a significant genetic basis, which may allow for adaptation, at least under conditions of artificial selection (e.g., supplementation programs or aquaculture). There is some evidence that the sex of the fish also affects the ability to withstand or respond to high water temperatures. It has been suggested that females are relatively more vulnerable than males [87,88], and therefore, the timing or duration of behavioural thermoregulation may be affected by sex.

Table 1. A compilation of water temperatures (°C) when behavioural thermoregulation (as defined in the text) was observed to occur for various stream salmonids, stratified by life stage.

Species	Life Stage	Setting	Behavioural Thermoregulation	Study Location	Reference
Atlantic salmon *Salmo salar*	Adult	In situ	22 °C	Miramichi River, NB	[89]
	Adult	In situ	23 °C	Miramichi River, NB	[23]
	Adult	In situ	17–19 °C	Rivière Sainte-Marguerite Nord-Est, QC	[90]
	Adult	In situ	19–22 °C	Miramichi River, NB	[74]
	Parr (1+, 2+)	In situ	>24 °C	Miramichi River, NB	[23]
	Parr (1+, 2+)	In situ	27.3 °C	Miramichi River, NB	[13]
	Parr (1+, 2+)	In situ	28.3 °C	Ouelle River, QC	[13]
	Parr (1+, 2+)	In situ	25 °C, 27 °C	Miramichi River, NB	[25]
	Juvenile	In situ	24.2–27.1 °C	Miramichi River, NB	[72]
	YOY	In situ	29.8 °C; 30.8 °C	Miramichi River, NB	[62]
	YOY	In situ	30.1 °C	Miramichi River, NB	[66]
Brown trout *Salmo trutta* (L., 1758)	Adult	In situ	>22 °C, <26 °C	MI and WI, USA	[91]
	Adult	Lab	20 °C	Fish from hatchery	[92]
	Adult	In situ	24–25 °C	Firehole River, WY, USA	[93]
Cutthroat trout *Oncorhynchus clarkii* (Richardson, 1836)	Adult	In situ	21.7 °C	South Fork Clearwater River, ID, USA	[94]
	Adult	In situ	>22 °C	Willamette River, OR, USA	[95]
	Adult	In situ	>22–23 °C	Bear River, ID and WY, USA	[96]
	Adult	In situ	~22 °C	Henry's Lake, ID, USA	[97]
Rainbow trout *Oncorhynchus mykiss*	Adult	In situ	18–25 °C	Grande Ronde River and Pine Creek, OR, USA	[37]
(Walbaum, 1792)	Adult	In situ	24–25 °C	Firehole River, WY, USA	[93]
	Adult	In situ	~19 °C	Columbia River, WA, USA	[98]
	Adult	In situ	≥19 °C	Columbia River, WA, USA	[99]
	Juvenile	In situ	50% at 22–23 °C 100% at 25 °C	Klamath River, CA, USA	[100]
	Juvenile	In situ	>21 °C	South Fork Eel River, CA, USA	[53]
	Juvenile	In situ	22–23 °C	Klamath River, CA, USA	[33]
Chinook salmon *Oncorhynchus tschawytscha* (Walbaum, 1792)	Adult	In situ	>18 °C	Willamette River, OR USA	[101]
	Adult	In situ	20 °C	Columbia River, OR, USA	[102]
	Adult	In situ	≥20–21 °C	Columbia River, WA, USA	[99]
	Adult	In situ	>25 °C	John Day River, OR, USA	[31]

Table 1. *Cont.*

Species	Life Stage	Setting	Behavioural Thermoregulation	Study Location	Reference
	Juvenile	In situ	22–23 °C	Klamath River, CA, USA	[33]
Coho salmon *Oncorhynchus kisutch* (Walbaum, 1792)	Juvenile	In situ	22–23 °C	Klamath River, CA, USA	[33]
Sockeye salmon *Oncorhynchus nerka* (Walbaum, 1792)	Adult	In situ	>15 °C	Wood River, AK, USA	[103]
Brook trout *Salvelinus fontinalis*	Adult	In situ	>21 °C	Miramichi River, NB	[25]
	Adult	Lab	20 °C	Fish from hatchery	[92]
	Adult	In situ	>22 °C, <26 °C	MI and WI, USA	[91]
	YOY	In situ	~20 °C	Charles Lake, ON	[104]
Arctic char *Salvelinus alpinus* (L., 1758)	Juvenile	Exp. lake	>16 °C	Lake Ellingtjern, Norway	[105]
Lake trout *Salvelinus namaycush* (Walbaum, 1792)	Adult	In situ	<20 °C	Canadian Shield Lakes, ON	[106]
Bull trout *Salvelinus confluentus* (Suckley, 1859)	Adult	In situ	~21 °C	Lostine River, OR, USA	[107]
	Adult	In situ	>15 °C	Kinbasket Reservoir, BC	[108]
	Adult	In situ	>20 °C	Blackfoot River, MT, USA	[109]

Since the obligate behavioural thermoregulation is dependent on the species, life stage, and population, is there then "a fixed temperature threshold" that triggers an aggregation response into thermal refuges? Early work on the phenomenon provided some evidence to such fixed temperature thresholds [58] and, certainly, much of the current conservation management is based on simple water temperature thresholds targeting specific activities. The "fixed temperature threshold model" provides a relatively good approximation to the likely timing of the initial obligate behavioural thermoregulation events, for example, during the first thermal stress event of a summer or after a sufficiently long period of recovery, allowing for any previously accumulated physiological metabolites to dissipate. However, to explain the apparent variability in the water temperatures at which fish are aggregating into thermal refuges, O'Sullivan et al. [72] recently developed a model based on the concept of thermal hysteresis. As proposed in the previous studies, the accumulating physiological by-products due to anaerobic metabolism remain in the bodies of salmonids even after critically stressful water temperature conditions subside, and the presence of these by-products may compromise the ability to cope with subsequent heat events [25,110]. This has been termed "thermal fatigue", "thermal baggage", or "thermal hysteria" [25,72]. The existence of these accumulated anaerobic by-products is the proximate reason why the thermal threshold to aggregate into cold-water refuges seem to decline during subsequent thermal events and it can be sufficiently modelled by taking into account the time since the previous obligate behavioural thermoregulation event and the frequency of these events [72]. Therefore, the concept of thermal hysteresis captures the idea that the timing of obligate behavioural thermoregulation is inherently dependent on the history of previous obligate thermoregulation events resulting in a variable, rather than static, threshold temperature [72].

5. Behaviour Leading to, during, and after Behavioural Thermoregulation Event

5.1. Finding Refuges

Stream salmonids are known for their extreme spatial cognition as long-distance migrants [111]. However, their spatial capabilities as juveniles during their freshwater phase are also remarkable [112], with a sub-metre ability to return to prior territories after >5 month periods [11]. The searching ability of spatially sparse thermal refuges corroborate these earlier works that the spatial capabilities of at least juvenile Atlantic salmon appear excellent, and their movements to and from thermal refuges are not by chance. Breau [113] and Corey et al. [62] both examined individually tagged juvenile (age 1+ and 2+) Atlantic salmon leading to and during obligate behavioural thermoregulation events. Breau [113] designed a study with a maximum distance away from a known monitored cold-water refuge of 714 m; the tagged salmon parr were able to find the cold-water refuge from these distances. Continuing the previous work, Corey et al. [62] followed 636 passive integrated transponder-tagged Atlantic salmon parr and documented their movement of up to ~7.5 km to cold-water refuges. In both studies, the majority of the fishes' movement to refuges was relatively localized and occurred within the reach of study; however, the movement distances regularly exceeded hundreds of metres [62,113]. How far the 0+ juveniles came to the thermal refuges remains unknown. Adult salmonids are capable of moving longer distances in response to thermally stressful events. For example, Carrow [74] observed a rapid migration of adult Atlantic salmon through relatively warmer downstream reaches in the Northwest Miramichi River into generally cooler upstream locations (~50 km) before the water temperature reached conditions necessitating obligate behavioural thermoregulation. Facultative thermoregulation was observed in the refuges that the migrating salmon encountered on their way to the cooler upstream reaches. As the water temperature continued to rise, migration further upstream became limited by concomitantly low water levels that prevented a search for cold-water refuges further upstream, and the adult salmon then started searching for cold-water refuges in the downstream direction [74]. However, if adult salmonids identified reaches wherein cold-water refuges are located during the later phases of their migration, they engaged in a similar behaviour as the juvenile life stages wherein they remained "conveniently" close to the refuges that they repeatedly accessed as the water temperature increased [9,65].

The mechanisms that salmonids use to find cold-water refuges during water temperature-related events are not fully understood. Upstream-migrating adults presumably continue swimming upstream until they encounter a gradient of cooler water temperature where behavioural thermoregulation is possible. For territory holding Atlantic salmon parr, both Breau [113] and Corey et al. [62] observed that individually tagged juveniles that located cold-water refuges came both from up and downstream directions, and across a river stemming >30 m in width. The likelihood of finding refuges either in the up or downstream direction was dependent on the reach of the juvenile Atlantic salmon, and similarly the movement distances in either the up or downstream direction appeared to be study reach-specific [62]. Leading up to obligate thermoregulation events, Atlantic salmon parr have been observed to assume positions off the stream substratum, actively swimming in water columns within their territories (but not feeding), and then grouping to small clusters of few parr, and traveling towards presumed cold-water refuges in small aggregations [113]. The possible explanations for traveling towards cold-water refuges in groups were hypothesized as potential safety in numbers, migration with kin, or the possibility of "following the leader", wherein only some individuals knew location of the refuges, and others followed the conspecifics [113]. It has also been hypothesized that social cues, whether olfactory signals or visual cues of other aggregating fish, lead into an advertisement of the refuge locations [114]. Naturally, it is also possible that the refuge seeking fish may start aimlessly swimming in the direction where other conspecifics are moving, and they haphazardly encounter a cold-water refuge. The specific mechanisms during travel towards cold-water refuges warrant further research.

5.2. Behaviour in the Refuges and Post-Aggregation Movements

While the ultimate reason for the obligate use of thermal refuges is to survive past the thermal conditions that exceed the species/life stage/population-specific physiological criteria, it is clear that behaviour and spatial distribution of aggregating salmonids while within the thermal refuge is governed by a number of criteria. While it is recognized that the anadromous adult salmonids do not generally feed during their freshwater migration [115,116], the cessation of feeding and aggression has also been observed for many adult non-anadromous fish and the older juvenile life stages of anadromous salmonids when obligate behavioural thermoregulation takes place [23,113]. Such cessation results from the anaerobiosis and lack of energy available for specific dynamic action (i.e., metabolism related to digestion). All the available energy is used to maintain critical basal metabolism, and thus, to survive. Species-specific differences naturally occur and, for example, Brewitt et al. [117] showed that during facultative thermoregulation, the use of tributary confluence refuges allowed rainbow trout to move into the warmer main river to forage for prey and then retreat back to the cool thermal refuge.

During obligate thermoregulation events, it is notable that while energy preservation is at prime, the aggregating fish, and especially the juveniles, form their aggregations in the low(er) water column and continue to swim against the current as opposed to resting at the stream bottom or hiding within the substratum, as one would suspect from an energy conservation point of view (Figure 4c). The same behaviour has been observed in laboratory studies [69]. Breau et al. [69] hypothesized that the maintenance of swimming may have a role in recycling anaerobic waste products such as lactic acid in their muscles and could, therefore, have an active role in reducing physiological oxygen debt [118] However, the spatial distribution of salmonids in the cold-water refuge is not independent of energy conservation or other biotic interactions. The research points towards many additional physical habitat characteristics that define what one might term the "optimal thermal refuges". This is indicated by the fact that while the primary reason for salmonids to occupy cold-water refuges is indeed to encounter cooler temperatures than the ambient water temperature in the river, it is very common that the fish aggregations are not in the coolest part of the refuge [33,53,65,66,69,119]. In lieu of the coldest locations within the refuge, O'Sullivan et al. [65] observed adult Atlantic salmon consistently aggregating in specific hydraulic fields defined by a narrow range of a Froude number that was significantly smaller than the Froude number in the locations during non-thermally stressful events. In addition, they observed salmon aggregating in tight geometric formations—thermal-pelotons—modulating the hydrodynamics, especially for the individuals in the centre and back of the formations [65]. They proposed that this was likely linked to the interplay of the hydraulic and thermodynamic controls on the bioenergetic cost of holding differing positions in the thermal refuge [65]. Perhaps not surprisingly, the individuals in most energetically profitable positions (in the center of the peloton) were the large multi-sea winter salmon, while the smaller adults had to "sacrifice" themselves in higher energy-consuming positions in the front of the aggregation. This observation supports the notion that the spatial distribution within the cold-water refuges appears to be, at least in part, determined by intra-specific competition for the most profitable (energetic) positions. Similarly, under extreme water temperature conditions (>31.5 °C), hierarchical species and age-specific structuring within the cold-water refuge has been observed [66]. While still not necessarily occupying the coldest possible locations within the refuge, the species (brook trout and Atlantic salmon) and different age classes hierarchically structured themselves along a temperature gradient from relatively cool (21.8 °C) to warm (30.1 °C), corresponding to size and structured from largest (adults) to smallest (0+ juveniles), with 2+ and 1+ parr structured by age in between, further indicating that the more dominant (larger) individuals were able to exert control over spatial structuring within the refuge [66]. Aggregating salmonids may also remain in their typical tight clusters due to the physiological advantages of remaining in groups, i.e., the calming effect wherein aggregating may reduce the high oxygen demand associated with high temperature events [23]. The microhabitat parameters also affect the spatial distribution of aggregating salmonids, with access to relatively deeper locations in the refuge [23,119] or the availability of deeper microhabitats adjacent to the refuge [25] having been shown to be of particular importance. Why salmonids primarily aggregate in tributary-derived cold-water refuge plumes and do not generally appear to proceed upstream into the cold-water tributaries that generate these refuges may also be explained by the necessity to remain adjacent to deeper water or by other aforementioned energetic benefits that may be achieved by staying in the confluence of the cold-water tributaries and the main river.

When the obligate behavioural thermoregulation events dissipate, stream salmonids generally leave the refuge, albeit facultative behavioural thermoregulation may continue, and some individuals may find the refuge a physiologically better option than continuing into the ambient river. For territorial juveniles such as Atlantic salmon, the evidence suggests that the fish have the capability to "anticipate" the relative risk of future thermally stressing events, and if future thermal events can be foreseen, the fish may decide to establish territories in proximity to the areas where cold-water refuges are available. Breau [113] documented that while juvenile Atlantic salmon abandoned thermal refuges when the water temperature decreased below their physiological thresholds (e.g., during the nights between subsequent thermoregulation events), the fish did not return to their original tagging locations but remained on "standby" in some adjacent location where returning to the thermal refuge was facilitated. Corey et al. [62] observed that a very high proportion (96–100%) of juvenile Atlantic salmon remained in their original tagging reaches post-thermal events if they contained thermal refuges, but a lower proportion (62%) returned to a reach where no cold-water refuges were located. This finding led Corey et al. [120] to investigate the long-term effects of cold-water refuge availability on the relative abundance of juvenile Atlantic salmon along a 17-km segment of river and they found that while the relative abundance increased in the reaches with thermal refuges in summer, an equilibrium in the abundance was regained later in autumn when subsequent thermal events became unlikely. In the same study, Corey et al. [120] revealed that the tributary confluence thermal refuges with the highest temperature difference from the main river had the highest effect on the relative abundance of Atlantic salmon across five thermal refuges during the summer season. Similar observations regarding a preference towards a larger temperature differential between the mainstem and tributary has also been seen for juvenile rainbow (steelhead) trout [100].

5.3. Consequences of Behavioural Thermoregulation

The consequences of using thermal refuges are presumably clear. Salmonids that find and use cold-water refuges during thermally stressful events are able to survive through the bottleneck periods and use physiologically more benign environments in comparison to conspecifics that are not thermoregulating [69,87,121,122]. The consequence of a failure to find a thermal refuge during thermal events that necessitate obligate behavioural thermoregulation is also clear. Widespread mortality events of both adult and juvenile salmonids have been widely reported for decades [36], with numerous contemporary reports of mortalities resulting in a direct response to high water temperature events in combination with other human activities [62,121,123,124] (Figure 6). The consequences of refuge use on somatic growth are presumably small as behavioural thermoregulation occurs during conditions when there is no longer an aerobic scope for growth, feeding has ceased, and therefore, growth is arrested even during the time spent in the thermal refuges. We did not find published work examining the direct growth effect of refuge use by stream salmonids. Many others who have reported biological consequences have examined situations where the stream salmonid performance was compared between warm vs. relatively cooler water exposure, and therefore, the results can be inferred to apply to a situation wherein fish would have used cold-water refuges. Crossin et al. [87] noted a lower migration success in (chronically) thermally challenged female sockeye salmon, and similarly, Hinch et al. [88] reported a higher mortality in female Pacific salmonids relative to males, which they associated, in part, to high temperatures. This indicates that, at least for adults, females may benefit relatively more from using cold-water refuges and this may be explained by the fact that the stresses related to a higher investment of the gonadal development that occurs during migration [125] may be partially offset by cold-water refuge use. Gametes are also less viable in thermally challenged adult female salmonids [126], and similarly, negative effects of warm acclimation have been observed in male brown trout sperm, albeit the effects manifested only early in the spawning season [127]. The use of cold-water refuges have benefits related to energy conservation. For example, Berman and Quinn [128] estimated a 12 to 20% decrease in the basal metabolic demand for chinook salmon that exhibited facultative behavioural thermoregulation. Such energy conservation from the use of cold-water refuges is presumably beneficial as it allows for saved energy for spawning, and in the case of iteroparous salmonids, may allow for better survival post-spawning. However, the simulation work by Snyder et al. [129] in a heavily hydropower-regulated Columbia River system did not indicate significant contributions of cold-water refuge use to energy savings by migrating adults, and therefore, the energetic consequences of behavioural thermoregulation appear to be context specific.

While using thermal refuges is generally thought to provide benefits for salmonids, there also appear to be some negative consequences. The delay of migration for anadromous adult salmonids have been reported in numerous studies [41,129]. Babin et al. [41] followed acoustically tagged adult Atlantic salmon in a deep and large thermally stratified hydropower reservoir in the Wolastoq I Saint John River and observed that upon leaving the cold hyporheic water of the reservoir for a warmer riverine section, some Atlantic salmon turned back when the river temperatures exceeded 22 °C. They repeated such behaviour multiple times over the course of a month. Snyder et al. [129] similarly showed through a simulation exercise in the Columbia River system that behavioural thermoregulation contributed significantly to the total migration time (56% of total migration time) in summer-run steelhead. However, the effect was much more benign for fall-run Chinook (6% of total migration time). While these observations may be interpreted as a migration delay due to an "ecological trap" phenomenon [130], with a potential further consequence that these fish may not be able to reach their purported upstream spawning grounds, the delay may also ensure that the fish arrive in appropriate locations only after the temperature regime of the putative spawning grounds have declined into a suitable range [129]. Aside from being delayed in their migration, the salmonids aggregating in a cold-water refuge may also become exposed to high fishing pressure as many refuges are locally well-known fishing locations [119]. For example, Keefer et al. [98] documented a high rate for harvesting steelhead trout in the Columbia River in refuge tributaries. It is also plausible that large aggregations of salmonids in cold-water refuges attract predators, thus increasing the natural predation rate. However, while a plausible possibility, to our surprise we have in fact never personally documented either avian or mammalian predators at cold-water refuges even during the densest aggregation events despite having collectively observed a large number of these events. The only documented direct observations have been those of American crows (*Corvus brachyrhynchos*) that have been feeding on moribund or already dead juvenile Atlantic salmon floating downstream in the Miramichi River during water temperature-related events.

6. Management and Conservation Actions in Response to a Warming Climate, Thermally Challenging Conditions and Cold-Water Refuges–Future of Stream Salmonids

Since thermally stressful events are common and may become increasingly so in future, the management of the cold-water patches and refuges is becoming increasing important. Mejia et al. [28] recently outlined advice for bridging the gap between science and management in this field. Many rivers that sustain fisheries for stream salmonids have implemented management strategies to protect salmonid populations during these warm events. The purpose of the management actions with respect to warm-water events is commonly geared towards the protection of the adult life stages from exposure to recreational fishery [124,131], although actions also exist to protect specific cold-water refuges overall [26]. The evidence that the post-release survival of adult salmonids decreases as a function of increasing water temperature after species/population-specific thresholds are surpassed is convincing [131–134]. Depending on what management measures are implemented to protect the adults, the juvenile life stages may also benefit from these measures. For example, temporary closures of known cold-water refuges for recreational fishing will also protect the juveniles that aggregate there from additional "foot traffic" in these locations during thermally stressful times. Such protections may be necessary. During obligate behavioural thermoregulation, the aggregating juveniles are often physiologically stressed to a point that any additional stressor (such as tactile exhibitions by anglers who may "pick up" or poke juvenile fishes out of curiosity) may be detrimental. The protection of aggregating juveniles in key cold-water refuges may, therefore, be necessary in their own right. We have observed, for example, events where recreational vehicles (such as four-wheelers) are driven across dense juvenile aggregations in shallow refuges by unsuspecting recreational fishermen. However, extreme care must be practiced when protecting cold-water refuges so that the protective actions would not have unintended consequences. Indicating where cold-water refuges are by their closure (or by publicly sharing thermal imagery) may "backfire" by attracting audiences with non-laudable intentions. With thermal imagery becoming more readily available and shareable through open-source publications, thought must be directed as to whether or not the locations of the cold-water refuges should be made publicly known. Through autonomous underwater video camera imagery, we have indirectly observed poaching events of anadromous brook trout when the fish have been aggregating in cold-water refuges. In our example, a large number of anadromous brook trout were repeatedly observed in subsequent frames of underwater camera imagery (for >13 h) when the water temperature conditions exceeded the conditions for obligate behavioural thermoregulation (28 °C) only to abruptly, completely, and permanently disappear from the refuge even though the water temperature conditions remained unchanged. Concurrently, the "surface foam" covering the cold-water refuge also disappeared between the photograph frames. It is conceivable that the fish were abruptly removed by a poacher's gill net maneuvered through the cold-water refuge. While such poaching events or individuals engaged in such illegal activity may be sparse, it is nevertheless important to carefully consider whether the spatial location of critical cold-water refuges should be publicly advertised as they may attract unwanted attention to these important locations that may otherwise stay hidden to the public eye, and especially to the individuals who have affinity for illegal harvests. Alternatively, local river guardians, conservation officers, or automated (infrared) security cameras may be required to ensure the protection of cold-water refuges from the "two-legged predator".

VanLeeuwen et al. [131] recently reviewed when and how water temperature-related fishery closures are best implemented. The trade-offs in these management strategies are clear. Since salmonid fishing may represent one of the most socio-economically significant activities in rural communities, the closure of recreational angling that is too lenient may have a severe financial consequence on the local economy. However, on the reverse, the non-closure may have a drastic consequence on the salmonid population due to exposure to said fishing during an exceedingly stressful time. Various protocols are used to implement water temperature-related protections, and common strategies include closures of selected cold-water pools, closures of whole or larger sections of river, or restricting the timing of fishing activities within the day after pre-determined temperature thresholds are exceeded [131]. One of the strategies that appear to offer a promising balance between continued fishing vs. temperature-related closure is the implementation of a "morning only" fishery that would safeguard fishing opportunities for anglers who may have traveled a long distance for their fishing opportunity, while protecting the salmonid population during the most stressful time of the day [131]. Hourly water temperature consistently reaches its minimum at a predictable morning hour, e.g., 9:00 AM in the Miramichi River [135]. The best protection/management strategy depends, however, on the status of the salmonid stock, fishing pressure, and the overall temperature regime of the river that may remain too high throughout the diel period for any fishery in certain locations [131]. To help manage water temperature-related fishery closures in the future, it will be useful to better understand the population-specific plasticity with regard to their requirements to behaviourally thermoregulate [13]. The current temperature thresholds when fisheries closures are contemplated have often been derived from research that took place in relatively warmer rivers where populations may have slightly higher tolerance for temperature-induced behavioural thermoregulation. However, in northern latitudes or relatively cooler rivers, the physiological processes in response to water temperature may start earlier in such cold-evolved populations. The hysteresis curves may take a drastically different shape or the inflection points where stress begins to accumulate may occur much earlier in these rivers, necessitating earlier management intervention.

The opportunities for cold-water refuge management/manipulation is also presented by hydropower-regulated rivers through hypolimnetic water releases or other cold-water outflows released into the tailrace. On one hand, such artificial cold-water releases can create an opportunity for salmonid fisheries in rivers that would otherwise not be thermally suitable [43]. On the other hand, such cold-water plumes may become locations where upstream-migrating salmonids spend excessive time; therefore, significantly delaying or even stopping further upstream migration. These cases have been aptly named thermal pollution [42] or thermal traps [41]. The management of the cold-water regime in hydropower-regulated rivers can be particularly complex as the quantity of water flow is often tied to the energy demands and market conditions, and the protection/enhancement of salmonid populations via these mechanisms adds another layer of complexity. It also has to be recognized that toying with the natural temperature regime of a river using cold hypolimnetic releases is likely to have cascading repercussions throughout other aquatic biota and the ecosystem, and the overall impacts of changes in the thermal regime should be fully evaluated before cold-water releases are used to benefit salmonid fisheries. The trade-offs in tailwater thermal habitat management are often complex [43].

Management decisions aside, another increasingly common conservation measure aimed at protecting salmonid populations from stressful warmwater events is cold-water refuge restoration [21]. This practice is becoming particularly popular in Atlantic Canada among non-government environmental organizations, First Nation conservation groups, and fishing clubs, and can fall under the categories of improving/enhancing existing cold-water refuges or potentially creating new ones. While promising, and possibly a generally useful and desired activity, there are a few important considerations that need to be taken into account when cold-water refuge restoration in contemplated.

First, if the addition of new cold-water refuges (e.g., via groundwater pumping [21]) is contemplated, a very careful and multi-faceted consideration of the consequences will be necessary. For creating new cold-water refuges, the first important question regards the location of the proposed refuge. Presumably, new cold-water refuges could be contemplated in reaches where a river network analysis (e.g., thermal imagery) indicates a near-complete lack of current refuges. Creating a new refuge could, under these circumstances, have an important protective effect if situated in the reaches where the current nearest refuge is beyond the movement ability of e.g., juvenile life stages when obligate behavioural thermoregulation is triggered. In this context, the approach utilized by Snyder et al. [129] appears particularly useful for assessing different cold-water refuge addition scenarios. Such simulation tools allow for an assessment of the contribution of thermal refuges on the migration timing, distance, energy conservation of migrating/moving salmonids, and their eventual success, and therefore, allows for a determination of the utility of the added (or removed) thermal refuges on fish survival and health [129].

The second important consideration when contemplating the creation of new cold-water refuges is why a refuge would be created in a particular location. If the location is determined based on a large-scale strategic assessment and is determined to be a location in a reach where no other refuges exist, as above, then the location may be justified as a conservation measure. However, perhaps the creation of a new cold-water refuge is to retain more fish adjacent to a fishing lodge for the purpose of improved recreational angling opportunities? In such cases, the intent would be counterintuitive from the conservation perspective, even when the intent would be to practice catch-and-release fishing since mortality rates do increase with increasing water temperature [133,134]. If creating new cold-water refuges increasingly becomes a practice for the purpose of retaining fish close to a recreational fishing lodge, there is a real risk of such practice further exploding into an arms race between fishing establishments with the "most fish going to the person with the largest pump" and should generally be discouraged since groundwater sources are a finite resource. In fact, the third important consideration with regard to creating new cold-water refuges is whether such practice is advisable at all, as the use of groundwater to create a new refuge will explicitly mean that the groundwater will be "away" from naturally percolating into the river in another location. The action of pumping groundwater also takes water away from the terrestrial ecosystem. At periods when water levels are low and temperatures are high, water-induced forest stress is also likely [136]. Further, the impacts may resonate to domestic water supplies in regions where groundwater wells provide potable water. This is an aspect with no easy solution; even all the authors of the current article do not fully agree on the advice of groundwater pumping. Kurylyk et al. [21] presented different scenarios for creating new cold-water refuges through groundwater pumping and the simulated effects on river depletion. It is important to emphasize that the intent of this work was to suggest the creation of cold-water refuges only on an extreme temporary basis, i.e., activated automatically but temporarily when water temperature exceeds conditions that are extremely likely to cause obligate behavioural thermoregulation events and in areas where no other refuges are present. Implemented in this way, a new refuge would be temporarily created during the hours of day when the alternative scenario for stream salmonids is likely to be physiologically induced death but would minimize the overall effect on the amount of groundwater and river depletion.

With regard to projects that aim to enhance or restore cold-water refuges, multiple interesting avenues exist for the future. As our understanding increases regarding what characteristics conform a "goldilocks refuge", some cold-water refuges can possibly be made significantly more suitable through even small changes [21]. However, conservation groups should also keep in mind that the old adage of "fix it only if it is broken" may prove useful in this context. One potentially promising avenue for restoration/enhancement work may regard cold-water alcove refuges, which may become disconnected from the main rivers during low discharge conditions but may hold large volumes of cold water (Figure 1f). The cold-water alcoves may also be oxygen-depleted due to high groundwater concentrations and a very slow flow, but artificial oxygenation through an air pump may alleviate these situations or modifications in the alcove could induce drops that entrain air and thereby provide natural oxygenation. These possibilities are ripe for exploration in the future.

If the emphasis of cold-water restoration is on maximizing the net benefits for aggregating salmonids, then one must implicitly understand what the fish are looking for in a thermal refuge. While the secondary physical habitat characteristics may temporarily become less important in certain cases, e.g., during obligate behavioural thermoregulation events ("life-or-death situation") when access to some cold water is better than no access at all, we documented through numerous examples that a cold-water patch does not necessarily mean cold-water refuge, let alone an ideal refuge. As outlined earlier, Wilbur et al. [25] showed that access to relatively deeper water adjacent to the cold water was a major determinant separating a cold-water patch from a refuge for large-bodied brook trout. O'Sullivan et al. [65] demonstrated a specificity towards unique hydraulic conditions, and not necessarily the coolest water for adult Atlantic salmon, further indicating that knowledge of the thermal refuge habitat and hydraulic requirements is necessary for most successful cold-water refuge restoration. We propose that it may be necessary to mandate before–after monitoring protocols for cold-water refuge restoration/enhancement, which could include documenting both the tangible benefits in terms of the gained physical cold-water habitat but possibly also show that the enhancement activity at least does not reduce the density or absolute number of salmonids using the refuges post-enhancement. While these projects almost always proceed with the best future for the salmonid population in mind, the outcome is not necessarily guaranteed if the restorative work is conducted by untrained local groups without expertise in hydraulics, habitats, or salmonid ecology.

Particularly, if the restorative work is conducted without the knowledge of what the criteria for the salmonid population/life stage in question even are, the restorative work may prove to be simply "feel good" projects for participating organizations with limited tangible benefits. If public funds are dedicated to these restoration projects, the groups should involve individuals with appropriate expertise (hydrologists, hydromorphologists, ecologists) to minimize/avoid negative impacts, and there should be accountability for any tangible benefits.

At a larger (river network) scale, temperature network models—either spatial statistical network or random forest models, depending on the hydrogeological setup combined with other physical and biological variables and correlates—allow for species distribution modelling, and therefore, predictions of the species occurrence or abundance in future climates [27]. While such models ignore the heterogeneity of the thermal habitats at the micro-scale (i.e., the presence or absence and frequency of thermal refuges), they allow resource managers to predict "mean responses" and allow for the identification of the likelihood of aquatic species disappearing from parts of a river network as a function of a warming climate [27]. Such exercises have been shown to be useful in conservation as they will allow for the identification of sections of rivers (e.g., tributaries, sub-catchments) that are primary targets for added conservation or mitigation measures, or areas that will be likely candidates for species resiliency if rivers get warmer over time. On the other hand, the information derived from such models can also potentially be used in the future to identify areas of rivers that can be "sacrificed" if certain sections are predicted to become uninhabitable for a species due to climate warming-induced water temperature increases alone. Then, such areas may be prioritized for anthropogenic development while allowing the more resilient areas for the species presence be left as conservation reaches. Such prioritization vs. sacrificing may sound apocalyptic but may similarly provide a balance between the development and conservation in catchments where increasing anthropogenic pressures are inevitable, where the overall thermal habitat has become extremely limited, and therefore, where "everything cannot be saved".

The warming climate and the consequently warming freshwater is predicted to make life more difficult for stream salmonids in the southern extreme or lower elevational part of their range. Much attention, or even blame, has been directed toward landscape activities (e.g., agriculture, forestry), which may often be the only prominent anthropogenic activities in otherwise pristine catchments with salmonid populations, for their potential role in the warming of freshwater, and therefore, their potential role of being the reason or at least further exaggerating the negative trend on thermal habitat suitability [137,138]. It has been long recognized that such landscape activities, when poorly conducted and/or especially when occurring without appropriate mitigations such as appropriate riparian zones, can indeed have a significant and long-term warming effect on freshwater streams [139,140]. On the contrary, when landscape activities are undertaken on a moderate scale with mitigations (e.g., harvesting of 7% of a stream basin in the Catamaran Brook catchment, NB, Canada), the effects of freshwater warming have not been observed [141,142]. At least in Atlantic Canada, it is currently difficult to find unregulated landscape activities in areas where salmonids occur. Therefore, it may be more prudent to shift the focus of pressure from landscape users to landscape regulators (often government agencies stipulating laws, regulations, and permits) and ask whether there should be a concern regarding the effect of the landscape activity on warming of freshwater. It is the regulators who can establish the "bowling lane" of what (and how) the landscape users can do in agro-forested landscapes, and then ensure through a process of adaptive management that the regulations are effective in protecting cold-water habitats by also taking into account the cumulative effects in the interconnected waterways. Ideally, both landscape users and regulators would work together in the best interest of the fish.

It will also be crucially important to thoroughly quantify the role of freshwater warming that is attributable to climate patterns vs. landscape activities, especially in catchments that have regionally significant salmonid populations. There continues to be a prevailing misconception in many jurisdictions that landscape activities have a similar effect on the water temperature across the catchment's landscape [138], and therefore, that the management of landscape activities can follow similar regulations and mitigations irrespective of the spatial location of such activities within or between catchments. Examples of such erroneous views are, e.g., the fixed percentage of an allowable harvest of the catchment surface area or fixed width buffer zones across whole jurisdictions (e.g., Canadian Atlantic provinces). The "one-size-fits-all" approach is false; large river catchments are most likely to have different hydrological response areas and hydrological units that respond differently to landscape activities [52,136,143–146]. If the potential effect of landscape activities is of concern to fisheries management due to the warming effects on freshwater, then it is imperative that there is a thorough understanding of the science and the pathways of the potential warming effect that is stratified by appropriate hydrological units. It may turn out that the landscape activity has nothing to do with the purported or observed warming. Conversely, it may become apparent that landscape activities are driving broad shifts in hydrological processes. It is our belief that if these pathways are understood and the potential effects can be quantified and effectively communicated to both landscape managers, landscape operators, and the general public, a positive outcome can be achieved, and environmentally conscientious practices can be pursued while minimizing and/or limiting the impacts on freshwater resources. As for the effects of climate change on the warming of freshwater, only time will tell whether the current warming patterns can be curbed by implementing intergovernmental policies and whether stream salmonids have sufficient time and physiological/genetic scope to adapt to changing thermalscapes.

7. Conclusions

As the climate is predicted to continue to warm due to global climate change, it is very likely that high water temperature-related behavioural thermoregulation events, whether facultative or obligate, will continue to become a more common behaviour for stream salmonids. While this behaviour is already common in the southern part of the range for most stream salmonids, it is likely that the more northernly situated or higher elevation populations will also stat experiencing events that demand aggregation into cold-water refuges and the physiologically determined timing of these events may occur earlier than expected in these populations that have evolved in relatively colder water. It is imperative for fisheries management to carefully consider the locations and availability of current thermal refuges, determine management responses that are necessary in each location to protect salmonid populations, and work closely with land users to ensure the least possible impacts of high water temperature events on already stressed populations.

Author Contributions: Conceptualization, T.L.; investigation, all authors; data curation, T.L., A.M.O., E.M.C., E.N.C. and C.B.; writing—original draft preparation, T.L., A.M.O. and E.N.C.; writing—review and editing, T.L., A.M.O. and C.B.; visualization, T.L., A.M.O. and E.N.C.; funding acquisition, T.L., A.M.O., R.A.C. (R. Allen Curry) and R.A.C. (Richard A. Cunjak). All authors have read and agreed to the published version of the manuscript.

Funding: Over the years of research into cold-water refuge use by salmonids, our research programs have received funding from numerous organizations, including, but not limited to, the Natural Sciences and Engineering Research Council of Canada, the Foundation for Conservation of Atlantic Salmon, Fisheries and Oceans Canada, the New Brunswick Innovation Foundation, the Province of New Brunswick, J.D. Irving Ltd., the Miramichi Salmon Association (including Jack T.H. Fenety Scholarships for five of the authors over the years), and Mitacs.

Data Availability Statement: No new data were created or analyzed in this study. Data sharing is not applicable to this article.

Acknowledgments: The authors of this article represent "three academic generations" of scientists who have, all combined, worked on thermal refuges for a better part of last two decades in various combinations of tutelage. As such, numerous field technicians, both junior and senior, have greatly facilitated our understanding of thermal refuge use by stream salmonids and we are thankful for their hard work. We also thank Dan Isaak from the US Forest Service for helping to ensure the appropriate coverage of Pacific salmonid temperature thresholds, and we thank the three anonymous reviewers for their improvements to the earlier draft of this manuscript.

Conflicts of Interest: The authors declare no conflict of interest.

References

1. IPCC. *Synthesis Report of the IPCC Sixth Assessment Report (AR6)*; IPCC: Geneva, Switzerland, 2023; 85p.
2. Jonsson, B. Thermal Effects on Ecological Traits of Salmonids. *Fishes* **2023**, *8*, 337. [CrossRef]
3. Jonsson, B.; Jonsson, N. A Review of the Likely Effects of Climate Change on Anadromous Atlantic Salmon *Salmo salar* and Brown Trout *Salmo trutta*, with Particular Reference to Water Temperature and Flow. *J. Fish Biol.* **2009**, *75*, 2381–2447. [CrossRef]
4. Linnansaari, T.; Cunjak, R.A. Fish: Freshwater Ecosystems. In *Temperature Adaptation in a Changing Climate: Nature at Risk*; Storey, K.B., Tanino, K.K., Eds.; CABI: Wallingford, UK, 2012; pp. 80–97, ISBN 978-1-84593-822-2.
5. Isaak, D.J.; Wollrab, S.; Horan, D.; Chandler, G. Climate Change Effects on Stream and River Temperatures across the Northwest U.S. from 1980–2009 and Implications for Salmonid Fishes. *Clim. Chang.* **2012**, *113*, 499–524. [CrossRef]
6. Heino, J.; Virkkala, R.; Toivonen, H. Climate Change and Freshwater Biodiversity: Detected Patterns, Future Trends and Adaptations in Northern Regions. *Biol. Rev.* **2009**, *84*, 39–54. [CrossRef]
7. Caissie, D. The Thermal Regime of Rivers: A Review. *Freshw. Biol.* **2006**, *51*, 1389–1406. [CrossRef]
8. Dahms, C.; Killen, S.S. Temperature Change Effects on Marine Fish Range Shifts: A Meta-analysis of Ecological and Methodological Predictors. *Glob. Chang. Biol.* **2023**, *29*, 4459–4479. [CrossRef]
9. Moore, A.; Bendall, B.; Barry, J.; Waring, C.; Crooks, N.; Crooks, L. River Temperature and Adult Anadromous Atlantic Salmon, *Salmo salar*, and Brown Trout, *Salmo trutta*. *Fish. Manag. Ecol.* **2012**, *19*, 518–526. [CrossRef]
10. Finstad, A.G.; Ugedal, O.; Forseth, T.; Næsje, T.F. Energy-Related Juvenile Winter Mortality in a Northern Population of Atlantic Salmon (*Salmo salar*). *Can. J. Fish. Aquat. Sci.* **2004**, *61*, 2358–2368. [CrossRef]
11. Linnansaari, T.; Alfredsen, K.; Stickler, M.; Arnekleiv, J.V.; Harby, A.; Cunjak, R.A. Does Ice Matter? Site Fidelity and Movements by Atlantic Salmon (*Salmo salar* L.) Parr during Winter in a Substrate Enhanced River Reach. *River Res. Appl.* **2009**, *25*, 773–787. [CrossRef]
12. Prowse, T.D.; Bonsal, B.R.; Duguay, C.R.; Lacroix, M.P. River-Ice Break-up/Freeze-up: A Review of Climatic Drivers, Historical Trends and Future Predictions. *Ann. Glaciol.* **2007**, *46*, 443–451. [CrossRef]
13. Corey, E.; Linnansaari, T.; Dugdale, S.J.; Bergeron, N.; Gendron, J.; Lapointe, M.; Cunjak, R.A. Comparing the Behavioural Thermoregulation Response to Heat Stress by Atlantic Salmon Parr (*Salmo salar*) in Two Rivers. *Ecol. Freshw. Fish* **2020**, *29*, 50–62. [CrossRef]
14. Wenger, S.J.; Isaak, D.J.; Luce, C.H.; Neville, H.M.; Fausch, K.D.; Dunham, J.B.; Dauwalter, D.C.; Young, M.K.; Elsner, M.M.; Rieman, B.E.; et al. Flow Regime, Temperature, and Biotic Interactions Drive Differential Declines of Trout Species under Climate Change. *Proc. Natl. Acad. Sci. USA* **2011**, *108*, 14175–14180. [CrossRef]
15. Isaak, D.J.; Rieman, B.E. Stream Isotherm Shifts from Climate Change and Implications for Distributions of Ectothermic Organisms. *Glob. Chang. Biol.* **2013**, *19*, 742–751. [CrossRef]
16. O'Sullivan, A.M.; Corey, E.; Cunjak, R.A.; Linnansaari, T.; Curry, R.A. Salmonid Thermal Habitat Contraction in a Hydrogeologically Complex Setting. *Ecosphere* **2021**, *12*, e03797. [CrossRef]
17. Svenning, M.; Falkegård, M.; Dempson, J.B.; Power, M.; Bårdsen, B.; Guðbergsson, G.; Fauchald, P. Temporal Changes in the Relative Abundance of Anadromous Arctic Charr, Brown Trout, and Atlantic Salmon in Northern Europe: Do They Reflect Changing Climates? *Freshw. Biol.* **2022**, *67*, 64–77. [CrossRef]
18. Ebersole, J.L.; Liss, W.J.; Frissell, C.A. Thermal Heterogeneity, Stream Channel Morphology, and Salmonid Abundance in Northeastern Oregon Streams. *Can. J. Fish. Aquat. Sci.* **2003**, *60*, 1266–1280. [CrossRef]
19. Ebersole, J.L.; Liss, W.J.; Frissell, C.A. Cold Water Patches in Warm Streams: Physicochemical Characteristics and the Influence of Shading. *J. Am. Water Resour. Assoc.* **2003**, *39*, 355–368. [CrossRef]
20. Torgersen, C.E.; Ebersole, J.L.; Keenan, D.M. *Primer for Identifying Cold-Water Refuges to Protect and Restore Thermal Diversity in Riverine Landscapes*; U.S. Environmental Protection Agency: Washington, DC, USA, 2012; p. 91.
21. Kurylyk, B.L.; MacQuarrie, K.T.B.; Linnansaari, T.; Cunjak, R.A.; Curry, R.A. Preserving, Augmenting, and Creating Cold-water Thermal Refugia in Rivers: Concepts Derived from Research on the Miramichi River, New Brunswick (Canada). *Ecohydrology* **2015**, *8*, 1095–1108. [CrossRef]
22. Mejia, F.H.; Torgersen, C.E.; Berntsen, E.K.; Maroney, J.R.; Connor, J.M.; Fullerton, A.H.; Ebersole, J.L.; Lorang, M.S. Longitudinal, Lateral, Vertical, and Temporal Thermal Heterogeneity in a Large Impounded River: Implications for Cold-Water Refuges. *Remote Sens.* **2020**, *12*, 1386. [CrossRef]
23. Breau, C.; Cunjak, R.A.; Bremset, G. Age-Specific Aggregation of Wild Juvenile Atlantic Salmon *Salmo salar* at Cool Water Sources during High Temperature Events. *J. Fish Biol.* **2007**, *71*, 1179–1191. [CrossRef]
24. Dugdale, S.J.; Franssen, J.; Corey, E.; Bergeron, N.; Lapointe, M.; Cunjak, R.A. Main Stem Movement of Atlantic Salmon Parr in Response to High Temperatures. *Ecol. Freshw. Fish* **2016**, *25*, 429–445. [CrossRef]

25. Wilbur, N.M.; O'Sullivan, A.M.; MacQuarrie, K.T.B.; Linnansaari, T.; Curry, R.A. Characterizing Physical Habitat Preferences and Thermal Refuge Occupancy of Brook Trout (*Salvelinus fontinalis*) and Atlantic Salmon (*Salmo salar*) at High River Temperatures. *River Res. Appl.* **2020**, *36*, 769–783. [CrossRef]

26. Sullivan, C.J.; Vokoun, J.C.; Helton, A.M.; Briggs, M.A.; Kurylyk, B.L. An Ecohydrological Typology for Thermal Refuges in Streams and Rivers. *Ecohydrology* **2021**, *14*, e2295. [CrossRef]

27. Isaak, D.J.; Young, M.K. Cold-Water Habitats, Climate Refugia, and Their Utility for Conserving Salmonid Fishes. *Can. J. Fish. Aquat. Sci.* **2023**, *80*, 1187–1206. [CrossRef]

28. Mejia, F.H.; Ouellet, V.; Briggs, M.A.; Carlson, S.M.; Casas-Mulet, R.; Chapman, M.; Collins, M.J.; Dugdale, S.J.; Ebersole, J.L.; Frechette, D.M.; et al. Closing the Gap between Science and Management of Cold-water Refuges in Rivers and Streams. *Glob. Chang. Biol.* **2023**, *29*, 5482–5508. [CrossRef]

29. Cunjak, R.A. Winter Habitat of Selected Stream Fishes and Potential Impacts from Land-Use Activity. *Can. J. Fish. Aquat. Sci.* **1996**, *53* (Suppl. S1), 267–282. [CrossRef]

30. Cunjak, R.A.; Prowse, T.D.; Parrish, D.L. Atlantic Salmon (*Salmo salar*) in Winter: "The Season of Parr Discontent". *Can. J. Fish. Aquat. Sci.* **1998**, *55* (Suppl. S1), 161–180. [CrossRef]

31. Torgersen, C.E.; Price, D.M.; Li, H.W.; Mcintosh, B.A. Multiscale Thermal Refugia and Stream Habitat Associations of Chinook Salmon in Northeastern Oregon. *Ecol. Appl.* **1999**, *9*, 301–319. [CrossRef]

32. Dugdale, S.J.; Bergeron, N.; St-Hilaire, A. Temporal Variability of Thermal Refuges and Water Temperature Patterns in an Atlantic Salmon River. *Remote Sens. Environ.* **2013**, *136*, 358–373. [CrossRef]

33. Sutton, R.J.; Deas, M.L.; Tanaka, S.K.; Soto, T.; Corum, R.A. Salmonid Observations at a Klamath River Thermal Refuge under Various Hydrological and Meteorological Conditions. *River Res. Appl.* **2007**, *23*, 775–785. [CrossRef]

34. Gendron, J. Physical Controls on Summer Thermal Refuges for Salmonids in Two Gravel-Cobble Salmon Rivers with Contrasting Thermal Regimes: The Ouelle and Ste-Marguerite Rivers. Master's Thesis, McGill University, Montreal, QC, Canada, 2013.

35. Wondzell, S.M.; Gooseff, M. 9.13 Geomorphic Controls on Hyporheic Exchange across Scales: Watersheds to Particles. In *Treatise on Geomorphology*; Shroder, J.F., Ed.; Academic Press: Cambridge, MA, USA, 2013; Volume 9, pp. 203–218, ISBN 978-0-08-088522-3.

36. Huntsman, A.G. Death of Salmon and Trout with High Temperature. *J. Fish. Res. Board Can.* **1942**, *5*, 485–501. [CrossRef]

37. Ebersole, J.L.; Liss, W.J.; Frissell, C.A. Relationship between Stream Temperature, Thermal Refugia and Rainbow Trout Oncorhynchus mykiss Abundance in Arid-Land Streams in the Northwestern United States. *Ecol. Freshw. Fish* **2001**, *10*, 1–10. [CrossRef]

38. Nielsen, J.L.; Lisle, T.E.; Ozaki, V. Thermally Stratified Pools and Their Use by Steelhead in Northern California Streams. *Trans. Am. Fish. Soc.* **1994**, *123*, 613–626. [CrossRef]

39. Wondzell, S.M.; Ward, A.S. The Channel-Source Hypothesis: Empirical Evidence for in-Channel Sourcing of Dissolved Organic Carbon to Explain Hysteresis in a Headwater Mountain Stream. *Hydrol. Process.* **2022**, *36*, e14570. [CrossRef]

40. Enders, E.C.; Boisclair, D.; Roy, A.G. The Effect of Turbulence on the Cost of Swimming for Juvenile Atlantic Salmon (*Salmo salar*). *Can. J. Fish. Aquat. Sci.* **2003**, *60*, 1149–1160. [CrossRef]

41. Babin, A.B.; Peake, S.; Linnansaari, T.; Curry, R.A.; Ndong, M.; Haralampides, K.; Jones, R. Atlantic Salmon Upstream Migration Delay in a Large Hydropower Reservoir. *N. Am. J. Fish. Manag.* **2021**, *41*, 158–175. [CrossRef]

42. Olden, J.D.; Naiman, R.J. Incorporating Thermal Regimes into Environmental Flows Assessments: Modifying Dam Operations to Restore Freshwater Ecosystem Integrity: Incorporating Thermal Regimes in Environmental Flows Assessments. *Freshw. Biol.* **2010**, *55*, 86–107. [CrossRef]

43. Krause, C.W.; Newcomb, T.J.; Orth, D.J. Thermal Habitat Assessment of Alternative Flow Scenarios in a Tailwater Fishery. *River Res. Appl.* **2005**, *21*, 581–593. [CrossRef]

44. Hill, C.R.; Harrison, P.; University of New Brunswick, Fredericton, NB, Canada. Personal Communication, 2023.

45. Alfredsen, K.; Harby, A.; Linnansaari, T.; Ugedal, O. Development of an Inflow-Controlled Environmental Flow Regime for a Norwegian River: Inflow-Controlled Environmental Flow Regime. *River Res. Appl.* **2012**, *28*, 731–739. [CrossRef]

46. Cunjak, R.A.; Caissie, D.; El-Jabi, N.; Hardie, P.; Conlon, J.H.; Pollock, T.L.; Giberson, D.J.; Komadina-Douthwright, S.M. The Catamaran Brook (New Brunswick) Habitat Research Project: Biological, Physical and Chemical Conditions (1990–1992). *Can. Tech. Rep. Fish. Aquat. Sci.* **1993**, *1914*, 81.

47. Isaak, D.J.; Young, M.K.; Nagel, D.E.; Horan, D.L.; Groce, M.C. The Cold-Water Climate Shield: Delineating Refugia for Preserving Salmonid Fishes through the 21st Century. *Glob. Chang. Biol.* **2015**, *21*, 2540–2553. [CrossRef]

48. O'Sullivan, A.M.; Wegscheider, B.; Helminen, J.; Cormier, J.G.; Linnansaari, T.; Wilson, D.A.; Curry, R.A. Catchment-Scale, High-Resolution, Hydraulic Models and Habitat Maps–a Salmonid's Perspective. *J. Ecohydraulics* **2021**, *6*, 53–68. [CrossRef]

49. Boyer, C.; St-Hilaire, A.; Bergeron, N.; Daigle, A.; Curry, R.A.; Caissie, D.; Gillis, C.-A. RivTemp: A Water Temperature Network for Atlantic Salmon Rivers in Eastern Canada. *Water News* **2016**, *35*, 10–15. [CrossRef]

50. Isaak, D.J.; Wenger, S.J.; Peterson, E.E.; Ver Hoef, J.M.; Nagel, D.E.; Luce, C.H.; Hostetler, S.W.; Dunham, J.B.; Roper, B.B.; Wollrab, S.P.; et al. The NorWeST Summer Stream Temperature Model and Scenarios for the Western U.S.: A Crowd-Sourced Database and New Geospatial Tools Foster a User Community and Predict Broad Climate Warming of Rivers and Streams. *Water Resour. Res.* **2017**, *53*, 9181–9205. [CrossRef]

51. Isaak, D.J.; Peterson, E.E.; Ver Hoef, J.M.; Wenger, S.J.; Falke, J.A.; Torgersen, C.E.; Sowder, C.; Steel, E.A.; Fortin, M.-J.; Jordan, C.E.; et al. Applications of Spatial Statistical Network Models to Stream Data: Spatial Statistical Network Models for Stream Data. *Wiley Interdiscip. Rev. Water* **2014**, *1*, 277–294. [CrossRef]

52. O'Sullivan, A.M.; Devito, K.J.; Ogilvie, J.; Linnansaari, T.; Pronk, T.; Allard, S.; Curry, R.A. Effects of Topographic Resolution and Geologic Setting on Spatial Statistical River Temperature Models. *Water Resour. Res.* **2020**, *56*, e2020WR028122. [CrossRef]

53. Wang, T.; Kelson, S.J.; Greer, G.; Thompson, S.E.; Carlson, S.M. Tributary Confluences Are Dynamic Thermal Refuges for a Juvenile Salmonid in a Warming River Network. *River Res. Appl.* **2020**, *36*, 1076–1086. [CrossRef]

54. Ritter, T.D.; Zale, A.V.; Grisak, G.; Lance, M.J. Groundwater Upwelling Regulates Thermal Hydrodynamics and Salmonid Movements during High-temperature Events at a Montane Tributary Confluence. *Trans. Am. Fish. Soc.* **2020**, *149*, 600–619. [CrossRef]

55. Handcock, R.N.; Torgersen, C.E.; Cherkauer, K.A.; Gillespie, A.R.; Tockner, K.; Faux, R.N.; Tan, J. Thermal Infrared Remote Sensing of Water Temperature in Riverine Landscapes. In *Fluvial Remote Sensing for Science and Management*; Carbonneau, P.E., Piegay, H., Eds.; John Wiley & Sons: Hoboken, NJ, USA, 2012; pp. 85–113, ISBN 978-0-470-71427-0.

56. Roy, D.P.; Wulder, M.A.; Loveland, T.R.; Woodcock, C.E.; Allen, R.G.; Anderson, M.C.; Helder, D.; Irons, J.R.; Johnson, D.M.; Kennedy, R.; et al. Landsat-8: Science and Product Vision for Terrestrial Global Change Research. *Remote Sens. Environ.* **2014**, *145*, 154–172. [CrossRef]

57. Atwell, B.H.; MacDonald, R.B.; Bartolucci, L.A. Thermal Mapping of Streams from Airborne Radiometric Scanning. *J. Am. Water Resour. Assoc.* **1971**, *7*, 228–243. [CrossRef]

58. Dugdale, S.J.; Bergeron, N.E.; St-Hilaire, A. Spatial Distribution of Thermal Refuges Analysed in Relation to Riverscape Hydro-morphology Using Airborne Thermal Infrared Imagery. *Remote Sens. Environ.* **2015**, *160*, 43–55. [CrossRef]

59. Monk, W.A.; Wilbur, N.M.; Curry, R.A.; Gagnon, R.; Faux, R.N. Linking Landscape Variables to Cold Water Refugia in Rivers. *J. Environ. Manag.* **2013**, *118*, 170–176. [CrossRef] [PubMed]

60. Fullerton, A.H.; Torgersen, C.E.; Lawler, J.J.; Faux, R.N.; Steel, E.A.; Beechie, T.J.; Ebersole, J.L.; Leibowitz, S.G. Rethinking the Longitudinal Stream Temperature Paradigm: Region-Wide Comparison of Thermal Infrared Imagery Reveals Unexpected Complexity of River Temperatures. *Hydrol. Process.* **2015**, *29*, 4719–4737. [CrossRef]

61. Fuller, M.R.; Ebersole, J.L.; Detenbeck, N.E.; Labiosa, R.; Leinenbach, P.; Torgersen, C.E. Integrating Thermal Infrared Stream Temperature Imagery and Spatial Stream Network Models to Understand Natural Spatial Thermal Variability in Streams. *J. Therm. Biol.* **2021**, *100*, 103028. [CrossRef] [PubMed]

62. Corey, E.M.; Linnansaari, T.; O'Sullivan, A.M.; Curry, R.A.; Cunjak, R.A. Quantifying Movement Patterns of Wild Juvenile Atlantic Salmon (*Salmo salar*) in Relation to High Water Temperature and Proximity to Thermal Refuges. Ph.D. Thesis, University of New Brunswick, Fredericton, Canada, 2023.

63. Dugdale, S.J.; Kelleher, C.A.; Malcolm, I.A.; Caldwell, S.; Hannah, D.M. Assessing the Potential of Drone-based Thermal Infrared Imagery for Quantifying River Temperature Heterogeneity. *Hydrol. Process.* **2019**, *33*, 1152–1163. [CrossRef]

64. Casas-Mulet, R.; Pander, J.; Ryu, D.; Stewardson, M.J.; Geist, J. Unmanned Aerial Vehicle (UAV)-Based Thermal Infra-Red (TIR) and Optical Imagery Reveals Multi-Spatial Scale Controls of Cold-Water Areas over a Groundwater-Dominated Riverscape. *Front. Environ. Sci.* **2020**, *8*, 64. [CrossRef]

65. O'Sullivan, A.M.; Linnansaari, T.; Leavitt, J.; Samways, K.M.; Kurylyk, B.L.; Curry, R.A. The Salmon-peloton: Hydraulic Habitat Shifts of Adult Atlantic Salmon (*Salmo salar*) Due to Behavioural Thermoregulation. *River Res. Appl.* **2022**, *38*, 107–118. [CrossRef]

66. Morgan, A.M.; O'Sullivan, A.M. Cooler, Bigger; Warmer, Smaller: Fine-Scale Thermal Heterogeneity Maps Age Class and Species Distribution in Behaviourally Thermoregulating Salmonids. *River Res. Appl.* **2023**, *39*, 163–176. [CrossRef]

67. O'Sullivan, A.M.; Kurylyk, B.L. Limiting External Absorptivity of UAV-Based Uncooled Thermal Infrared Sensors Increases Water Temperature Measurement Accuracy. *Remote Sens.* **2022**, *14*, 6356. [CrossRef]

68. O'Sullivan, A.M.; Linnansaari, T.; Curry, R.A. Ice Cover Exists: A Quick Method to Delineate Groundwater Inputs in Running Waters for Cold and Temperate Regions. *Hydrol. Process.* **2019**, *33*, 3297–3309. [CrossRef]

69. Breau, C.; Cunjak, R.A.; Peake, S.J. Behaviour during Elevated Water Temperatures: Can Physiology Explain Movement of Juvenile Atlantic Salmon to Cool Water?: Temperature and Physiology of Juvenile Salmon. *J. Anim. Ecol.* **2011**, *80*, 844–853. [CrossRef]

70. Elliott, J.M.; Elliott, J.A. Temperature Requirements of Atlantic Salmon *Salmo salar*, Brown Trout *Salmo trutta* and Arctic Charr *Salvelinus alpinus*: Predicting the Effects of Climate Change. *J. Fish Biol.* **2010**, *77*, 1793–1817. [CrossRef] [PubMed]

71. Grant, J.W.A.; Kramer, D.L. Territory Size as a Predictor of the Upper Limit to Population Density of Juvenile Salmonids in Streams. *Can. J. Fish. Aquat. Sci.* **1990**, *47*, 1724–1737. [CrossRef]

72. O'Sullivan, A.M.; Corey, E.M.; Collet, E.N.; Helminen, J.; Curry, R.A.; MacIntyre, C.; Linnansaari, T. Timing and Frequency of High Temperature Events Bend the Onset of Behavioural Thermoregulation in Atlantic Salmon (*Salmo salar*). *Conserv. Physiol.* **2023**, *11*, coac079. [CrossRef]

73. Økland, F.; Erkinaro, J.; Moen, K.; Niemela, E.; Fiske, P.; McKinley, R.S.; Thorstad, E.B. Return Migration of Atlantic Salmon in the River Tana: Phases of Migratory Behaviour. *J. Fish Biol.* **2001**, *59*, 862–874. [CrossRef]

74. Carrow, R.M. Freshwater Migration and Behaviour of Wild Adult Atlantic Salmon (*Salmo salar*) in the Miramichi River, New Brunswick, Canada. Master's Thesis, University of New Brunswick, Department of Biology, Fredericton, NB, Canada, 2021.

75. Farrell, A.P.; Eliason, E.J.; Sandblom, E.; Clark, T.D. Fish Cardiorespiratory Physiology in an Era of Climate Change. *Can. J. Zool.* **2009**, *87*, 835–851. [CrossRef]
76. Eliason, E.J.; Clark, T.D.; Hinch, S.G.; Farrell, A.P. Cardiorespiratory Collapse at High Temperature in Swimming Adult Sockeye Salmon. *Conserv. Physiol.* **2013**, *1*, cot008. [CrossRef]
77. Ern, R.; Andreassen, A.H.; Jutfelt, F. Physiological Mechanisms of Acute Upper Thermal Tolerance in Fish. *Physiology* **2023**, *38*, 141–158. [CrossRef]
78. Allen, J.A. The Influence of Physical Conditions in the Genesis of Species. *Radic. Rev.* **1877**, *1*, 108–140.
79. Angilletta, M.J. *Thermal Adaptation: A Theoretical and Empirical Synthesis*; Oxford University Press: Oxford, UK, 2009.
80. Eliason, E.J.; Clark, T.D.; Hague, M.J.; Hanson, L.M.; Gallagher, Z.S.; Jeffries, K.M.; Gale, M.K.; Patterson, D.A.; Hinch, S.G.; Farrell, A.P. Differences in Thermal Tolerance among Sockeye Salmon Populations. *Science* **2011**, *332*, 109–112. [CrossRef]
81. Anlauf-Dunn, K.; Kraskura, K.; Eliason, E.J. Intraspecific Variability in Thermal Tolerance: A Case Study with Coastal Cutthroat Trout. *Conserv. Physiol.* **2022**, *10*, coac029. [CrossRef]
82. Breau, C. *Knowledge of Fish Physiology Used to Set Water Temperature Thresholds for In-Season Closures of Atlantic Salmon (Salmo salar) Recreational Fisheries*; Canadian Science Advisory Secretariat Research Document; Fisheries and Oceans Canada: Ottawa, ON, Canada, 2013; p. 24.
83. Collet, E.; O'Sullivan, A.M.; University of New Brunswick, Fredericton, NB, Canada. Personal Communication, 2023.
84. Pörtner, H.O.; Bennett, A.F.; Bozinovic, F.; Clarke, A.; Lardies, M.A.; Lucassen, M.; Pelster, B.; Schiemer, F.; Stillman, J.H. Trade-Offs in Thermal Adaptation: The Need for a Molecular to Ecological Integration. *Physiol. Biochem. Zool.* **2006**, *79*, 295–313. [CrossRef] [PubMed]
85. Anttila, K.; Couturier, C.S.; Øverli, Ø.; Johnsen, A.; Marthinsen, G.; Nilsson, G.E.; Farrell, A.P. Atlantic Salmon Show Capability for Cardiac Acclimation to Warm Temperatures. *Nat. Commun.* **2014**, *5*, 4252. [CrossRef] [PubMed]
86. Anttila, K.; Dhillon, R.S.; Boulding, E.G.; Farrell, A.P.; Glebe, B.D.; Elliott, J.A.K.; Wolters, W.R.; Schulte, P.M. Variation in Temperature Tolerance among Families of Atlantic Salmon (*Salmo salar*) Is Associated with Hypoxia Tolerance, Ventricle Size and Myoglobin Level. *J. Exp. Biol.* **2013**, *216*, 1183–1190. [CrossRef]
87. Crossin, G.T.; Hinch, S.G.; Cooke, S.J.; Welch, D.W.; Patterson, D.A.; Jones, S.R.M.; Lotto, A.G.; Leggatt, R.A.; Mathes, M.T.; Shrimpton, J.M.; et al. Exposure to High Temperature Influences the Behaviour, Physiology, and Survival of Sockeye Salmon during Spawning Migration. *Can. J. Zool.* **2008**, *86*, 127–140. [CrossRef]
88. Hinch, S.G.; Bett, N.N.; Eliason, E.J.; Farrell, A.P.; Cooke, S.J.; Patterson, D.A. Exceptionally High Mortality of Adult Female Salmon: A Large-Scale Pattern and a Conservation Concern. *Can. J. Fish. Aquat. Sci.* **2021**, *78*, 639–654. [CrossRef]
89. Gibson, R.J. Some Factors Influencing the Distributions of Brook Trout and Young Atlantic Salmon. *J. Fish. Res. Board Can.* **1966**, *23*, 1977–1980. [CrossRef]
90. Frechette, D.M.; Dugdale, S.J.; Dodson, J.J.; Bergeron, N.E. Understanding Summertime Thermal Refuge Use by Adult Atlantic Salmon Using Remote Sensing, River Temperature Monitoring, and Acoustic Telemetry. *Can. J. Fish. Aquat. Sci.* **2018**, *75*, 1999–2010. [CrossRef]
91. Wehrly, K.E.; Wang, L.; Mitro, M. Field-Based Estimates of Thermal Tolerance Limits for Trout: Incorporating Exposure Time and Temperature Fluctuation. *Trans. Am. Fish. Soc.* **2007**, *136*, 365–374. [CrossRef]
92. Hitt, N.P.; Snook, E.L.; Massie, D.L. Brook Trout Use of Thermal Refugia and Foraging Habitat Influenced by Brown Trout. *Can. J. Fish. Aquat. Sci.* **2017**, *74*, 406–418. [CrossRef]
93. Kaya, C.M.; Kaeding, L.R.; Burkhalter, D.E. Use of a Cold-Water Refuge by Rainbow and Brown Trout in a Geothermally Heated Stream. *Progress. Fish-Cult.* **1977**, *39*, 37–39. [CrossRef]
94. Dobos, M.E.; Corsi, M.P.; Schill, D.J.; DuPont, J.M.; Quist, M.C. Influences of Summer Water Temperatures on the Movement, Distribution, and Resource Use of Fluvial Westslope Cutthroat Trout in the South Fork Clearwater River Basin. *N. Am. J. Fish. Manag.* **2016**, *36*, 549–567. [CrossRef]
95. Barrett, H.S.; Armstrong, J.B. Move, Migrate, or Tolerate: Quantifying Three Tactics for Coldwater Fish Coping with Warm Summers in a Large River. *Ecosphere* **2022**, *13*, e4095. [CrossRef]
96. Hillyard, R.W.; Keeley, E.R. Temperature-Related Changes in Habitat Quality and Use by Bonneville Cutthroat Trout in Regulated and Unregulated River Segments. *Trans. Am. Fish. Soc.* **2012**, *141*, 1649–1663. [CrossRef]
97. McCarrick, D.K.; Dillon, J.C.; High, B.; Quist, M.C. Spatial and Temporal Distribution and Habitat Selection of Native Yellowstone Cutthroat Trout and Nonnative Utah Chub. *N. Am. J. Fish. Manag.* **2022**, *42*, 939–951. [CrossRef]
98. Keefer, M.L.; Peery, C.A.; High, B. Behavioral Thermoregulation and Associated Mortality Trade-Offs in Migrating Adult Steelhead (*Oncorhynchus mykiss*): Variability among Sympatric Populations. *Can. J. Fish. Aquat. Sci.* **2009**, *66*, 1734–1747. [CrossRef]
99. Keefer, M.L.; Clabough, T.S.; Jepson, M.A.; Johnson, E.L.; Peery, C.A.; Caudill, C.C. Thermal Exposure of Adult Chinook Salmon and Steelhead: Diverse Behavioral Strategies in a Large and Warming River System. *PLoS ONE* **2018**, *13*, e0204274. [CrossRef]
100. Brewitt, K.S.; Danner, E.M. Spatio-Temporal Temperature Variation Influences Juvenile Steelhead (*Oncorhynchus mykiss*) Use of Thermal Refuges. *Ecosphere* **2014**, *5*, art92. [CrossRef]
101. Keefer, M.L.; Clabough, T.S.; Jepson, M.A.; Naughton, G.P.; Blubaugh, T.J.; Joosten, D.C.; Caudill, C.C. Thermal Exposure of Adult Chinook Salmon in the Willamette River Basin. *J. Therm. Biol.* **2015**, *48*, 11–20. [CrossRef]

102. Goniea, T.M.; Keefer, M.L.; Bjornn, T.C.; Peery, C.A.; Bennett, D.H.; Stuehrenberg, L.C. Behavioral Thermoregulation and Slowed Migration by Adult Fall Chinook Salmon in Response to High Columbia River Water Temperatures. *Trans. Am. Fish. Soc.* **2006**, *135*, 408–419. [CrossRef]
103. Armstrong, J.B.; Ward, E.J.; Schindler, D.E.; Lisi, P.J. Adaptive Capacity at the Northern Front: Sockeye Salmon Behaviourally Thermoregulate during Novel Exposure to Warm Temperatures. *Conserv. Physiol.* **2016**, *4*, cow039. [CrossRef]
104. Biro, P.A. Staying Cool: Behavioral Thermoregulation during Summer by Young-of-Year Brook Trout in a Lake. *Trans. Am. Fish. Soc.* **1998**, *127*, 212–222. [CrossRef]
105. Langeland, A.; L'Abee-Lund, J.H. An Experimental Test of the Genetic Component of the Ontogenetic Habitat Shift in Arctic Charr (*Salvelinus alpinus*). *Ecol. Freshw. Fish* **1998**, *7*, 200–207. [CrossRef]
106. Sellers, T.J.; Parker, B.R.; Schindler, D.W.; Tonn, W.M. Pelagic Distribution of Lake Trout (*Salvelinus namaycush*) in Small Canadian Shield Lakes with Respect to Temperature, Dissolved Oxygen, and Light. *Can. J. Fish. Aquat. Sci.* **1998**, *55*, 170–179. [CrossRef]
107. Howell, P.J.; Dunham, J.B.; Sankovich, P.M. Relationships between Water Temperatures and Upstream Migration, Cold Water Refuge Use, and Spawning of Adult Bull Trout from the Lostine River, Oregon, USA. *Ecol. Freshw. Fish* **2010**, *19*, 96–106. [CrossRef]
108. Gutowsky, L.F.G.; Harrison, P.M.; Martins, E.G.; Leake, A.; Patterson, D.A.; Zhu, D.Z.; Power, M.; Cooke, S.J. Daily Temperature Experience and Selection by Adfluvial Bull Trout (*Salvelinus confluentus*). *Environ. Biol. Fishes* **2017**, *100*, 1167–1180. [CrossRef]
109. Swanberg, T.R. Movements of and Habitat Use by Fluvial Bull Trout in the Blackfoot River, Montana. *Trans. Am. Fish. Soc.* **1997**, *126*, 735–746. [CrossRef]
110. Gallant, M.J.; LeBlanc, S.; MacCormack, T.J.; Currie, S. Physiological Responses to a Short-Term, Environmentally Realistic, Acute Heat Stress in Atlantic Salmon, *Salmo salar*. *Facets* **2017**, *2*, 330–341. [CrossRef]
111. Thorstad, E.B.; Whoriskey, F.G.; Rikardsen, A.H.; Aarestrup, K. Aquatic Nomads: The Life and Migrations of the Atlantic Salmon. In *Atlantic Salmon Ecology*; Aas, Ø., Einum, S., Klemetsen, A., Skurdal, J., Eds.; Wiley-Blackwell: Oxford, UK, 2011; pp. 1–32.
112. Braithwaite, V.A.; Armstrong, J.D.; McAdam, H.M.; Huntingford, F.A. Can Juvenile Atlantic Salmon Use Multiple Cue Systems in Spatial Learning? *Anim. Behav.* **1996**, *51*, 1409–1415. [CrossRef]
113. Breau, C. The Ecophysiology, Behaviour and Movement Patterns of Juvenile Atlantic Salmon (*Salmo salar*) during Elevated Water Temperatures and the Importance of Cool Water Sites. Ph.D. Thesis, University of New Brunswick, Department of Biology, Fredericton, NB, Canada, 2011.
114. Elvidge, C.K.; Cooke, E.L.L.; Cunjak, R.A.; Cooke, S.J. Social Cues May Advertise Habitat Quality to Refuge-Seeking Conspecifics. *Can. J. Zool.* **2017**, *95*, 1–5. [CrossRef]
115. Johansen, M.; Erkinaro, J.; Amundsen, P.-A. The When What and Where of Freshwater Feeding. In *Atlantic Salmon Ecology*; Aas, Ø., Einum, S., Klemetsen, A., Skurdal, J., Eds.; Wiley-Blackwell: Oxford, UK, 2011; pp. 89–114.
116. Johansen, M. Evidence of Freshwater Feeding by Adult Salmon in the Tana River, Northern Norway. *J. Fish Biol.* **2001**, *59*, 1405–1407. [CrossRef]
117. Brewitt, K.S.; Danner, E.M.; Moore, J.W. Hot Eats and Cool Creeks: Juvenile Pacific Salmonids Use Mainstem Prey While in Thermal Refuges. *Can. J. Fish. Aquat. Sci.* **2017**, *74*, 1588–1602. [CrossRef]
118. Milligan, C.L.; Hooke, G.B.; Johnson, C. Sustained Swimming at Low Velocity Following a Bout of Exhaustive Exercise Enhances Metabolic Recovery in Rainbow Trout. *J. Exp. Biol.* **2000**, *203*, 921–926. [CrossRef]
119. Baird, O.E.; Krueger, C.C. Behavioral Thermoregulation of Brook and Rainbow Trout: Comparison of Summer Habitat Use in an Adirondack River, New York. *Trans. Am. Fish. Soc.* **2003**, *132*, 1194–1206. [CrossRef]
120. Corey, E.; Linnansaari, T.; Cunjak, R.A. High Temperature Events Shape the Broadscale Distribution of Juvenile Atlantic Salmon (*Salmo salar*). *Freshw. Biol.* **2023**, *68*, 534–545. [CrossRef]
121. Farrell, A.P.; Hinch, S.G.; Cooke, S.J.; Patterson, D.A.; Crossin, G.T.; Lapointe, M.; Mathes, M.T. Pacific Salmon in Hot Water: Applying Aerobic Scope Models and Biotelemetry to Predict the Success of Spawning Migrations. *Physiol. Biochem. Zool.* **2008**, *81*, 697–709. [CrossRef]
122. Corey, E.; Linnansaari, T.; Cunjak, R.A.; Currie, S. Physiological Effects of Environmentally Relevant, Multi-Day Thermal Stress on Wild Juvenile Atlantic Salmon (*Salmo salar*). *Conserv. Physiol.* **2017**, *5*, cox014. [CrossRef]
123. Bowerman, T.; Roumasset, A.; Keefer, M.L.; Sharpe, C.S.; Caudill, C.C. Prespawn Mortality of Female Chinook Salmon Increases with Water Temperature and Percent Hatchery Origin. *Trans. Am. Fish. Soc.* **2018**, *147*, 31–42. [CrossRef]
124. Fisheries and Oceans Canada. Temperature Threshold to Define Management Strategies for Atlantic Salmon (*Salmo salar*) Fisheries under Environmentally Stressful Conditions. *Can. Sci. Adv. Sec. Sci. Adv. Rep.* **2012**, *19*, 17.
125. Crossin, G.T.; Hinch, S.G.; Farrell, A.P.; Higgs, D.A.; Lotto, A.G.; Oakes, J.D.; Healey, M.C. Energetics and Morphology of Sockeye Salmon: Effects of Upriver Migratory Distance and Elevation. *J. Fish Biol.* **2004**, *65*, 788–810. [CrossRef]
126. Fenkes, M.; Shiels, H.A.; Fitzpatrick, J.L.; Nudds, R.L. The Potential Impacts of Migratory Difficulty, Including Warmer Waters and Altered Flow Conditions, on the Reproductive Success of Salmonid Fishes. *Comp. Biochem. Physiol. A Mol. Integr. Physiol.* **2016**, *193*, 11–21. [CrossRef]
127. Fenkes, M.; Fitzpatrick, J.L.; Ozolina, K.; Shiels, H.A.; Nudds, R.L. Sperm in Hot Water: Direct and Indirect Thermal Challenges Interact to Impact on Brown Trout Sperm Quality. *J. Exp. Biol.* **2017**, *220*, 2513–2520. [CrossRef]
128. Berman, C.H.; Quinn, T.P. Behavioural Thermoregulation and Homing by Spring Chinook Salmon, *Oncorhynchus tshawytscha* (Walbaum), in the Yakima River. *J. Fish Biol.* **1991**, *39*, 301–312. [CrossRef]

129. Snyder, M.N.; Schumaker, N.H.; Dunham, J.B.; Keefer, M.L.; Leinenbach, P.; Brookes, A.; Palmer, J.; Wu, J.; Keenan, D.; Ebersole, J.L. Assessing Contributions of Cold-Water Refuges to Reproductive Migration Corridor Conditions for Adult Salmon and Steelhead Trout in the Columbia River, USA. *J. Ecohydraulics* **2022**, *7*, 111–123. [CrossRef]
130. Robertson, B.A.; Hutto, R.L. A Framework for Understanding Ecological Traps and an Evaluation of Existing Evidence. *Ecology* **2006**, *87*, 1075–1085. [CrossRef] [PubMed]
131. Van Leeuwen, T.E.; Lehnert, S.J.; Breau, C.; Fitzsimmons, M.; Kelly, N.I.; Dempson, J.B.; Neville, V.M.; Young, M.; Keefe, D.; Bird, T.J.; et al. Considerations for Water Temperature-Related Fishery Closures in Recreational Atlantic Salmon (*Salmo salar*) Catch and Release Fisheries: A Case Study from Eastern Canada. *Rev. Fish. Sci. Aquac.* **2023**, *31*, 598–619. [CrossRef]
132. Gale, M.K.; Hinch, S.G.; Donaldson, M.R. The Role of Temperature in the Capture and Release of Fish: Temperature Effects on Capture-Release. *Fish Fish.* **2013**, *14*, 1–33. [CrossRef]
133. Lennox, R.J.; Cooke, S.J.; Davis, C.R.; Gargan, P.; Hawkins, L.A.; Havn, T.B.; Johansen, M.R.; Kennedy, R.J.; Richard, A.; Svenning, M.-A.; et al. Pan-Holarctic Assessment of Post-Release Mortality of Angled Atlantic Salmon *Salmo salar*. *Biol. Conserv.* **2017**, *209*, 150–158. [CrossRef]
134. Keefe, D.; Young, M.; Van Leeuwen, T.E.; Adams, B. Long-term Survival of Atlantic Salmon Following Catch and Release: Considerations for Anglers, Scientists and Resource Managers. *Fish. Manag. Ecol.* **2022**, *29*, 286–297. [CrossRef]
135. Caissie, D.; Breau, C.; Hayward, J.; Cameron, P. *Water Temperature Characteristics within the Miramichi and Restigouche Rivers*; Canadian Science Advisory Secretariat Research Document; Fisheries and Oceans Canada: Ottawa, ON, Canada, 2013; p. vii + 31.
136. O'Sullivan, A.M.; Devito, K.J.; D'Orangeville, L.; Curry, R.A. The Waterscape Continuum Concept: Rethinking Boundaries in Ecosystems. *WIREs Water* **2022**, *9*, e1598. [CrossRef]
137. Lynch, J.A.; Rishel, G.B.; Corbett, E.S. Thermal Alteration of Streams Draining Clearcut Watersheds: Quantification and Biological Implications. *Hydrobiologia* **1984**, *111*, 161–169. [CrossRef]
138. Moore, R.D.; Spittlehouse, D.L.; Story, A. Riparian Microclimate and Stream Temperature Response to Forest Harvesting: A Review. *J. Am. Water Resour. Assoc.* **2005**, *41*, 813–834. [CrossRef]
139. Johnson, S.L.; Jones, J.A. Stream Temperature Responses to Forest Harvest and Debris Flows in Western Cascades, Oregon. *Can. J. Fish. Aquat. Sci.* **2000**, *57*, 30–39. [CrossRef]
140. Hall, J.D.; Cederholm, C.J.; Murphy, M.L.; Koski, K.V. Fish-Forestry Interactions in Oregon, Washington and Alaska, USA. In *Fishes and Forestry; Worldwide Watershed Interactions and Management*; Northcote, T.G., Hartman, G.F., Eds.; Blackwell Publishing: Oxford, UK, 2004; pp. 363–388.
141. Cunjak, R.A.; Curry, R.A.; Scruton, D.A.; Clarke, K.D. Fish-Forestry Interactions in Freshwaters of Atlantic Canada. In *Fishes and Forestry; Worldwide Watershed Interactions and Management*; Northcote, T.G., Hartman, G.F., Eds.; Blackwell Publishing: Oxford, UK, 2004; pp. 439–462.
142. Alexander, M.D.; MacQuarrie, K.T.B.; Caissie, D.; Butler, K.E. The Thermal Regime of Shallow Groundwater and a Small Atlantic Salmon Stream Bordering a Clearcut with a Forested Streamside Buffer. In Proceedings of the Annual Conference of the Canadian Society for Civil Engineering, Moncton, NB, Canada, 4–7 June 2003; p. 10.
143. Winter, T.C.; Harvey, J.W.; Franke, O.L.; Alley, W.M. *Ground Water and Surface Water a Single Resource*; Diane Publishing: Collingdale, PA, USA, 1998; p. 88.
144. Devito, K.J.; Creed, I.F.; Fraser, C.J.D. Controls on Runoff from a Partially Harvested Aspen-Forested Headwater Catchment, Boreal Plain, Canada. *Hydrol. Process.* **2005**, *19*, 3–25. [CrossRef]
145. Devito, K.J.; Hokanson, K.J.; Moore, P.A.; Kettridge, N.; Anderson, A.E.; Chasmer, L.; Hopkinson, C.; Lukenbach, M.C.; Mendoza, C.A.; Morissette, J.; et al. Landscape Controls on Long-Term Runoff in Subhumid Heterogeneous Boreal Plains Catchments. *Hydrol. Process.* **2017**, *31*, 2737–2751. [CrossRef]
146. Briggs, M.A.; Lane, J.W.; Snyder, C.D.; White, E.A.; Johnson, Z.C.; Nelms, D.L.; Hitt, N.P. Shallow Bedrock Limits Groundwater Seepage-Based Headwater Climate Refugia. *Limnologica* **2018**, *68*, 142–156. [CrossRef]

Article

Expected Climate Change in the High Arctic—Good or Bad for Arctic Charr?

Martin A. Svenning [1,*], Eigil T. Bjørvik [2], Jane A. Godiksen [3], Johan Hammar [4], Jack Kohler [5], Reidar Borgstrøm [6] and Nigel G. Yoccoz [7]

[1] Norwegian Institute for Nature Research (NINA-Tromsø), Department of Arctic Ecology, Fram Centre, NO-9296 Tromsø, Norway
[2] CFEED, Verkstedveien 4, NO-7125 Vanvikan, Norway; eigiltb@hotmail.com
[3] Institute of Marine Research, Nordnes, P.O. Box 1870, NO-5817 Bergen, Norway; jane.godiksen@hi.no
[4] Institute of Freshwater Research, SE-178 93 Drottningholm, Sweden; jhammar.isacf@gmail.com
[5] Norwegian Polar Institute, Fram Centre, NO-9296 Tromsø, Norway; jack.kohler@npolar.no
[6] Faculty of Environmental Sciences and Natural Resource Management, Norwegian University of Life Sciences, P.O. Box 5003, NO-1432 Ås, Norway; reidar.borgstrom@nmbu.no
[7] Department of Arctic and Marine Biology, The University of Tromsø, NO-9037 Tromsø, Norway; nigel.yoccoz@uit.no
* Correspondence: martin.svenning@nina.no

Abstract: Lakes in the High Arctic are characterized by their low water temperature, long-term ice cover, low levels of nutrients, and low biodiversity. These conditions mean that minor climatic changes may be of great importance to Arctic freshwater organisms, including fish, by influencing vital life history parameters such as individual growth rates. In this study, Arctic charr sampled from two Svalbard lakes (78–79° N) over the period 1960–2008 provided back-calculated length-at-age information extending over six decades, covering both warm and cold spells. The estimated annual growth in young-of-the-year (YOY) Arctic charr correlated positively with an increasing air temperature in summer. This increase is likely due to the higher water temperature during the ice-free period, and also to some extent, due to the winter air temperature; this is probably due to thinner ice being formed in mild winters and the subsequent earlier ice break-up. However, years with higher snow accumulation correlated with slower growth rates, which may be due to delayed ice break-up and thus a shorter summer growing season. More than 30% of the growth in YOY charr could be explained specifically by air temperature and snow accumulation in the two Arctic charr populations. This indicated that juvenile Svalbard Arctic charr may experience increased growth rates in a future warmer climate, although future increases in precipitation may contradict the positive effects of higher temperatures to some extent. In the longer term, a warmer climate may lead to the complete loss of many glaciers in western Svalbard; therefore, rivers may dry out, thus hindering migration between salt water and fresh water for migratory fish. In the worst-case scenario, the highly valuable and attractive anadromous Arctic charr populations could eventually disappear from the Svalbard lake systems.

Keywords: High Arctic; Svalbard lake systems; climate impact; Arctic charr; growth rate; anadromy

Key Contribution: Annual growth in YOY Svalbard Arctic charr correlated positively with increasing air temperature, while years with higher snow accumulation correlated with slower growth rates.

Citation: Svenning, M.A.; Bjørvik, E.T.; Godiksen, J.A.; Hammar, J.; Kohler, J.; Borgstrøm, R.; Yoccoz, N.G. Expected Climate Change in the High Arctic—Good or Bad for Arctic Charr? *Fishes* **2024**, *9*, 8. https://doi.org/10.3390/fishes9010008

Received: 13 November 2023
Revised: 14 December 2023
Accepted: 20 December 2023
Published: 23 December 2023

1. Introduction

The salmonid Arctic charr (*Salvelinus alpinus*) has a Holarctic distribution and is the only freshwater fish species in most High Arctic regions [1], including Svalbard, an archipelago in the Arctic Ocean between a latitude of 74 and 81° N. Arctic charr populations demonstrate remarkable ecological plasticity, showing various life-history adaptations to harsh northern environments. In addition, it is the only freshwater fish species on Svalbard, forming both anadromous, resident, and landlocked stocks [2].

Lakes in the High Arctic are characterized by low water temperatures, long-term ice cover, low levels of nutrients, and low biodiversity [3]. Depending on the location, most Svalbard lakes are ice-covered for 9–12 months a year [2], with maximum water temperatures usually reaching 6–8 °C in summer–autumn [4]. The thickness of the ice, the timing of ice break-up/cover, and water temperatures during the ice-free season, however, vary among lake systems, due to local differences in air temperature and precipitation [5,6]. Studies during the early 1980s revealed that year-class strength correlates positively with mean air temperature during the two summers preceding the spawning of Arctic charr in Svalbard [7].

Temperature, which is known to influence both ingestion and metabolism, will thus influence the somatic growth rate in fishes, although the effects on growth are also influenced by interactions between temperature and food supply [8]. Although [9] suggested that as the temperature rises, somatic growth of Arctic charr would increase in high-latitude lakes, ref. [10] showed that the knowledge of fish growth in response to climate change in general remains incomplete, and that some findings are even contradictory. Svalbard Arctic charr usually experience water temperatures in June–August in the range of 1–8 °C, such that even small increases in water temperature (0.5–1 °C) may lead to significant changes in the growth of younger fish [11]. In the subalpine Lake Øvre Heimdalsvatn (>1000 m a.s.l.), an increase of 1 °C in the June air temperature resulted in a 10% growth increase in brown trout (*Salmo trutta*) [12]. In Greenland, ref. [13] recorded a positive correlation between air temperature and the growth of landlocked Arctic charr in one lake, whilst the opposite was found in another. This was probably due to differences in energy demand and food supply. In the open lake systems on Svalbard, i.e., systems containing anadromous Arctic charr with additional access to marine prey, fish density seems rather low [2,4]. As noted above, even small increases in thermal conditions may increase the growth rate in fish significantly.

Recent information on global warming shows a higher increase in air temperatures at higher northern latitudes [14,15]. In Longyearbyen, Svalbard, the annual mean temperature has increased significantly from 1912 to 2008 [6], corresponding to +0.22 °C per decade, with the highest increase shown in spring (+0.45 °C). In northeastern and southwestern Svalbard, an increase in air temperature of 8 and 3 °C, respectively, are suggested for the next 100 years [6,14]. So far, annual precipitation in the Svalbard region has been low, with a gradient from higher values in the southwest to lower values in the northeast. Precipitation, however, is also expected to increase in Svalbard during the next 100 years, especially in winter. This is predicted to lead to a larger accumulation of snow in spring, which in turn, will likely postpone the timing of ice break-up and shorten the ice-free season [16]. On the other hand, both an increasing air temperature in winter, implying thinner ice [17], and a higher temperature in spring inducing earlier ice-melt [2,18], will prolong the ice-free period. The sun's radiation will then affect the water more efficiently and this may lead to both a higher maximum water temperature as well as a higher number of degree days [5,19]. An increase in accumulated snow cover in spring has been shown by [16] in the alpine Lake Litlos (>1100 m a.s.l.) in Hardangervidda, Norway, and by [5] in Greenland lakes. This has been shown to have a strong effect on the time of ice break-up and thus a negative effect on the growth and the survival in young-of-the-year (YOY) brown trout and Arctic charr, respectively. In High Arctic and alpine lakes, a change in climate that results in higher temperatures is therefore assumed to increase growth in juvenile charr; more snowfall, on the other hand, probably induces decreased growth.

To study these opposite effects of temperature and precipitation on life history parameters in fish, access to long-term survey data covering both warm and cold periods/years is essential [20]. On Svalbard, two weather stations that have been in operation for more than 100 years [21,22] showed mainly positive air temperature trends before the 1930s, and a warm period during the 1930s and 1940s. This was followed by a negative temperature trend in the 1950s and 1960s. From the 1960s onwards there has been a general temperature increase in all seasons (see [6,14] for more details). From around 1960 to 2008, samples of Arctic charr otoliths have been collected from several populations, thus providing the possibility to use Arctic charr otoliths as flight recorders to back-calculate fish growth in juvenile Arctic charr back to the early 1950s (Gullestad, Hammar and Svenning unpublished).

Accordingly, in this study, we compare back-calculated fish growth in YOY Arctic charr from otoliths with climate indices, specifically air temperature and the accumulation of snow, in two charr populations on Svalbard, situated between 78 and 79° N. We hypothesize that both warm winters and warm summers will result in an improved growth rate in YOY Arctic charr, which may be due to thinner ice, earlier ice break-up, and a higher water temperature during the ice-free season. In years where high snowfall has occurred, however, ice break-up may be postponed, resulting in decreased growth of juvenile charr.

2. Materials and Methods

2.1. Study Sites

The Archipelago of Svalbard is situated between 74 to 81° N and 10 to 35° E, and consists of four large islands, Spitsbergen, Nordaustlandet, Barentsøya, and Edgeøya, in addition to many smaller islands (Figure 1). The climate is highly arctic, with mean July air temperatures normally being between 4 and 5 °C on western Spitsbergen, with temperatures being 2–3 °C lower on Nordaustlandet, the northernmost island [23]. The difference in temperatures between the western and northern coasts is mainly due to the influence of the Gulf Stream on the western coast of Spitsbergen, while Nordaustlandet and the eastern part of Spitsbergen are more influenced by cold, easterly air masses [24]. Annual precipitation on Svalbard is low, with 300–400 mm mostly falling as snow [23], but there are large local variations.

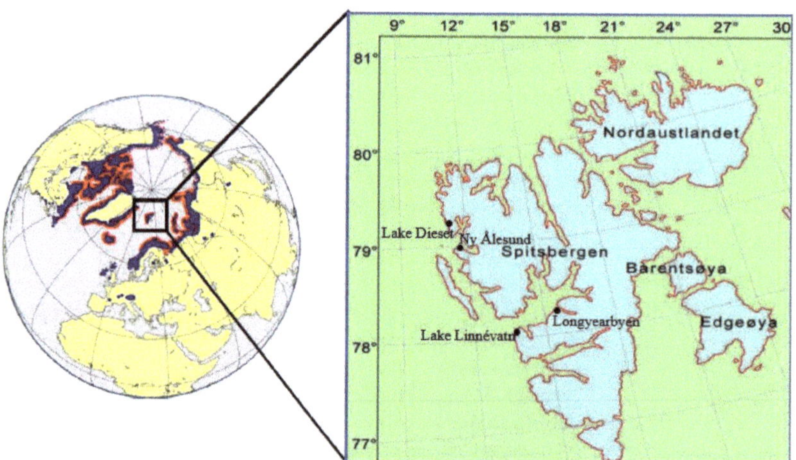

Figure 1. Map showing the circumpolar distribution area of resident (blue) and anadromous (red) Arctic charr (**left**), and the localities of Lake Dieset, Lake Linnévatn, and the meteorological stations in Longyearbyen and Ny-Ålesund (**right**).

Arctic charr is the only freshwater fish on Svalbard, with populations in approximately 200 lakes; of these, probably less than 20 lake systems contain stable anadromous populations.

Sampling has been carried out in Lake Linnévatn in the outermost part of Isfjorden, and Lake Dieset on the Mitra Peninsula, both situated on the western side of Spitsbergen (Figure 1). The lake areas are in the range of 4–4.5 km^2, and both support anadromous Arctic charr in addition to resident charr [2,4]. Lake Linnévatn (78° N, 13° E) is the southernmost of the studied lakes (Figure 1). Water temperatures are low even during summer, due to the inflow of glacier meltwater as well as being surrounded by high mountains that reduce the solar radiation of the lake [25]. Owing to glacial silting, light transmission, as characterized by the Secchi disk transparency values during summer at 0.3 m, is relatively low [4,25]. The lake is ice-covered for approximately 9–10 months of the year, but during the ice-free season, the 2.5 km outlet river to the Isfjorden Bay has a relatively stable discharge [4]. In the summers of 2008 and 2017, a fish trap was mounted in the outlet river, and approximately 2500 anadromous Arctic charr ascended the watercourse in both seasons (Svenning unpublished).

Dieset (79° N, 11° E) comprises a system of two lakes situated approximately 140 km to the north of Lake Linnévatn (Figure 1). The lakes are ice-covered for approximately 10 months of the year, but with large annual variations [2]. The discharge in the outlet river may vary considerably, and in some years Arctic charr were prevented from ascending upriver to the spawning areas due to low water flow [2]. Monitoring upstream migrations using fish traps in the 1970s [2] and in the early 1990s [26] showed that 500 to 900 individual Arctic charr ascended the water course annually. In the last 10 years, the anadromous part of the population has increased, and probably up to 2000 Arctic charr now ascend the lakes each year.

2.2. Meteorological Data

Both lakes Linnévatn and Diesetvatn are located near two meteorological stations (Figure 1). The meteorological station at Isfjord Radio was situated 4 km to the west of Lake Linnévatn, while the meteorological station at Ny-Ålesund in the Kongsfjorden Bay is situated 35 km southeast of Lake Dieset. Monthly means of air temperature and precipitation are available at Isfjord Radio from 1935 to 1976, except for the period 1941–1947, due to the evacuation of Svalbard during World War II [21]. Missing data from this latter period have been estimated using observations at Jan Mayen and the Russian station Bukta Tikhaya, on Franz Josef Land [21]. After 1976, air temperature data for Lake Linnévatn have been estimated from observations made at the Svalbard Airport and Longyearbyen stations, using regression analyses of the overlapping data with Isfjord Radio during the period 1957–1976. Monthly mean values of air temperature and precipitation were available for Ny-Ålesund from 1969 to the present, with data for the period before 1969 estimated from the Isfjord Radio records [21].

Until recently, there have been no consistent and regular measurements of snow depths made in Svalbard, in general, and only sporadic measurements of snow depth were made at the meteorological sites in Ny-Ålesund and Isfjord Radio. Long-term annual snow depth measurements are available, however, from the winter glacier mass balance records of two glaciers, Midtre Lovénbreen and Austre Brøggerbreen, in the vicinity of Ny-Ålesund [27]. Winter balance is obtained by snow-depth soundings over the glacier at the end of the winter accumulation period. The mass balance data used here cover the period 1967–2008 and comprise averages of snow thickness measurements made on the two glaciers in late April to early May, at the lowermost glacier elevations (50–250 m). While restricted spatially to these two glaciers, the data provide good proxies for spring snowfall in western Svalbard. This has been demonstrated by parallel measurements made over shorter periods from 2004 to 2008 on the Ny-Ålesund glaciers and Linnébreen near Lake Linnévatn (Jack Kohler, unpublished data).

2.3. Fish Data and Otolith Analyses

A total of 804 sagittal otoliths of Arctic charr from Lake Dieset and 289 from Lake Linnévatn have been used in this study. Arctic charr from Lake Dieset were sampled between 1970 and 2008 using a variety of equipment, including gillnetting in the lake (n = 360), a trap catching ascending fish from seawater (n = 257), electrofishing in the outlet river and in the lake (n = 182), and rod fishing in the lake (n = 5). In Lake Linnévatn, Arctic charr were sampled periodically from 1968 to 2006, and most fish were sampled by gillnetting in the lake (n = 174), while 115 were sampled by electrofishing in the lake and the outlet river. The material covers year classes of charr from 1948 to 2006 in Lake Linnévatn and from 1951 to 2008 in Lake Dieset (Figure 2). During the three periods with consistent data from YOY in both lakes (1959–1963, 1987–1991, and 1998–2006), we found no significant differences in the length of YOY between the two populations (in either period).

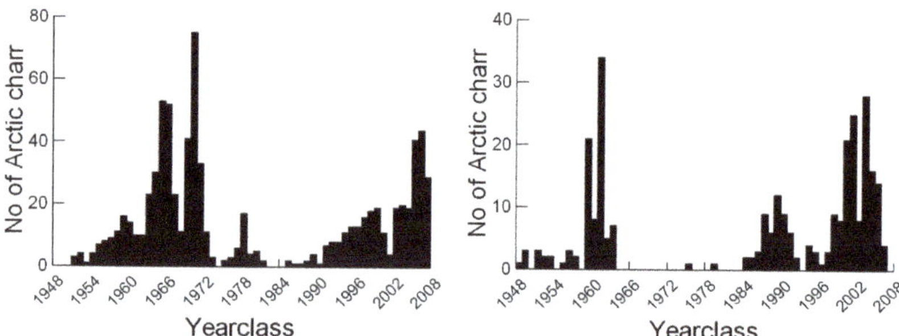

Figure 2. The number of aged Arctic charr in different year-classes in 1951–2008 from Lake Dieset (**left**; n = 804) and 1948–2006 from Lake Linnévatn (**right**; n = 289). Samplings of Arctic charr used for back-calculating the body length in young of the year (YOY) were conducted during the periods 1970–2008 (Lake Dieset) and 1968–2006 (Lake Linnévatn).

Arctic charr otoliths from Svalbard have a distinctive zonal differentiation between summer and winter increments, even among old individuals. They are much easier to age compared to otoliths sampled from Arctic charr populations further south, which is likely due to their slow annual growth rates, giving a clear distinction between summer and winter zones (Figure 3). Thus, age determination was carried out without any additional treatment or preparation of the otoliths, except placing them in glycerol and then viewing them under a binocular microscope.

Figure 3. The two sagittal otoliths from a 10-year-old (winter) Arctic charr captured in Lake Linnévatn in September 1998.

Otolith increment width measurements were performed as described in [28]. The annuli were clearly visible along the rostrum, and in old fish rostrum counts have been shown to give the highest age estimates in Arctic charr [29]. Consequently, rostral radii were used for both age determination and otolith measurements; hence, relationships between fish fork length and otolith radius were evaluated.

Otoliths were photographed using a digital camera (DS-5M; Nikon Instruments Europe B.B., Kingston, Surrey, England) connected to a Leica Wild MZ8 stereo microscope and computer. Images were captured using NIS-Element software (Nikon Instruments Europe B.V., Amstelveen, The Netherlands) (https://www.microscope.healthcare.nikon.com/en_EU/products/software/nis-elements, accessed on 1 January 2023), while the width of the otolith increments and total otolith length was measured using the image analysis program ImageJ (Version 1.41o; US National Institutes of Health, Bethesda, MD, USA). The center of each otolith was identified as precisely as possible by sight. The larval zone of the otolith is being formed when the fish larva is still in the egg and is defined as the innermost, darker part of the 0^+ otolith when observed with reflected light. Difficulties associated with differentiating between the end of the larval increment and the end of the first summer increment necessitated that the distance from the otolith center to the end of the first opaque (summer) increment was measured as a relative measure of the fish's first summer growth. Identified otolith growth increments of the first summer were assigned to calendar years based on the estimated age of the fish at catch.

For fish younger than 5 years of age and showing no records of anadromy, strong positive and linear correlation was observed between fish length and otolith size from both Lake Dieset (r^2 = 0.88; n = 175) and Lake Linnévatn (r^2 = 0.90; n = 73) (Figure 4). Thus, otolith increment width is a suitable predictor of somatic growth of the fish. See details in [28].

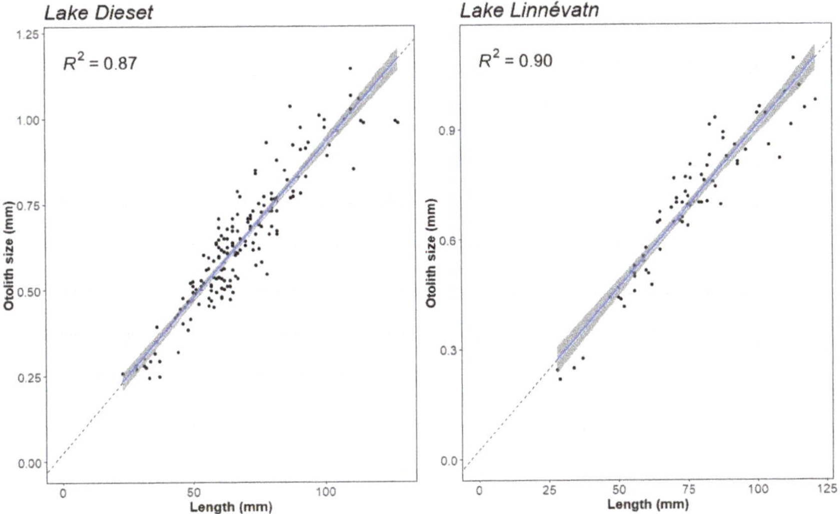

Figure 4. The relationship between fish length and otolith size for Arctic charr younger than 5 years of age and less than 130 mm, captured in Lake Dieset (**left**; n = 175; Y = 0.028 + 0.0090 X; *p* < 0.001) and Lake Linnévatn (**right**; n = 73; Y = 0.025 + 0.0088 X; *p* < 0.001).

2.4. Statistics

We used additive models in the R-library mgcv [30] to investigate temporal trends, effects of climatic variables, and the linearity of these effects (see [31,32] for examples). Additive models use flexible nonlinear functions to describe the relationships between response and predictor variables; here, we used the default thin-plate regression splines of the gam() function to model these nonlinear relationships [30]. The effective degrees of freedom (edf) for the model, measure the degree of smoothness, with edf = 1 representing a linear relationship and an edf value above 1 representing increased nonlinearity. Effective degrees of freedom are estimated using general cross-validation. However, as we were interested in the main patterns, we restricted the dimension of the spline basis to 4 so that edf would not be too large. To describe temporal trends in the first summer growth, we used the year of hatching as a predictor variable. For climatic effects, we considered mass balance (as a measure of spring snow accumulation), average winter (December to February), and summer (June to August) air temperatures as the predictor variables. All three variables were included as simple additive terms (i.e., without interaction terms such as products), and we fitted all models with one, two, and three smooth terms. We used adjusted r^2 as a measure of the predictive power of each model, and together with a measure based on cross-validation (general cross-validation score), we selected the best model for predicting growth [30]. For the models with climatic variables as predictors, the figures show the partial residuals and the corresponding smoothed effect, that is, when other variables in the model are adjusted for.

3. Results

During the period when the individual Arctic charr were recruited to lakes Dieset and Linnévatn, the average winter (January–March) air temperatures (1950–2008) varied from -21.6 to -5.4 °C and from -18.6 to -3.5 °C at Isfjord Radio (Lake Linnévatn). In contrast, the average summer (June–August) air temperatures varied from 1.7 to 5.4 and 2.0 to 5.7 °C at Ny-Ålesund (Lake Dieset) and Isfjord Radio (Lake Linnévatn), respectively (Figure 5).

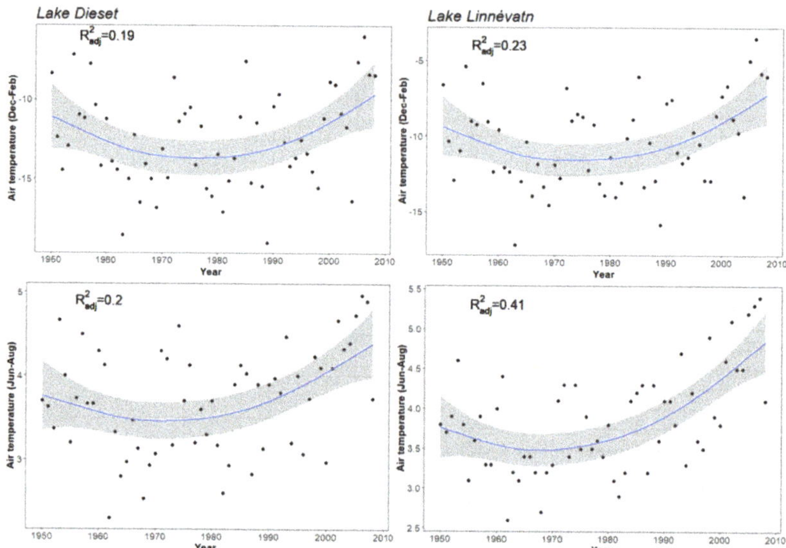

Figure 5. Estimated average winter (above, December–February) and summer (lower, June–August) air temperatures (°C) at Ny-Ålesund (**left**; Lake Dieset area) and Isfjord Radio (**right**; Lake Linnévatn area) in the period 1950 to 2008. The lines are smoothed by GAM.

Both winter and summer air temperatures significantly co-varied between the two meteorological locations (r = 0.83 and 0.90, respectively; df = 59; $p < 0.001$). Both weather stations also recorded a decreasing air temperature trend (both winter and summer) from 1950 to 1978 and an increasing temperature from 1979 to 2008 (Figure 5).

The temporal trend in the back-calculated length based on measured otolith increments in YOY Arctic charr was positively associated with the shift in temperature trends related to year class (Figure 6; Dieset: edf = 2.95. $p < 0.001$; Linnévatn: edf = 2.93, $p < 0.001$). Further, the estimated growth of YOY Arctic charr from Lake Dieset and Lake Linnévatn were both positively associated with increasing summer and winter air temperatures, and negatively correlated with the accumulated snow depth (Figure 7, Table 1). For Lake Dieset, the additive model showed an association with summer T (edf = 1.93, F = 23.45, $p < 0.0001$), winter T (edf = 2.74, F = 6.15, $p = 0.0004$), and snow depth (edf = 2.89, F = 8.12, $p < 0.0001$). For Lake Linnévatn, the additive model showed an association with summer T (edf = 2.92, F = 13.29, $p < 0.0001$), winter T (edf = 1.28, F = 4.46, $p = 0.018$), and snow depth (edf = 1, F = 5.94, $p = 0.016$). Based on the generalized cross-validation score, the inclusion of all three predictors (summer and winter air temperatures, and snow accumulation) explained more than 30% and 25% of the variation in YOY growth in the two lake systems of Lake Dieset and Lake Linnévatn, respectively (Table 1).

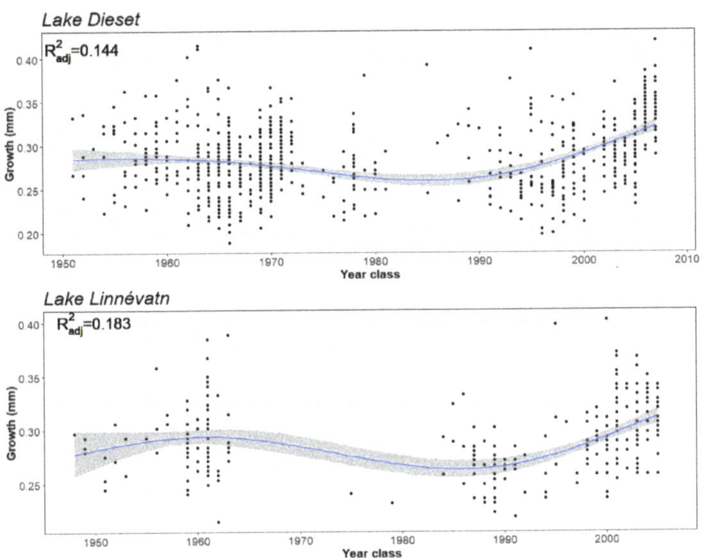

Figure 6. Estimated trends in fish growth in the young-of-the-year (YOY) Arctic charr from Lake Dieset (**above**) and Lake Linnévatn (**lower**), Svalbard, based on back-calculation from sagittal otoliths. The smooth lines are based on an additive model with year class as a predictor variable, and a spline basis of dimension of 4 to reduce overfitting.

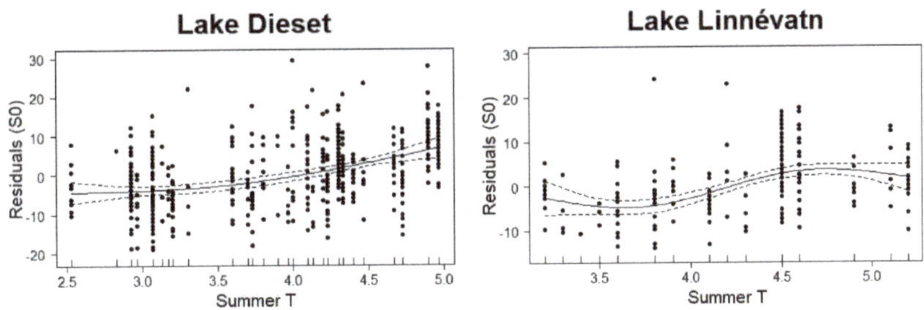

Figure 7. *Cont.*

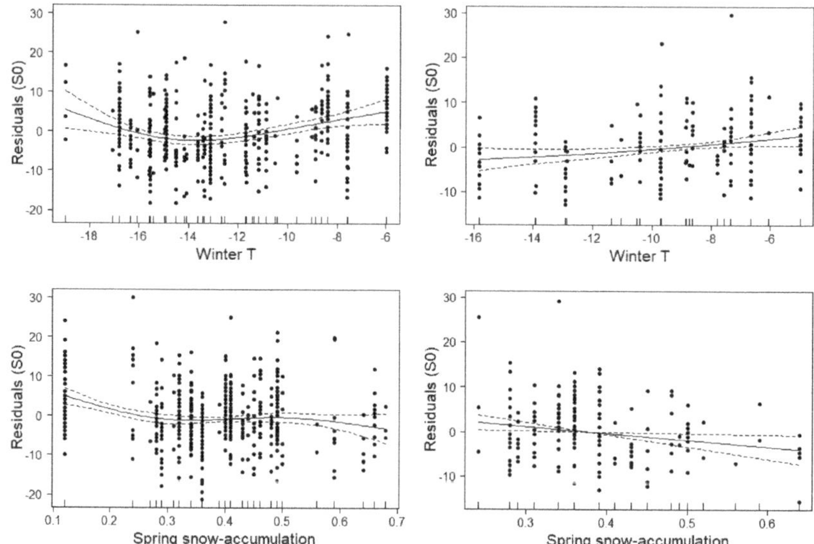

Figure 7. Residual plot for the first summer growth (S0) in the young-of-the-year (YOY) Arctic charr from Lake Dieset (**left**) and Lake Linnévatn (**right**), Svalbard, based on back-calculation from otoliths, and in relation to summer air temperature (above), winter air temperature (middle) and snow accumulations (lower).

Table 1. Influence on YOY growth explained by the different parameters (in %); summer (June–August) temperature (summer T), winter (December–February) temperature (winter T), and snow = mass balance. Percentage of variation explained (adjusted R^2) with generalized cross-validation score (in parenthesis) for all models used to predict the first summer growth. The best models (in bold, based on GCV score) for both lakes included all three predictors. Note that the sample sizes given as [N=] differ as mass balance was not available for all years, and that R^2 and GCV are not strictly comparable.

Lake	Lake Dieset	Lake Linnévatn
Year of Birth	14.4 (67.1) [N = 772]	18.3 (47.6) [N =285]
Summer T	19.6 (63.0) [N = 772]	10.9 (51.8) [N =285]
Winter T	16.0 (65.8) [N = 772]	1.9 (57.0) [N = 285]
Snow	0.00 (77.4) [N = 504]	3.9 (59.0) [N = 190]
Summer T + Winter T	20.1 (62.7) [N = 772]	10.6 (52.2) [N = 285]
Summer T + Snow	30.3 (54.6) [N = 504]	22.5 (48.4) [N = 190]
Winter T + Snow	23.9 (59.4) [N = 504]	11.6 (55.1) [N = 190]
Summer T + Winter T + Snow	31.6 (53.9) [N = 504]	24.9 (47.1) [N = 190]

4. Discussion

The estimated first summer growth of young-of-the-year (YOY) Arctic charr in the two lakes Linnévatn and Dieset, Svalbard, was found to be positively correlated with the summer air temperature. To some extent, YOY growth was also positively correlated with winter air temperature. This was most probably due to thinner ice formed in mild winters, and thereby an earlier ice break-up. In contrast, we found a negative relationship between the accumulated snow depth and growth in YOY Arctic charr.

Ref. [33] found that otolith-derived water temperatures estimated from YOY Arctic charr sampled in Lake Dieset were consistent with temperatures found in the shallowest part of the littoral areas in Svalbard lakes during summer. Further, they also found that the otolith-derived temperatures differed significantly from the monitored water temperatures recorded at the outlet river. Ref. [34] captured all age classes of juveniles in Lake Dieset as well as in the outlet river, with the exception of YOY Arctic charr that were never found in the river. This corresponds well with our sampling in both Lake Dieset and Lake Linnévatn (this study), i.e., that all YOY Arctic charr were caught close to the lake shore and at depths of less than 20 cm. These shallow areas of the littoral zone are highly influenced by solar radiation during the summer, thus explaining our findings demonstrating that summer growth of YOY Arctic charr in Like Dieset and Lake Linnévatn was positively correlated with the summer air temperature.

Otoliths continue to form annular zones even when the growth in body length has ceased [35,36], and back-calculation of length of such fish based on annuli widths is not possible. Likewise, in periods with rapid somatic growth, for instance, in Arctic charr during the sea residence in summer [28] or when Arctic charr shift to cannibalistic behavior (Svenning, unpublished), this is not manifested in a similar marked increase in otoliths (see details in [28]). As a consequence, otoliths are highly reliable with regard to age determination. However, the uncoupling of body and otolith growth reported for many species (see, for instance [37,38], causes the general back-calculation of body length of older fish based on annuli widths in otoliths to be problematic. The sampled otolith material from lakes Dieset and Linnévatn, covering a period of more than 50 years, has made it possible to relate the annual growth of Arctic charr to the fluctuation in climatic conditions on Svalbard during this long period. We have used the size of YOY in this study, because this age class seems to stay in the shallowest littoral area. This is contrary to the other juvenile age classes and adult fish, who use both the deeper part of the littoral, profundal, and river habitats with variable temperature regimes [33]. On the other hand, it is likely that all size classes respond similarly to variations in temperature [12,16,39], and the positive growth effect of YOY Arctic charr due to increased temperature is suggested to be representative for all age groups.

The lake water temperature during the ice-free season is mainly influenced by the timing of ice break-up and the following summer air temperature. Earlier studies have indicated that ice break-up may be explained by air temperature alone [40,41]. Meanwhile, more recent studies, such as [42], found that regional variation in ice-off dates in northwestern Canada were also driven by relationships between lake size, snow thickness, and ice thickness. The annual mean air temperature in Svalbard has increased by 3–5 °C over the last four to five decades [24]. This should have led to earlier ice break-ups in Svalbard lakes. However, ref. [43] found that although the ice-free season in Lake Linnévatn has increased by 1.5 days per year in the last ten years, the time of ice break-up has been relatively constant. Further, water temperature has increased by an average of 0.06 °C per year, and stayed warm for a longer period each autumn, thus indicating a further positive effect on Arctic charr annual growth.

Mean winter and spring air temperatures have also shown an increasing trend in mainland Norway since the 1990s [44], with a significant and accelerating trend of earlier ice break-up and delayed freeze-up dates after 1991 [45]. A corresponding development has also been described for lowland and subalpine lakes elsewhere in Europe, North America, and in most of the northern hemisphere during the last decades, with a reduction in lowland areas covered by snow, earlier ice break-ups, and increasing surface water temperatures [46–48]. In the subarctic lake Takvatn, Norway, a positive effect of water temperature and a negative effect of fish abundance on somatic growth were found in individual juvenile Arctic charr [9]. This suggests that, as temperatures rise, the somatic growth of Arctic charr will increase in high-latitude lakes. Our findings suggest a similar development for YOY Arctic charr in the two Svalbard lakes studied.

Ref. [24] showed that in addition to air temperature, precipitation (both rain and snow) in Svalbard has increased in the last decades. Increasing amounts of snow on top of the ice in Svalbard lakes may compensate for increasing air temperature. Thus, this may explain the timing of ice break-up showing no change over the last 10–15 years for Lake Linnévatn [43], while dates of ice-cover have occurred later. The positive correlation between air temperature, and ice break-up and water temperature found in Lake Linnévatn [43], plus a longer period with elevated temperatures in warmer years, support the findings from our study, i.e., the clear trend of a positive correlation between air temperature and YOY growth in Arctic charr, both in Lake Linnévatn and Lake Dieset. Further, the more or less constant annual timing of ice cover found for Lake Linnévatn [43] supports our second finding, i.e., a negative relationship between accumulated snow depth and growth in YOY Arctic charr.

Ice thickness in High Arctic lakes typically reaches its maximum in May [49]. Although winter air temperatures in Svalbard show an increasing trend, the December–February temperature is still well below zero. However, higher air temperatures may nevertheless affect ice thickness on the lakes, i.e., ice becomes thinner, resulting in an earlier ice break-up in summer, and a longer open-water season. In Lake Dieset, where the drainage area is 57 km^2, of which more than 30% is covered by glaciers, the authors in [2], found a positive correlation between the water level in the outlet river and the air temperature. Thus, increasing air temperature obviously increases the melting of the glaciers, and depending on the summer air temperature, may have a cooling effect on the water in the lake. With future glacier loss due to melting, lakes fed by cool meltwater will more or less disappear, and lake water temperatures may become considerably higher than today. Likewise, the water flow in outlet rivers will be reduced; thereby, the access for migratory fish to feed in seawater during summer, or to swim upriver from the sea may be hindered, i.e., leading to a risk for collapse of anadromous Arctic charr populations in Svalbard and other High Arctic Lake systems.

Temperature has pervasive effects on recruitment, year-class strength, somatic growth, and other life history and behavioral traits in salmonid fishes, especially through metabolic rates and food consumption [50]. When YOY and older juveniles of anadromous Arctic charr obtain a better annual growth in length, both the survival and length of the freshwater fish remain constant until smoltification and migration to salt water may be influenced. A shorter time period in fresh water before smoltification may increase the number of smolts due to reduced predation, thereby resulting in a higher density of migratory fish. An increase in the length of YOY resident Arctic charr may also result in a higher annual survival, causing larger year classes and resulting in an increased population density. This, in turn, would lead to increased predation pressure on food resources, a lower food intake by individual fish, and stagnation in growth at a smaller size [51]. A corresponding development has been observed for brown trout on the Hardangervidda mountain plateau (>1100 m a.s.l.), i.e., low accumulation of snow in spring, early ice-free lakes and streams, and higher summer temperatures all contribute to larger and more numerous YOY [16].

As in most of the Arctic, the anadromous populations of Arctic charr in Svalbard are far more attractive for commercial fishing than resident populations, mainly due to the size and higher quality of the migratory fish. As a consequence, some of the Svalbard populations were previously heavily exploited, resulting in a strong decrease in large and spawning migratory fish. As a result of this, the Governor of Svalbard introduced a sanctuary for some of the anadromous populations in 1993. This lasted until 1997, when gillnetting was re-opened in four lakes. This was followed by a strict harvest quota for the same populations after 2008. The anadromous Arctic charr population in Lake Linnévatn has returned to a higher number than 30 years ago, but this is far less than before the 1950s.

Global warming is widespread, but it is occurring much faster in higher northern latitudes [6]. Sea surface temperatures in the Barents Sea as well as on the western coast of Svalbard have increased significantly during the last decades [52]. Thus, climate change now facilitates the northward movement of many fishes previously constrained by low temperatures from dispersal to High Arctic environments. Moreover, during the summers of 2001 and 2006, the first recorded three-spined sticklebacks, *Gasterosteus aculeatus*, were captured in Lake Linnévatn and Lake Straumsjøen, in Isfjorden, Svalbard [53]. These mature sticklebacks, observed in two Svalbard lake systems, are the only evidence of a vagrant occurrence of the species and not a verification of its establishment. Although an established population of sticklebacks may lead to food competition with juvenile Arctic charr and/or a new vital prey for piscivorous Arctic charr, it is difficult to predict whether a potential establishment of stickleback would have negative or positive consequences for the production, life history, and harvest potential of Svalbard Arctic charr.

Even if climate warming leads to an increased somatic growth of Arctic charr in High Arctic lakes, it is still uncertain how this will influence the life history and size-structured interactions in Arctic charr populations [9]. In lakes with migratory Arctic charr in Svalbard, improved annual somatic growth could potentially lead to larger year classes and an increasing fraction of anadromous versus resident fish This implies a potential for larger catch quotas and higher catches. However, in the longer term, warmer climates are projected to eventually lead to complete loss of many Svalbard glaciers [54,55]. With the loss of glaciers due to melting in lake catchments, rivers could potentially dry out in late summer, thus hindering migration between freshwater and saltwater environments for anadromous Arctic charr (see [28]). In the worst-case scenario, the highly valuable and attractive anadromous Arctic charr populations could eventually disappear from Svalbard lake systems.

5. Conclusions

The Arctic charr species complex dominates the fish communities in High Arctic rivers and lakes. Besides being an essential component of the diversity of many High Arctic fish communities, Arctic charr also comprise a significant economic resource as food for northern people. Despite being adapted to extreme environmental conditions, the distribution of Arctic charr is limited in the far north by abiotic barriers. This paper exemplifies the possible impact of changes in climate on the growth of juvenile Arctic charr in two populations in Svalbard. We further conclude that, in the longer term, a warmer climate may lead to the complete loss of most glaciers along the western coast of Svalbard, i.e., outlet rivers may dry out in summer/autumn, thus hindering migration between saltwater and freshwater environments for migratory fish. Thus, in the worst-case scenario the highly valuable and attractive anadromous Arctic charr populations could eventually disappear from Svalbard lake systems.

Author Contributions: Conceptualization, M.A.S.; Methodology, E.T.B., M.A.S., N.G.Y. and J.A.G.; Software, N.G.Y.; Validation, M.A.S. and R.B., Formal Analysis, N.G.Y.; Investigation, M.A.S., J.A.G. and E.T.B.; Resources; Data Curation, M.A.S., E.T.B. and J.A.G.; Writing—Original Draft Preparation, M.A.S., E.T.B. and J.A.G.; Writing—Review and Editing, M.A.S., R.B., J.H., J.K. and N.G.Y.; Visualization, M.A.S., N.G.Y. and R.B.; Supervision, M.A.S.; Project Administration, M.A.S. and J.A.G.; Funding Acquisition, M.A.S. All authors have read and agreed to the published version of the manuscript.

Funding: This research received no external funding.

Institutional Review Board Statement: In accordance with Norwegian and European legislation related to animal research, formal approval of the experimental protocol by the Norwegian Animal Research Authority (NARA) was not required, because the regulation only applies to procedures on live animals. All samples (otoliths) from this study were taken from animals that were already dead.

Data Availability Statement: The data that support the findings of this study are available from the corresponding author upon reasonable request.

Acknowledgments: This study is based on the sampling of Arctic charr in two Svalbard lake systems over the period of 1960–2008. We would like to dedicate this work to Nils Gullestad, who started collecting Svalbard Arctic charr otoliths with great enthusiasm in the early 1960s, continuing for almost two decades, but who sadly passed away last autumn before the project was completed. We would also like to thank Sigmund Spjelkavik, Thomas Olsen, and the staff at Isfjord Radio for collecting Arctic charr from Lake Linnévatn, and to Hanne Hegseth who sampled Arctic charr in both Lake Linnévatn and Lake Dieset. We also thank Eirik Førland, Knut Langeland, Bård Jørgen Bårdsen, Brian Dempson, and Mike Power for their valuable input during the project period.

Conflicts of Interest: Author Eigil T. Bjørvik was employed by the company CFEED. The remaining authors declare that the research was conducted in the absence of any commercial or financial relationships that could be construed as a potential conflict of interest.

References

1. Johnson, L. *The Arctic Charr, Salvelinus alpinus*; Balon, E.K., Ed.; W. Junk Publishers: Hague, The Netherlands, 1980.
2. Svenning, M.-A.; Gullestad, N. Adaptations to Stochastic Environmental Variations: The Effects of Seasonal Temperatures on the Migratory Window of Svalbard Arctic Charr. *Environ. Biol. Fish* **2002**, *64*, 165–174. [CrossRef]
3. Hammar, J. Freshwater Ecosystems of Polar Regions: Vulnerable Resources. *Ambio* **1989**, *18*, 6–22.
4. Svenning, M.-A.; Klemetsen, A.; Olsen, T. Habitat and Food Choice of Arctic Charr in Linnévatn on Spitsbergen, Svalbard: The First Year-Round Investigation in a High Arctic Lake. *Ecol. Freshw. Fish* **2007**, *16*, 70–77. [CrossRef]
5. Christoffersen, K.S.; Amsinck, S.L.; Landkildehus, F.; Lauridsen, T.L.; Jeppesen, E. Lake Flora and Fauna in Relation to Ice-Melt, Water Temperature and Chemistry at Zackenberg. *Adv. Ecol. Res.* **2008**, *40*, 371–389. [CrossRef]
6. Førland, E.J.; Benestad, R.E.; Hanssen-Bauer, I.; Haugen, J.E.; Skaugen, T.E. Temperature and Precipitation Development at Svalbard 1900–2100. *Adv. Meteorol.* **2011**, *2011*, 893790. [CrossRef]
7. Hammar, J. Natural Selection at the Edge of Life: Allelic Polymorphism and Recruitment in High Latitude Arctic Char (*Salvelinus alpinus*) Generated and Maintained by Environmental Extremes. *Diversity* **2023**, *15*, 74. [CrossRef]
8. Jobling, M. The Influence of Environmental Manipulations on Inter- and Intra-Individual Variation in Food Acquisition and Growth Performance of Arctic Charr, *Salvelinus alpinus*. *J. Fish. Biol.* **1994**, *44*, 1069–1087. [CrossRef]
9. Kotowych, N.; Smalås, A.; Amundsen, P.A.; Primicerio, R. Climate Warming Accelerates Somatic Growth of an Arctic Fish Species in High-Latitude Lakes. *Sci. Rep.* **2023**, *13*, 16749. [CrossRef]
10. Huang, M.; Ding, L.; Wang, J.; Ding, C.; Tao, J. The Impacts of Climate Change on Fish Growth: A Summary of Conducted Studies and Current Knowledge. *Ecol. Indic.* **2021**, *121*, 106976. [CrossRef]
11. Jobling, M. Influence of Body Weight and Temperature on Growth Rates of Arctic Charr, *Salvelinus alpinus* (L.). *J. Fish Biol.* **1983**, *22*, 471–475. [CrossRef]
12. Jensen, K.W. On the Dynamics and Exploitation of the Population of Brown Trout, Salmo Trutta, in Lake Øvre Heimdalsvatn, Southern Norway. *Rep. Inst. Freshw. Res. Drottningholm* **1977**, *56*, 18–69.
13. Kristensen, D.M.; Jørgensen, T.R.; Larsen, R.K.; Forchhammer, M.C.; Christoffersen, K.S. Inter-Annual Growth of Arctic Charr (*Salvelinus alpinus* L.) in Relation to Climate Variation. *BMC Ecol.* **2006**, *6*, 10. [CrossRef] [PubMed]
14. Isaksen, K.; Nordli, Ø.; Ivanov, B.; Køltzow, M.A.; Aaboe, S.; Gjelten, H.M.; Mezghani, A.; Eastwood, S.; Førland, E.; Benestad, R.E.; et al. Exceptional Warming over the Barents Area. *Sci. Rep.* **2022**, *12*, 9371. [CrossRef] [PubMed]
15. Rantanen, M.; Karpechko, A.Y.; Lipponen, A.; Nordling, K.; Hyvärinen, O.; Ruosteenoja, K.; Laaksonen, A. The Arctic Has Warmed Nearly Four Times Faster than the Globe since 1979. *Commun. Earth Environ.* **2022**, *3*, 168. [CrossRef]
16. Borgstrøm, R. Annual Growth of Brown Trout in Alpine Lakes Is Highly Influenced by Spring Snow Depth and Ice-out Day. *Fauna Nor.* **2023**, *42*, 37–46. [CrossRef]
17. Fang, X.; Stefan, H.G. Potential Climate Warming Effects on Ice Covers of Small Lakes in the Contiguous U.S. *Cold Reg. Sci. Technol.* **1998**, *27*, 119–140. [CrossRef]
18. Duguay, C.R.; Prowse, T.D.; Bonsal, B.R.; Brown, R.D.; Lacroix, M.P.; Ménard, P. Recent Trends in Canadian Lake Ice Cover. *Hydrol. Process* **2006**, *20*, 781–801. [CrossRef]
19. Stefan, H.G.; Fang, X.; Hondzo, M. Simulated Climate Change Effects on Year-Round Water Temperatures in Temperate Zone Lakes. *Clim. Change* **1998**, *40*, 547–576. [CrossRef]
20. Wrona, F.J.; Prowse, T.D.; Reist, J.; Hobbie, J.E.; Levesque, L.M.; Vincent, W.F. Climate Change Effects on Aquatic Biota, Ecosystem Structure and Function. *Ambio* **2006**, *35*, 359–369. [CrossRef]
21. Nordli, Ø.; Hanssen-Bauer, I.; Førland, E.J. Homogeneity Analyses of Temperature and Precipitation Series from Svalbard and Jan Mayen. *Klima* **1996**, *16*, 41.
22. Nordli, Ø.; Przybylak, R.; Ogilvie, A.E.J.; Isaksen, K. Long-Term Temperature Trends and Variability on Spitsbergen: The Extended Svalbard Airport Temperature Series, 1898–2012. *Polar Res* **2014**, *33*, 21349. [CrossRef]
23. Hisdal, V.S. Svalbard Nature History. In *Polarhåndbok nr 11*; Norsk Polarinstitutt: Tromsø, Norway, 1998.
24. Hanssen-Bauer, I. *Climate in Svalbard 2100—A Knowledge Base for Climate Adaptation*; NCCS Report No. 1/2019; Norwegian Centre for Climate Services (NCCS): Oslo, Norway, 2019; p. 207.
25. Bøyum, A.; Kjensmo, J. Physiography of Lake Linnévatn Western Spitsbergen. *Int. Ver. Für Theor. Angew. Limnol. Verhandlungen* **1978**, *20*, 609–614. [CrossRef]
26. Gulseth, O.A. Seawater Tolerance, Migratory Behavior and Growth of Charr, *Salvelinus alpinus*, with Emphasis on the High Arctic Dieset Charr on Spitsbergen, Svalbard. Ph.D. Thesis, Department of Zoology, Norwegian University of Science and Technology (NTNU), Trondheim, Norway, 2000.
27. Anon Norwegian Polar Institute. *Austre Brøggerbreen and Midtre Lovenbreen Mass Balance*; Environmental Monitoring of Svalbard and Jan Mayen (MOSJ): Tromsø, Norway, 2023.
28. Svenning, M.-A.; Smith-Nilsen, A.M.; Jobling, M. Sea Water Migration of Arctic Charr (*Salvelinus alpinus*). Correlation between Freshwater Growth and Seaward Migration, Based on Back-Calculation from Otoliths. *Nord. J. Freshw. Res.* **1992**, *67*, 18–26.

29. Kristoffersen, K.; Klemetsen, A. Age Determination of Arctic Charr (*Salvelinus alpinus*) from Surface and Cross Section of Otoliths Related to Otolith Growth. *Nord. J. Freshw. Res.* **1991**, *66*, 98–107.

30. Wood, S.N. *Generalized Additive Models: An Introduction with R*, 2nd ed.; Chapman Hall/CRC Press: Boca Raton, FL, USA, 2017. [CrossRef]

31. Bustnes, J.O.; Yoccoz, N.G.; Bangjord, G.; Polder, A.; Skaare, J.U. Temporal Trends (1986–2004) of Organochlorines and Brominated Flame Retardants in Tawny Owl Eggs from Northern Europe. *Environ. Sci. Technol.* **2007**, *41*, 8491–8497. [CrossRef]

32. Mysterud, A.; Stenseth, N.C.; Yoccoz, N.G.; Langvatn, R.; Steinheim, G. Nonlinear Effects of Large-Scale Climatic Variability on Wild and Domestic Herbivores. *Nature* **2001**, *410*, 1096–1099. [CrossRef]

33. Godiksen, J.A.; Svenning, M.A.; Sinnatamby, R.N.; Dempson, J.B.; Borgstrøm, R.; Power, M. Stable Isotope-Based Determinations of the Average Temperatures Experienced by Young-of-the-Year Svalbard Arctic Charr (*Salvelinus alpinus* (L.)). *Polar Biol.* **2011**, *34*, 591–596. [CrossRef]

34. Gulseth, O.A.; Nilssen, K.J. Growth Benefit from Habitat Change by Juvenile High-Arctic Char. *Trans. Am. Fish. Soc.* **1999**, *128*, 593–602. [CrossRef]

35. Nordeng, H. On the Biology of Char (Salmo Alpinus L.) in Salangen, North Norway. 1. Age and Spawning Frequency Determined from Scales and Otoliths. *Nytt Mag. Zool.* **1961**, *10*, 67–123.

36. Power, G. Fish Population Structure in Arctic Lakes. *J. Fish. Res. Bd. Can.* **1978**, *35*, 53–59. [CrossRef]

37. Reay, P.J. The Seasonal Pattern of otolith Growth and Its Application to Back-Calculation Studies in *Ammodytes Tobianus* L. *J. Cons. Int. Explor. Mer.* **1972**, *34*, 485–504. [CrossRef]

38. Casselman, J.M. Growth and Relative Size of Calcified Structure of Fish. *Trans. Amer. Fish. Soc.* **1990**, *119*, 673–688. [CrossRef]

39. Elliott, J.M. The Energetics of Feeding, Metabolism and Growth of Brown Trout (*Salmo trutta* L.) in Relation to Body Weight, Water Temperature and Ration Size. *J. Anim. Ecol.* **1976**, *45*, 923. [CrossRef]

40. Ruosteenoja, K. The Date of Break-up of Lake Ice as a Climate Index. *Geophysica* **1986**, *22*, 89–99.

41. Weyhenmeyer, G.A.; Meili, M.M.; Livingstone, D.M. Nonlinear Temperature Response of Lake Ice Breakup. *Geophys. Res. Lett.* **2004**, *31*, L07203. [CrossRef]

42. Higgins, S.N.; Desjardins, C.M.; Drouin, H.; Hrenchuk, L.E.; Sanden, J.J. The Role of Climate and Lake Size in Regulating the Ice Phenology of Boreal Lakes. *J. Geophys. Res. Biogeosci.* **2021**, *126*, e2020JG005898. [CrossRef]

43. Tuttle, S.E.; Roof, S.R.; Retelle, M.J.; Werner, A.; Gunn, G.E.; Bunting, E.L. Evaluation of satellite-derived estimates of lake ice cover timing on Linnévatnet, Kapp Linné, Svalbard using in-situ data. *Remote Sens.* **2022**, *14*, 1311. [CrossRef]

44. Rizzi, P.; Graziano, P.; Dallara, A. A Capacity Approach to Territorial Resilience: The Case of European Regions. *Ann. Reg. Sci.* **2018**, *60*, 285–328. [CrossRef]

45. L'Abée-Lund, J.H.; Vøllestad, A.; Brittain, J.E.; Kvambekk, S.; Solvang, T. Geographic Variation and Temporal Trends in Ice Phenology in Norwegian Lakes during the Period 1890–2020. *Cryosphere* **2021**, *15*, 2333–2356. [CrossRef]

46. Kainz, M.J.; Ptacnik, R.; Rasconi, S.; Hager, H.H. Irregular Changes in Lake Surface Water Temperature and Ice Cover in Subalpine Lake Lunz, Austria. *Inland. Waters* **2017**, *7*, 27–33. [CrossRef]

47. Lopez, L.S.; Hewitt, B.A.; Sharma, S. Reaching a Breaking Point: How Is Climate Change Influencing the Timing of Ice Breakup in Lakes across the Northern Hemisphere? *Limnol. Oceanogr.* **2019**, *64*, 2621–2631. [CrossRef]

48. Caldwell, P.V.; Kennen, J.G.; Hain, E.F.; Nelson, S.A.; Sun, G.; McNulty, S.G. *Hydrologic Modeling for Flow-Ecology Science in the Southeastern United States and Puerto Rico; e-Gen*; US Department of Agriculture Forest Service, Southern Research Station: Asheville, NC, USA, 2020; p. 77.

49. Rigler, F.H. Limnology in the High Arctic: A Case Study of Char Lake. *Verh. Internat. Verein. Limnol.* **1978**, *20*, 127–140. [CrossRef]

50. Jonsson, B. Thermal Effects on Ecological Traits of Salmonids. *Fishes* **2023**, *8*, 337. [CrossRef]

51. Amundsen, P.A.; Klemetsen, A. Diet, Gastric Evacuation Rates and Food Consumption in a Stunted Population of Arctic Charr, *Salvelinus alpinus* L., in Takvatn, Northern Norway. *J. Fish. Biol.* **1988**, *33*, 697–709. [CrossRef]

52. Berge, J.; Heggland, K.; Lønne, O.J.; Cottier, F.; Hop, H.; Gabrielsen, G.W.; Nøttestad, L.; Misund, O.A. First Records of Atlantic Mackerel (*Scomber scombrus*) from the Svalbard Archipelago, Norway, with Possible Explanations for the Extension of Its Distribution. *Arctic* **2015**, *68*, 54–61. [CrossRef]

53. Svenning, M.-A.; Aas, M.; Borgstrøm, R. First Records of Three-Spined Stickleback Gasterosteus Aculeatus in Svalbard Freshwaters: An Effect of Climate Change? *Polar Biol.* **2015**, *38*, 1937–1940. [CrossRef]

54. van Pelt, W.; Schuler, T.V.; Pohjola, V.; Pettersson, R. Accelerating Future Mass Loss of Svalbard Glaciers from a Multi-Model Ensemble. *J. Glaciol.* **2021**, *67*, 485–499. [CrossRef]

55. Geyman, E.; van Pelt, W.; Maloof, A.; Aas, H.F.; Kohler, J. Historical Record of Glacier Change on Svalbard Predicts Doubling of 21st Century Mass Loss. *Nature* **2022**, *601*, 374–379. [CrossRef]

Article

Winter Behavior of Juvenile Brown Trout in a Changing Climate: How Do Light and Ice Cover Affect Encounters with Instream Predators?

Karl Filipsson [1,2], Veronika Åsman [1], Larry Greenberg [1,*], Martin Österling [1], Johan Watz [1] and Eva Bergman [1]

[1] River Ecology and Management, Department of Environmental and Life Sciences, Karlstad University, Universitetsgatan 2, SE-651 88 Karlstad, Sweden; martin.osterling@kau.se (M.Ö.); johan.watz@kau.se (J.W.); eva.bergman.1868@kau.se (E.B.)

[2] Jakobi Sustainablility AB, SE-412 58 Gothenburg, Sweden

* Correspondence: larry.greenberg@kau.se; Tel.: +46-54-7001543

Abstract: During winter, stream fishes are vulnerable to semi-aquatic predators like mammals and birds and reduce encounters by being active in darkness or under surface ice. Less is known about the behavior of fishes towards instream piscivorous fishes. Here, we examined how surface ice and light affected the anti-predator behavior of juvenile brown trout (*Salmo trutta* Linnaeus, 1758) in relation to piscivorous burbot (*Lota lota* Linnaeus, 1758) and northern pike (*Esox lucius* Linnaeus, 1758) at 4 °C in experimental flumes. Trout had lower foraging and swimming activity and spent more time sheltering when predators were present than when absent. In daylight, trout's swimming activity was not affected by predators, whereas in darkness trout were less active when predators were present. Trout consumed more drifting prey during the day when ice was present, and they positioned themselves further upstream when under ice cover, regardless of light conditions. Trout stayed closer to conspecifics under ice, but only in the presence of pike. Piscivorous fishes thus constitute an essential part of the predatory landscape of juvenile trout in winter, and thus loss of ice cover caused by climate warming will likely affect trout's interactions with predators.

Keywords: anti-predator; global changer; diel behavior; foraging; piscivores; predators

Key Contribution: Most studies involving predation on fishes in northern temperate streams in winter have focused on prey behavior in relation to semi-aquatic predators like mammals and birds. Here, we show that the prey also modify their behavior in the presence of instream piscivorous fishes and how this interaction is affected by surface ice cover and light.

Citation: Filipsson, K.; Åsman, V.; Greenberg, L.; Österling, M.; Watz, J.; Bergman, E. Winter Behavior of Juvenile Brown Trout in a Changing Climate: How Do Light and Ice Cover Affect Encounters with Instream Predators? *Fishes* **2023**, *8*, 521. https://doi.org/10.3390/fishes8100521

Academic Editor: Assaf Barki

Received: 14 September 2023
Revised: 13 October 2023
Accepted: 16 October 2023
Published: 20 October 2023

1. Introduction

The global climate is gradually becoming warmer [1], with the magnitude of the effect dependent on location and time of year. In most northern latitudes, it is expected that temperature increases will be more pronounced during winter than during summer [2]. Hence, the effects of global warming during winter may be profound and have direct and indirect effects on organisms [3–5]. In aquatic systems situated in northern latitudes, the effects of warmer winters on the physical environment have already been documented. Warmer winters have resulted in reduced snow and ice cover, both temporally and spatially, as well as erratic ice cover formation and breakup throughout winter [6–10]. Surface ice cover is thought to protect ectothermic fish from many endothermic predators, and thus ongoing and expected further reductions of the period with intact surface ice cover may have negative effects for fish.

Many juvenile salmonids overwinter in streams and rivers, where they are vulnerable to predation [11–13]. Semi-aquatic predators, such as mammals and birds, both endotherms, are some of the main winter predators of stream salmonids [11,13]. As poikilotherms, salmonids and most other fishes have reduced predator detection and escape capabilities during winter, as a result of constrained physiological performance at low temperatures [14–17]. This thermal effect makes fish relatively easy to catch for these endothermic predators. In addition, the abundance of terrestrial prey, *e.g.*, reptiles, amphibians and insects, for semi-aquatic predators is generally low in winter, whereas fish provide a reliable food source.

Mammals and birds that prey on fish in winter are mainly active during daylight hours. Stream fishes can therefore reduce their risk of being preyed upon if they reduce their diurnal activity. During the day, juvenile salmonids seek shelter within dead wood, beaver dams, crevices in the streambed or along undercut streambanks [18–21]. Stream fishes are often nocturnal in winter as long as there is enough invertebrate prey to meet their energetic needs. In fact, this relationship between food availability and diel activity is not only evident during winter but in other seasons as well [22]. However, in winter, fish have low metabolic rates in cold water, and therefore a reduced need to forage. Therefore, fish often remain inactive during the day in winter, and this is not only true for salmonids [23–25], but also other stream fishes such as minnows and galaxiids [26,27].

Surface ice cover reduces the risk that stream fishes succumb to predation by mammals and birds [28]. Overwintering fish often have higher growth and survival rates when surface ice cover is present than when it is absent [12,29–34]. Ice cover leads to larger energy reserves, as fish increase the time they spend foraging and are less vigilant to predators [30,33,35]. Stream salmonids are active both in darkness and in daylight when surface ice cover is present [34,36]. In a winter field experiment, juvenile brown trout (*Salmo trutta* Linnaeus, 1758) in stream sections covered with simulated surface ice cover had higher growth rates, were more active and used more of the width of the stream during the day than trout in exposed stream sections. Instead, trout in exposed sections remained inactive and sheltered along the streambanks during the day [34]. Juvenile brown trout have also been shown to allocate more time to foraging and territoriality in winter when under ice cover [36,37]. Similarly, field studies of bull (*Salvelinus confluentus* Suckley, 1859) and cutthroat (*Oncorhynchus clarkii* Richardson, 1836) trout show that these fishes spend less time sheltering among instream structures if surface ice cover is present [38]. Salmonids also exhibit reduced stress levels when surface ice cover is present, which has been quantified as reductions in oxygen consumption [39], cortisol levels, opercular beat rate and lighter body coloration [36,40]. Surface ice cover thus plays a major role in behavioral trade-offs between foraging and predator avoidance of juvenile salmonids in winter [41,42].

In boreal regions, burbot (*Lota lota* Linnaeus, 1758) [43,44], northern pike (*Esox lucius* Linnaeus, 1758) [45,46] and salmonids [47] prey on juvenile salmonids in winter. As predicted, juvenile salmonids exhibit vigilance and anti-predator behaviors in the presence of these predators. Juvenile brown trout are both less active and increase the time they spend sheltering in winter when burbot are nearby [20,21]. Enefalk et al. [20] found that trout sheltered less within the streambed and more among dead wood when a burbot was present. This behavioral change has been interpreted as a response to the benthic burrowing lifestyle of burbot. Filipsson et al. [21] found that juvenile trout maintained positions further away from burbot in darkness, and in daylight trout sheltered under overhead cover. It thus seems plausible that the risk of predation from piscivorous fish is most pronounced in darkness (at night or under surface ice cover) when diurnal mammals and birds are less successful at foraging for fish.

Climate warming and altered ice regimes are likely to have major consequences for the winter ecology of stream salmonids. The aim of this study was to examine the anti-predator behavior of an overwintering juvenile salmonid in the presence and absence of surface ice cover. We observed the behavior of juvenile brown trout in an experimental setting in darkness and in daylight, both with and without artificial surface ice cover present. We studied the trout in the presence of either burbot or northern pike or in the absence of predators. We chose these two predatory fish species as they are sympatric with overwintering juvenile brown trout. In addition, burbot is active at low temperatures and mainly in darkness [48], whereas pike is a visual predator that has reduced physiological performance at winter temperatures [49]. We predicted that trout would exhibit anti-predator behaviors in the presence of burbot and pike. We also predicted that trout would become more active and forage more when under ice cover, at least in daylight, when the risk of predation from semi-aquatic, endothermic predators should be low, even though predatory fish should still pose a threat.

2. Materials and Methods

2.1. Study Fish

We captured 76 juvenile brown trout (12 yearlings and 64 one-year-old fish) in late September 2020, by electrofishing in the forest stream, Barlingshultsälven, Sweden (59°31.356′ N 12°18.728′ E). The fish were subsequently transported to the aquarium facility at Karlstad University and kept in four 200 L aquaria (~20 trout in each aquarium). Trout were acclimated to aquarium conditions for approximately 4 months before the experiment commenced. Water in the aquaria was constantly filtered (EHEIM 2217 Classic canister filter; Eheim GmbH & Co KG; sourced from Karlstad, Sweden) and cooled (Teco TK 2000, Teco, sourced from Fornace Zarattini, Italy), and 25% of the water in each aquarium was changed once a week. The water temperature in all aquaria was initially maintained at 11 °C, the temperature in the Barlingshultsälven when trout were captured. After one week of acclimatization, we reduced the water temperature by 1 °C/week, until temperatures reached 4 °C in the beginning of December. Photoperiod followed natural daylight cycles. During the acclimatization period, we fed trout approximately 1% of their body mass three times a week with thawed, previously frozen, red chironomid larvae. At the beginning of the experiment, the mean wet mass $\pm$ SD of the trout was 10.02 $\pm$ 4.66 g (min–max = 2.67–23.17 g). The mean total length $\pm$ SD was 109 $\pm$ 19 mm (min–max = 71–151 mm).

One burbot and one pike were used as predators. The burbot (53 cm and 1064 g) was captured using net cages during winter 2019 and was kept in a 4000 L flume until the start of this experiment. The pike (53 cm and 1179 g) was captured by angling in early October 2020 and kept in a second 4000 L flume. Burbot and pike were captured at the same location, at the mouth of the River Klarälven, close to Lake Vänern, Sweden (59°21.905′ N 13°33.075′ E). Light conditions and the water temperature regime for these fish followed the same protocol as for the brown trout. One week prior to starting the experiment, both burbot and pike were fed two thawed, previously frozen, 5 g brown trout.

2.2. Experimental Design

This experiment was conducted in the mid-sections of three 7-meter-long stream flumes (Figure 1). Glass windows on one side of the flumes enabled observations of fish during the behavioral trials. We demarcated the sections of the flumes used for the experiment with stainless steel mesh screens (mesh size 5.35 mm, thread 1 mm, 71% open area). Two experimental arenas were constructed in each flume, separated by green plastic net screens (mesh size 5.35 mm, thread 1 mm, 71% open area) attached to wood frames (95 cm wide and 60 cm high). Upstream and downstream arenas had areas of 95 × 50 cm and 95 × 130 cm, respectively. Burbot and pike were kept in the upstream sections, and placed there 48 h before the experiment started. The third flume was kept completely free from burbot and pike and functioned as a predator-free control. Downstream arenas served as experimental arenas for the trout. The water depth in all the flumes was 25 cm, and the average water velocity was 5 cm/s. Water temperature was kept constant at 4 $\pm$ 0.1 °C throughout the experiment, and the substrate consisted of 5–20 mm gravel. In the middle of each trout arena, we put one large stone (~10 × 5 × 3 cm) to provide a focal point for the trout when foraging. We also attached a measuring scale to the flume's glass panel, with tick marks to enable measurements of the upstream-downstream position of the trout.

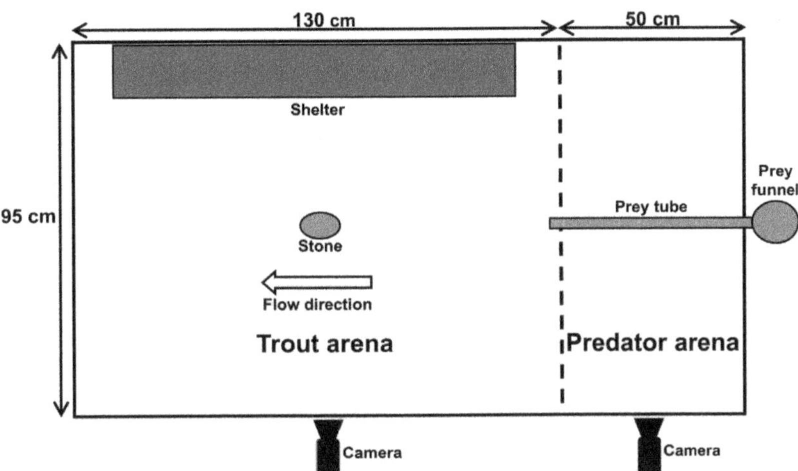

Figure 1. Birds-eye view of the experimental setup. Predators, i.e., burbot or pike, were kept in enclosed predator arenas upstream of the brown trout. We used net screens to separate trout from the predators. A plastic sheet 10 cm above the substrate provided overhead shelter for the trout, and a stone in the middle of the trout arenas provided a focal point when trout were foraging. Red chironomid larvae were used as food for the trout and were delivered to the trout through a funnel and tube, which entered the trout arenas through an opening in the net in the middle of the cross-section at 5 cm water depth. No predators were present in the control treatment.

We constructed one large shelter for the trout per stream by attaching a thin black opaque plastic sheet (975 × 190 × 5 mm) to three concrete blocks (10 × 16 × 3 cm). Four iron legs (7 cm) supported each concrete block. We positioned the concrete blocks at each end and in the middle of the shelters. The shelter structure was placed in an upstream-downstream direction along the side opposite the glass panels (Figure 1). Ice cover was simulated by using transparent multiwall polycarbonate boards (six layers, 30 mm thick in total, hereafter these plastic boards are referred to as ice cover), cut to fit snugly over the entire surface of the middle section of the stream flume. The ends of the boards were covered with duct tape to prevent water from entering the walled chambers, and thus ensuring that the boards would float on the water surface. The boards (ice cover) reduced illuminance by ~200 lx.

We conducted the experiment in January 2021. Prior to each trial, we removed six trout from the holding aquaria and sorted them into three size-matched pairs. Trout were anesthetized (benzocaine, 0.1 g/L), weighed and measured before the trials. Size differences within each pair did not exceed 15 mm or 3 g and trout sizes did not differ between the three predator treatments (One-way ANOVA, $p > 0.8$ for both wet mass and total length). We placed one pair of trout into each of the three flumes. Thereafter, trout were left in the flumes overnight for *c.* 12 h before observations of fish started. Trout were not fed during the 48 h prior to the experimental trials. During the experiment, we kept the light regime at 17 h darkness and 7 h daylight, which reflects the natural daylight cycle for January in the area from which the trout originated. Trout spent *c.* 20 h in the stream flumes during each experimental trial.

During the experimental trials, each pair of trout was video recorded (Canon XA10; Canon Inc.; sourced from Umeå, Sweden) during four 10-min-long recording sessions throughout the day. Two of the recording sessions were conducted in darkness (< 0.05 lx), and occurred early in the morning and late in the afternoon. The remaining two recording sessions were conducted in daylight (300 lx) in the morning and afternoon, between the two recordings in darkness. We used infrared illuminators (IR illuminator No. 40748, Kjell & Co Elektronik AB) and the infrared function on the camera to enable observations of fish in darkness. Four infrared illuminators were used for each flume, attached to wood posts 50 cm directly above the water surface. During two of the four recording sessions, one in darkness and one in daylight and either in the morning or in the afternoon, fish were subjected to simulated ice cover by placing the plastic boards directly on the water's surface. We placed or removed artificial ice cover at least 5 h before any observations of fish behavior. We randomized whether trout were subjected to surface ice cover in the morning or in the afternoon.

We fed trout one red chironomid larvae (> 10 mm long) every 15 s during the first three minutes of each recording session. Prey items were flushed with water through a funnel and delivered through a plastic tube, which entered the flume through an opening in the net screen in the middle of the upstream cross-section of the trout arenas at a water depth of 5 cm. During the last 7 min of each recording session, trout were not fed but behavioral observations still continued. We decided to feed trout for three minutes to prevent fish from becoming satiated and thus be less willing to forage during the remaining recording sessions throughout the day. When four 10-min recording sessions (during darkness/daylight and with ice cover/no ice cover) had been carried out for all three predator treatments (burbot, pike, control), trout were removed from the flumes and replaced with new pairs of trout. In total, 12 pairs of trout were tested for each of the three predator treatments, resulting in 72 trout being used for this experiment. When all experimental trials had been carried out, all fish, including the pike and burbot, were returned to the sites where they were caught.

2.3. Data Collection and Statistical Analyses

We examined the effects of predator presence, ice cover and light by quantifying data for seven response variables. These data were obtained by the first author from watching and scoring the following behaviors from the films: (1) whether trout foraged or not; (2) the number of consumed prey; (3) whether trout exhibited aggressive behaviors or not; (4) proportion of time that trout were active; (5) proportion of time that trout sheltered; (6) distance between the trout's anterior end and the upstream predator arenas (hereafter referred to as *distance from predator arena*, regardless of predator presence/absence) and (7) the average longitudinal distance between the two trout in each pair.

Foraging behavior was quantified both as a binomial response, if trout within a pair foraged or not, and as the total number of prey that trout consumed (0 to 13 prey per trial). Aggression was measured as a binomial response, whether trout pairs exhibited aggressive behaviors or not. Activity was measured as the proportion of time that trout were actively moving during each 10-min observation period, and shelter use was quantified as the proportion of time that trout spent under the overhead shelter. To meet the assumptions for statistical testing, we analyzed all variables expressed as proportions as arcsine transformed proportions [50], based on the arithmetic mean for each pair of trout. The position of each trout in the upstream-downstream (longitudinal) direction was measured every 15 s, and from these values, we could estimate the average trout distance from the predator demarcation and the average distance between trout in the upstream-downstream direction (longitudinal) during every recording session. Trout distances were measured in body lengths, based on the average length of the two trout in each pair.

We analyzed all data using generalized linear mixed models. The models included predator treatment as a between-subject factor and ice cover and light treatments as within-subject factors. We also included the 3-min session when trout received food and the subsequent 7-min session when trout did not receive food as two different levels in a within-subject factor, hereafter referred to as the "feeding" term. Statistical models used to analyze foraging behavior did not include this within-subject feeding term as an explanatory variable, as trout were only exposed to drifting prey (i.e., foraging) during the first three minutes of the 10-min observation period. All treatments were tested in full-factorial models that included all interaction terms. In addition, we added the mean body mass of each trout pair as a covariate to the models. We also conducted three pairwise contrasts (burbot vs.

control, pike vs. control, burbot vs. pike), which were determined *a priori*. For all behaviors except foraging and aggression, a linear distribution was used, as these datasets met assumptions of normality and homoscedasticity. Whether trout fed or exhibited aggressive behaviors or not (i.e., binary outcomes), was analyzed using a binomial distribution. We analyzed the number of consumed prey by using a negative binomial distribution. For the model on aggression, we excluded observations in darkness, as no trout exhibited aggressive behaviors in the presence of pike in darkness or in the presence of burbot under ice cover in darkness. We used compound symmetry covariance structures, as all models included repeated measures [51]. All statistical analyses were conducted in IBM SPSS Statistics 26 (IBM).

3. Results

3.1. Foraging Behavior

Only the presence of a predator ($F_{2,127} = 4.35$, $p = 0.02$) and light conditions ($F_{1,127} = 14.14$, $p < 0.001$) had significant effects on the number of trout pairs that foraged; none of the other factors were significant (Appendix A). Pairwise contrasts between predator treatments showed that the number of trout that foraged differed statistically between burbot and control treatments ($p = 0.017$), and pike and control treatments ($p = 0.001$), but not between burbot and pike treatments ($p = 0.25$). More trout foraged in daylight (86%) than in darkness (66%) (Figure 2A). In the darkness, fewer trout foraged in the presence of a pike (42%) and burbot (75%) than in the control treatments (83%) (Figure 2A). This difference was not as pronounced in daylight, where 75% of trout foraged in the presence of pike, 92% in the presence of burbot and 92% in the control treatments.

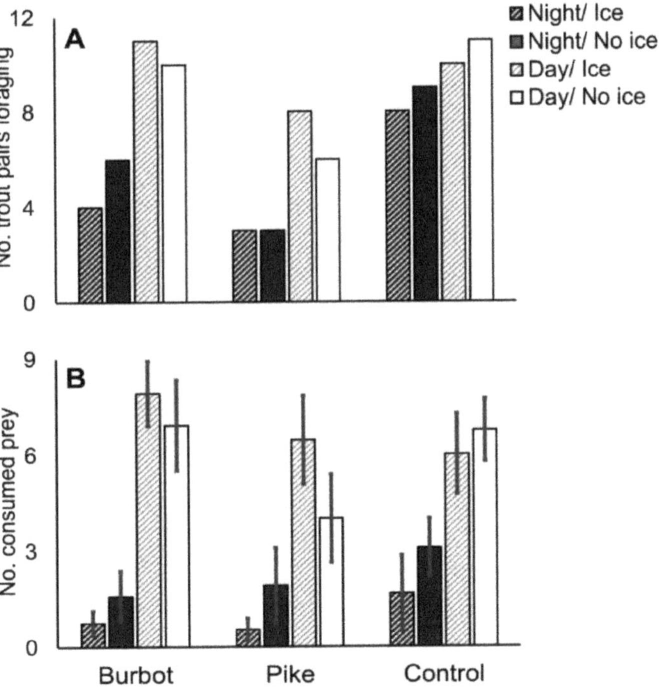

Figure 2. Effects of the presence of piscivorous fish, surface ice cover and light conditions on the (day/night) foraging behavior of juvenile brown trout. Foraging behavior is quantified as (**A**) the number of trout pairs (out of 12) that foraged and (**B**) the number of consumed prey (out of 13). Error bars in panel B indicate ± 1 SE.

Light conditions ($F_{1,127} = 53.55$, $p < 0.001$) and the light x ice cover interaction ($F_{1,127} = 6.25$, $p = 0.01$) affected the foraging rate of trout. Trout consumed almost four times as much prey in daylight compared to darkness. When ice cover was present, trout consumed more prey during the day but fewer prey in darkness (Figure 2B). The number of consumed prey also differed with trout size ($F_{1,127} = 4.49$, $p = 0.04$), as larger trout captured fewer prey. None of the other fixed terms or interactions had a significant effect on foraging behavior (Appendix A).

3.2. Aggression

No trout exhibited aggression in the presence of pike in darkness, or in the presence of burbot in darkness when surface ice cover was present (Figure 3). Only 8% of trout exhibited aggression in darkness during foraging trials, compared to 42% in daylight. During the 7-min observation periods after the foraging trials, 19% of trout exhibited aggression in darkness and 56% in daylight. In the presence of ice cover, 47% of trout exhibited aggression, compared to 39% when no ice cover was present. In the presence of burbot, 75% of trout exhibited aggression, similar to the control, whereas in the presence of pike only 33% of the trout exhibited aggression (Figure 3). Predator, ice cover and light treatments did not have significant effects on aggression (Appendix A). The only significant difference in the number of trout pairs that exhibited aggression was between the initial three minutes of foraging and the following seven minutes ($F_{1,127} = 6.53$, $p = 0.01$). In total, 47% of trout exhibited aggression during the 3-min long period when drifting prey were delivered, whereas 58% did so during the subsequent 7-min period without drifting prey.

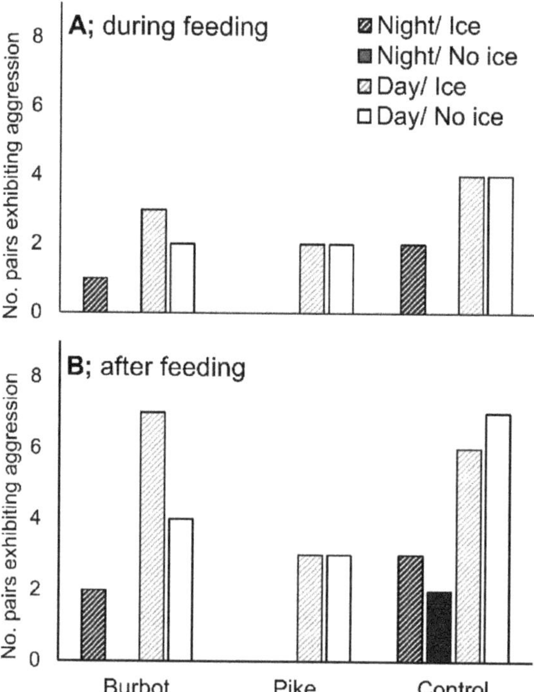

Figure 3. The number of trout pairs (out of 12) that exhibited aggression under different predator, ice cover and light treatments (day/night), both during the 3-min period when drifting prey were delivered (foraging trials) (**A**) and the 7-min period immediately after (**B**).

3.3. Activity and Shelter Use

Trout reduced their swimming activity both in the presence of a predator ($F_{2,271} = 7.29$, $p < 0.001$) and in darkness ($F_{1,271} = 25.81$, $p < 0.001$). Pairwise contrasts indicated statistical differences between burbot and control treatments ($p = 0.026$), and pike and control treatments ($p < 0.001$), but not between burbot and pike treatments ($p = 0.10$). There was a significant interaction effect between predator and light treatments ($F_{2,271} = 6.52$, $p = 0.002$). On average, trout were 29% less active in darkness than in daylight (Figure 4A,B), and this difference was almost entirely attributed to the behavioral change of trout in the presence of predators. Trout in the control treatment spent on average the same amount of time active (67%) regardless of light treatment. In the presence of burbot and pike, trout were 33 and 51% less active in darkness than in daylight, respectively (Figure 4A,B).

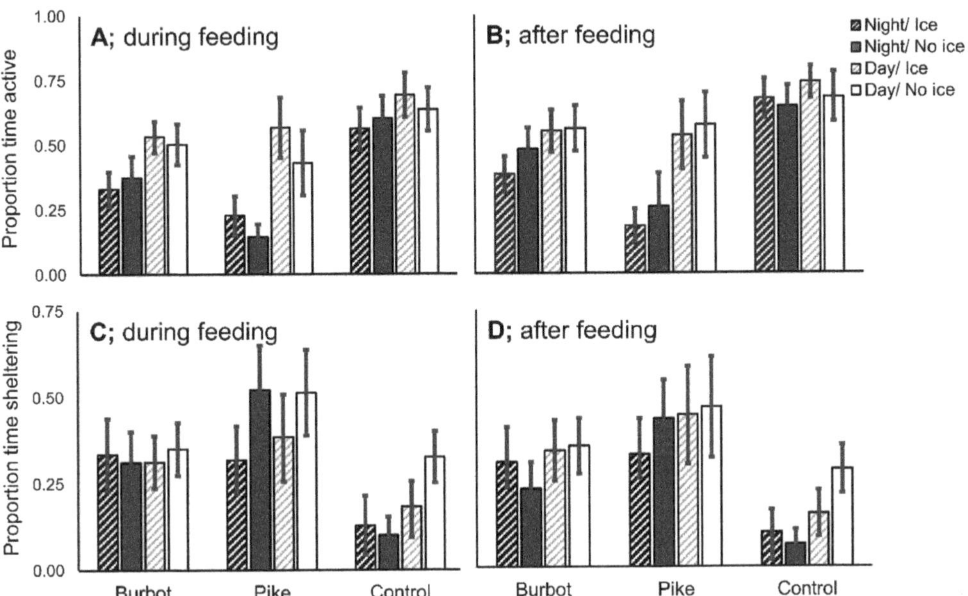

Figure 4. Effects of the presence of piscivorous fish, surface ice cover and light conditions on the (day/night) proportion of time that juvenile brown trout are (**A,B**) actively swimming and (**C,D**) seeking shelter under overhead cover, during and after the feeding period, i.e., during the 3-min period when drifting prey were delivered and the 7-min period immediately after. Error bars indicate ± 1 SE.

Both predator presence ($F_{2,271} = 3.70$, $p = 0.03$) and light treatment ($F_{1,271} = 11.35$, $p < 0.001$) had significant effects on the amount of time that trout spent sheltering. Pairwise contrasts between predator treatments indicated that only the difference between pike and control treatments was statistically significant ($p = 0.007$), not differences between burbot and control treatments ($p = 0.11$) or burbot and pike treatments ($p = 0.25$). Trout spent more time sheltering when pike was present and in daylight (Figure 4C,D). On average, trout spent 30% of their time sheltering in the presence of a burbot, 37% in the presence of a pike and 16% when no predators were present. The amount of time that trout spent sheltering was, on average, 35% higher in daylight than in darkness. None of the other fixed terms or interactions had significant effects on the proportion of time that trout spent actively swimming or sheltering (Appendix A). Trout mass had a significant effect on both the proportion of time that trout spent actively swimming ($F_{2,271} = 8.72$, $p = 0.003$) and sheltering ($F_{2,271} = 8.94$, $p = 0.003$) (Appendix A). Time that trout spent being active correlated negatively with body size, whereas the time they spent sheltering correlated positively with size.

3.4. Position of Trout in the Flumes

Ice cover ($F_{1,267}$ = 13.48, p < 0.001) and light treatments ($F_{1,267}$ = 8.16, p = 0.01) had significant effects on the distance the trout were from the predator arena. On average, trout positioned themselves further upstream when ice cover was present and in daylight (Figure 5A,B). Both the predator x ice cover interaction ($F_{1,265}$ = 6.60, p < 0.001) and ice cover x light interaction ($F_{1,265}$ = 4.72, p = 0.03) had significant effects on the longitudinal distance between trout within pairs. Trout positioned themselves closer to one another when surface ice cover was absent, but only in the presence of a pike. When both surface ice cover and pike were present, trout on average kept a longitudinal distance of 4.3 body lengths from each other. In pike treatments without surface ice cover, trout kept an average distance of 2.7 body lengths from each other (Figure 5C,D). In general, trout kept a greater distance from each other when ice cover was present in darkness, whereas in daylight this pattern was reversed. No other fixed factors or interactions had a significant effect on the position of trout in the flumes (Appendix A).

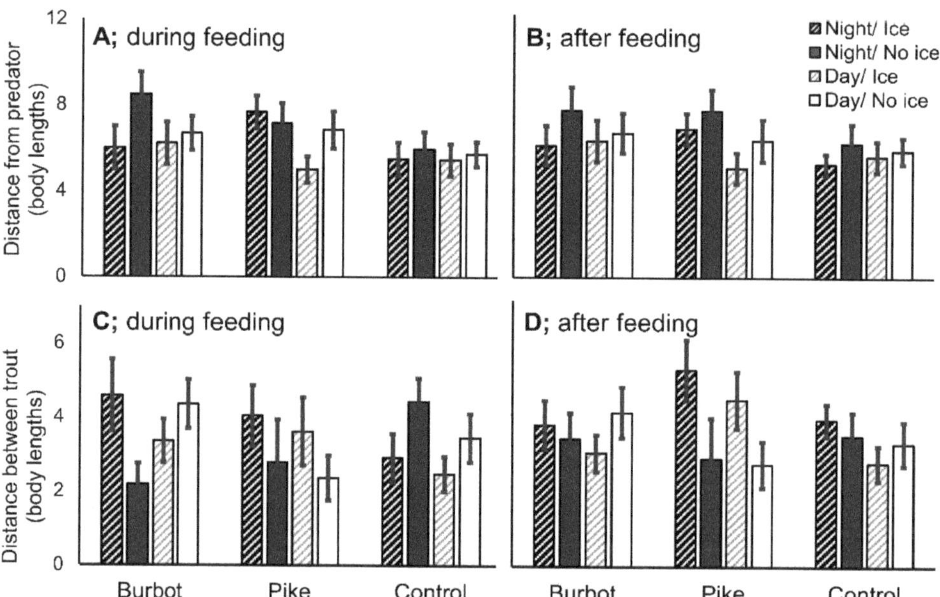

Figure 5. Effects of the presence of piscivorous fish, surface ice cover and light conditions (day/night) on the longitudinal distance between the trout pairs and between the trout and the predator arena (trout positions along the upstream-downstream axis) in the experimental flumes, during and after the feeding period, i.e., during the 3-min period when drifting prey were delivered and the 7-min period immediately after. The panels show (**A,B**) how far the trout positioned themselves downstream of the demarcation lines of the predator arenas (no predators present in control treatments), and (**C,D**) the average distance between the two trout in each pair. Error bars indicate ± 1 SE.

4. Discussion

In this study, we found that juvenile brown trout exhibited anti-predator behaviors in the presence of piscivorous fish, e.g., reduced propensity to forage, lower activity and increased time spent sheltering. Piscivorous fish have been previously shown to affect the ecology and behavior of overwintering stream salmonids [11,12,20,21]. Diel variation in behavior during winter has however often been attributed to the diurnal activity of semi-aquatic mammals and birds [13,34,36]. Predation from these endothermic animals is likely one of the main contributors to fish mortality in winter, and a major driver of behavioral adaptation to avoid predators. However, piscivorous fish may also constitute an essential part of the predatory landscape of stream salmonids in winter [21,52], and light and ice cover conditions can affect how juvenile salmonids respond in the presence of piscivorous fish.

Trout in our study had similar activity levels in daylight, regardless of whether a predator was present or not, but were less active in darkness when predators were nearby. Considering the fact that piscivorous fish are targeted as prey by mammals and birds, these fishes are also likely to reduce their activity during the day in winter. This could explain the higher vigilance of juvenile salmonids in the presence of piscivorous fish at night. Likewise, as juvenile salmonids are more vigilant towards predatory birds and mammals in daylight they may not pay as much attention to piscivorous fish. In addition, some piscivorous fish species that occur in boreal streams are generally nocturnal, such as the winter-active burbot that we used in this experiment [53–56]. Stream salmonids are mainly nocturnal in winter and tend to forage in darkness as long as their energetic needs can be fulfilled [22], but have higher foraging efficiency in daylight than in darkness [30,57]. A high energetic demand may explain why trout in this study mainly were active and foraged in daylight. Trout positioned themselves further upstream in daylight than in darkness, which also may indicate active foraging behavior during the day. Prior to each behavioral trial, we did not feed trout for 48 h, which may have resulted in the trout having a high propensity to forage during the trials. In addition, the water temperature during this experiment was relatively high (4 °C).

We used both burbot and northern pike as predators in our study. Burbot are mainly nocturnal, active at low temperatures, can locate their prey by olfactory cues, and are known to excavate the streambed in search of prey [48,53,55,56]. Pike is a diurnal visual predator that has reduced foraging success at low temperatures [49]. In our study, trout exhibited similar changes in foraging behavior and activity to both predators. Both burbot and pike may constitute a threat under low light conditions when semi-aquatic endothermic predators are not as active. Burbot and pike are common in the stream where we captured trout for this study (Karl Filipsson, personal observations during electrofishing), and it therefore seems plausible that trout from this stream would have anti-predator responses to both burbot and pike. In addition, fishes (including juvenile salmonids) elicit more pronounced anti-predator responses if the predator has been feeding on conspecifics [58,59]. Studies of European perch (*Perca fluviatilis* Linnaeus, 1758) show that fish can respond stronger to predators when they can both see and smell them [60], and if predators have consumed the prey prior to experimental testing, the prey may respond even stronger to the predators. Both burbot and pike were fed brown trout prior to the experimental trials, which may have increased the vigilance and resulted in more pronounced anti-predator responses by the trout in our study.

Trout consumed more prey when surface ice cover was present in daylight than at night. Earlier work shows that when ice cover is present juvenile brown trout may forage more during the day [37], although other studies show no effect of surface ice cover on foraging behavior [36]. In the study by Watz et al. [37] trout were tested in pairs, similar to in this study, whereas in the study by Watz et al. [36] trout were tested in groups of four. It is thus possible that the behavior of juvenile trout is affected by group dynamics and social interactions that depend on the number of fish present. Similarly, trout in the study by Watz et al. [36] exhibited higher swimming activity and more aggressive behaviors when surface ice cover was present, which we did not observe in our study. We used plastic boards to simulate ice cover, and it is possible that real surface ice has properties other than reducing light and functioning as a physical barrier and insulation, which may affect fish behavior. For example, surface ice cover in contact with stream surface water affects hydrology [61], and light conditions differ depending on the thickness of both the ice cover itself and the snow layer on top of the ice. Real ice cover has been used in some experimental studies on salmonid behavior [36,37], but plastic sheets [34] and reduced light [30] have been used to simulate ice cover as well. All of these studies have generated the predicted biological responses of salmonids under surface ice. We also found that trout positioned themselves further upstream (nearer to the predator arena) when ice cover was present, possibly at locations more beneficial for foraging. In treatments with pike, trout also positioned themselves further away from each other (longitudinal direction) when ice cover was present. This could be a response to the lunging foraging style of pike, assuming pike would forage more when ice cover is present, and thus trout attempt to reduce their risk of being preyed upon by keeping a greater distance from conspecifics. Current knowledge only provides limited insights into the behavioral effects of surface ice cover. Research on this topic thus seems especially timely considering the current loss of winter ice cover at northern latitudes.

5. Conclusions

Changes in snow and ice cover are some of the most evident effects of climate warming in temperate and boreal regions [2,9,62,63]. These changes have resulted in less ice cover spatially and temporally and in more erratic ice cover formation and breakup throughout winter [7,10], which can have a major impact on the ecology of overwintering fishes [30,33,34]. This study presents results on the diel winter behavior of a juvenile salmonid and shows that behavioral responses are affected by the presence of piscivorous fish and by surface ice cover. Ice cover and light conditions play an important role in behavioral trade-offs of overwintering stream fishes [12,34]. These trade-offs revolve around whether or not fish should allocate time to foraging or be vigilant to predators [21]. Behavioral decisions can affect overwinter survival, and thus population dynamics, but also how anti-predator responses are shaped through evolutionary adaptation [41,42]. Evident changes to the northern winter environment, i.e., loss of ice and snow, are already noticeable [2,7] and will likely have major effects on the dynamics of predator-prey interactions. Results from experimental studies like the one we present here thus provide knowledge on how overwintering stream fishes are adapted to encounters with predators, and the possible effects that global change has on these interactions.

Author Contributions: Conceptualization, methodology, K.F., V.Å., E.B., L.G., M.Ö. and J.W.; formal analysis, K.F.; investigation, K.F. and V.Å.; resources, Karlstad University; writing—original draft preparation, K.F.; writing—review and editing K.F., V.Å., E.B., L.G., M.Ö. and J.W.; supervision, E.B., L.G., M.Ö. and J.W.; project administration, E.B.; funding acquisition, Karlstad University. All authors have read and agreed to the published version of the manuscript.

Funding: This research received no external funding. Instead, it was funded by Karlstad University, through its special program for the university's strong research environments.

Institutional Review Board Statement: The Swedish Ethical Committee on Animal Welfare approved this study, both fish collection in the field (5.8.18-03819/2018; ID no. 001530) and the laboratory experiments (5.8.18-03818/2018; ID no. 001500).

Data Availability Statement: https://doi.org/10.6084/m9.figshare.24125931 (accessed on 15 October 2023).

Acknowledgments: The authors owe immense gratitude to Niclas Carlsson, who generously provided the pike for this study. Jeffery Marker, Roman Motyka and Sebastian Rock also provided valuable assistance during fish collection in the field. We are grateful to members of the Övre Gla fishing district and to the County Board in Värmland, who allowed us to capture trout in Barlingshultälven.

Conflicts of Interest: The authors declare no conflict of interest. The funders had no role in the design of the study; in the collection, analyses, or interpretation of data; in the writing of the manuscript; or in the decision to publish the results.

Appendix A

Table A1. Fixed effects from generalized linear mixed models, exploring the behavioral responses of juvenile brown trout under experimental winter conditions. The effects of predator presence (between-subject treatment), surface ice cover, light conditions and differences during and after the feeding period (i.e., during the 3-min period when drifting prey were delivered and the 7-min period immediately after, i.e., the within-subject factor "feeding"), as well as all interactions between these treatments, are included. Trout body mass is included as a covariate. *p*-values in bold denote statistically significant effects.

Variable	Source of Variation	*F*	*df*	*p*
Foraging (Y/N)	Predator	4.35	2, 127	**0.02**
binomial distribution	Ice cover	0.01	1, 127	0.91
	Light	14.14	1, 127	**<0.001**
	Predator x Ice cover	0.44	2, 127	0.65
	Predator x Light	0.61	2, 127	0.55
	Ice cover x Light	0.51	1, 127	0.48
	Predator x Ice cover x Light	0.34	2, 127	0.72
	Mass	0.79	1, 127	0.38
Foraging rate (continuous)	Predator	1.96	2, 127	0.15
linear distribution	Ice cover	3.12	1, 127	0.08
	Light	53.55	1, 127	**<0.001**
	Predator x Ice cover	0.07	2, 127	0.94
	Predator x Light	1.94	2, 127	0.15
	Ice cover x Light	6.25	1, 127	**0.01**
	Predator x Ice cover x Light	0.43	2, 127	0.65
	Mass	4.49	1, 127	**0.04**
Aggression (Y/N)	Predator	1.00	2, 131	0.37
binomial distribution	Ice cover	0.34	1, 131	0.56
	Feeding	6.53	1, 131	**0.01**
	Predator x Ice cover	0.79	2, 131	0.46
	Predator x Feeding	0.26	2, 131	0.77
	Ice cover x Feeding	0.01	1, 131	0.93
	Predator x Ice cover x Feeding	0.16	2, 131	0.85
	Mass	3.62	1, 131	0.06
Swimming activity (proportion)	Predator	7.29	2, 271	**<0.001**
linear distribution	Ice cover	0.01	1, 271	0.91
	Light	25.81	1, 271	**<0.001**
	Feeding	3.02	1, 271	0.08
	Predator x Ice cover	0.31	2, 271	0.73
	Predator x Light	6.52	2, 271	**0.002**
	Predator x Feeding	0.01	2, 271	0.96
	Ice cover x Light	0.86	1, 271	0.36
	Ice cover x Feeding	0.33	1, 271	0.57
	Light x Feeding	0.02	1, 271	0.88

Table A1. *Cont.*

Variable	Source of Variation	F	df	p
	Predator x Ice cover x Light	0.20	2, 271	0.99
	Predator x Light x Feeding	0.13	2, 271	0.88
	Ice cover x Light x Feeding	0.18	1, 271	0.67
	Predator x Ice cover x Feeding	0.78	2, 271	0.46
	Predator x Ice cover x Light x Feeding	0.08	2, 271	0.92
	Mass	8.72	1, 271	**0.003**
Shelter use (proportion)	Predator	3.70	2, 271	**0.03**
linear distribution	Ice cover	2.52	1, 271	0.11
	Light	11.35	1, 271	**<0.001**
	Feeding	0.72	1, 271	0.40
	Predator x Ice cover	1.88	2, 271	0.16
	Predator x Light	1.05	2, 271	0.35
	Predator x Feeding	0.13	2, 271	0.88
	Ice cover x Light	0.77	1, 271	0.38
	Ice cover x Feeding	0.23	1, 271	0.63
	Light x Feeding	0.33	1, 271	0.57
	Predator x Ice cover x Light	1.10	2, 271	0.34
	Predator x Light x Feeding	0.12	2, 271	0.88
	Ice cover x Light x Feeding	0.00	1, 271	0.97
	Predator x Ice cover x Feeding	0.27	2, 271	0.76
	Predator x Ice cover x Light x Feeding	0.03	2, 271	0.98
	Mass	8.94	1, 271	**0.003**
Distance from predator arena (continuous)	Predator	1.54	2, 267	0.22
linear distribution	Ice cover	13.48	1, 267	**<0.001**
	Light	8.16	1, 267	**0.01**
	Feeding	0.59	1, 267	0.44
	Predator x Ice cover	1.25	2, 267	0.29
	Predator x Light	2.55	2, 267	0.08
	Predator x Feeding	0.15	2, 267	0.86
	Ice cover x Light	0.01	1, 267	0.82
	Ice cover x Feeding	0.00	1, 267	0.98
	Light x Feeding	0.06	1, 267	0.81
	Predator x Ice cover x Light	2.84	2, 267	0.06
	Predator x Light x Feeding	0.09	2, 267	0.92
	Ice cover x Light x Feeding	0.36	1, 267	0.55
	Predator x Ice cover x Feeding	0.35	2, 267	0.70
	Predator x Ice cover x Light x Feeding	0.64	2, 267	0.53
	Mass	13.34	1, 267	**<0.001**
Distance between trout (continuous)	Predator	0.18	2, 265	0.83

Table A1. *Cont.*

Variable	Source of Variation	F	df	p
linear distribution	Ice cover	2.46	1, 265	0.12
	Light	2.76	1, 265	0.10
	Feeding	0.18	1, 265	0.67
	Predator x Ice cover	6.60	2, 265	**<0.001**
	Predator x Light	0.58	2, 265	0.57
	Predator x Feeding	0.29	2, 265	0.75
	Ice cover x Light	4.72	1, 265	**0.03**
	Ice cover x Feeding	0.16	1, 265	0.69
	Light x Feeding	0.01	1, 265	0.92
	Predator x Ice cover x Light	2.04	2, 265	0.13
	Predator x Light x Feeding	0.21	2, 265	0.81
	Ice cover x Light x Feeding	0.00	1, 265	0.98
	Predator x Ice cover x Feeding	1.78	2, 265	0.17
	Predator x Ice cover x Light x Feeding	0.75	2, 265	0.47
	Mass	1.21	1, 265	0.27

References

1. McKenzie, D.J.; Geffroy, B.; Farrell, A.P. Effects of global warming on fishes and fisheries. *J. Fish Biol.* **2021**, *98*, 1489–1492. [CrossRef] [PubMed]
2. IPCC. *Climate Change 2021: The Physical Science Basis. Contribution of Working Group I to the Sixth Assessment Report of the Intergovernmental Panel on Climate Change*; Masson-Delmotte, V., Zhai, P., Pirani, A., Connors, S.L., Péan, C., Berger, S., Caud, N., Chen, Y., Goldfarb, L., Gomis, M.I., et al., Eds.; Cambridge University Press: Cambridge, UK, 2021.
3. Heino, J.; Erkinaro, J.; Huusko, A.; Luoto, M. Climate change effects on freshwater fishes, conservation and management. In *Conservation of Freshwater Fishes*; Closs, G.P., Krkosek, M., Olden, J.D., Eds.; Cambridge University Press: Cambridge, UK, 2015; pp. 76–106. [CrossRef]
4. Falvo, C.A.; Koons, D.N.; Aubry, L.M. Seasonal climate effects on the survival of a hibernating mammal. *Ecol. Evol.* **2019**, *9*, 3756–3769. [CrossRef] [PubMed]
5. Clark, N.J.; Kerry, J.T.; Fraser, C.I. Rapid winter warming could disrupt coastal marine fish community structure. *Nat. Clim. Change* **2020**, *10*, 862–867. [CrossRef]
6. Brown, R.D. Northern hemisphere snow cover variability and change, 1915–97. *J. Clim.* **2000**, *13*, 2339–2355. [CrossRef]
7. Magnuson, J.J.; Robertson, D.M.; Benson, B.J.; Wynne, R.H.; Livingstone, D.M.; Arai, T.; Assel, R.A.; Barry, R.G.; Card, V.V.; Kuusisto, E.; et al. Historical trends in lake and river ice cover in the northern hemisphere. *Science* **2000**, *289*, 1743–1746. [CrossRef]
8. Post, E.; Forchhammer, M.C.; Bret-Harte, M.S.; Callaghan, T.V.; Christensen, T.R.; Elberling, B.; Fox, A.D.; Gilg, O.; Hik, D.S.; Høye, T.T.; et al. Ecological dynamics across the Arctic associated with recent climate change. *Science* **2009**, *325*, 1355–1358. [CrossRef]
9. Choi, G.; Robinson, D.A.; Kang, S. Changing northern hemisphere snow seasons. *J. Clim.* **2010**, *23*, 5305–5310. [CrossRef]
10. Callaghan, T.V.; Johansson, M.; Brown, R.D.; Groisman, P.; Labba, N.; Radionov, V.; Barry, R.; Bulygina, O.; Essery, R.; Frolov, D.; et al. The changing face of Arctic snow cover: A synthesis of observed and projected changes. *Ambio* **2011**, *40*, 17–31. [CrossRef]
11. Hurst, T.P. Causes and consequences of winter mortality in fishes. *J. Fish. Biol.* **2007**, *71*, 315–345. [CrossRef]
12. Huusko, A.; Greenberg, L.; Stickler, M.; Linnansaari, T.; Nykänen, M.; Vehanen, T.; Koljonen, S.; Louhi, P.; Alfredsen, K. Life in the ice lane: The winter ecology of stream salmonids. *River Res. Appl.* **2007**, *23*, 469–491. [CrossRef]
13. Harvey, B.C.; Nakamoto, R.J. Seasonal and among-stream variation in predator encounter rates for fish prey. *Trans. Am. Fish. Soc.* **2013**, *142*, 621–627. [CrossRef]
14. Webb, P.W. Temperature effects on acceleration of rainbow trout, *Salmo gairdneri*. *J. Fish. Res. Board Canada* **1978**, *35*, 1417–1422. [CrossRef]
15. Bennett, A.F. Thermal dependence of locomotor capacity. *Am. J Physiol. Regul. Integr. Comp. Physiol.* **1990**, *259*, 253–258. [CrossRef] [PubMed]
16. Guderley, H. Metabolic responses to low temperature in fish muscle. *Biol. Rev. Camb. Philos. Soc.* **2004**, *79*, 409–427. [CrossRef] [PubMed]
17. Johnston, I.A. Environment and plasticity of myogenesis in teleost fish. *J. Exp. Biol.* **2006**, *209*, 2249–2264. [CrossRef]

18. Valdimarsson, S.K.; Metcalfe, N.B. Shelter selection in juvenile Atlantic salmon, or why do salmon seek shelter in winter? *J. Fish Biol.* **1998**, *52*, 42–49. [CrossRef]
19. Jonsson, B.; Jonsson, N. *Ecology of Atlantic Salmon and Brown Trout: Habitat as a Template for Life Histories*; Springer: Dordrecht, The Netherlands, 2011; ISBN 978-94-007-1188-4.
20. Enefalk, Å.; Watz, J.; Greenberg, L.; Bergman, E. Winter sheltering by juvenile brown trout (*Salmo trutta*)—Effects of stream wood and an instream ectothermic predator. *Freshw. Biol.* **2017**, *62*, 111–118. [CrossRef]
21. Filipsson, K.; Bergman, E.; Österling, M.; Erlandsson, A.; Greenberg, L.; Watz, J. Effects of temperature and a piscivorous fish on diel winter behaviour of juvenile brown trout (*Salmo trutta*). *Freshw. Biol.* **2019**, *64*, 1797–1805. [CrossRef]
22. Orpwood, J.E.; Griffiths, S.W.; Armstrong, J.D. Effects of food availability on temporal activity patterns and growth of Atlantic salmon. *J. Anim. Ecol.* **2006**, *75*, 677–685. [CrossRef]
23. Heggenes, J.; Krog, O.M.W.; Lindås, O.R.; Dokk, J.G. Homeostatic behavioural responses in a changing environment: Brown trout (*Salmo trutta*) become nocturnal during winter. *J. Anim. Ecol.* **1993**, *62*, 295–308. [CrossRef]
24. Metcalfe, N.B.; Fraser, N.H.C.; Burns, M.D. Food availability and the nocturnal vs. diurnal foraging trade-off in juvenile salmon. *J. Anim. Ecol.* **1999**, *68*, 371–381. [CrossRef]
25. Nykänen, M.; Huusko, A.; Lahti, M. Changes in movement, range and habitat preferences of adult grayling from late summer to early winter. *J. Fish Biol.* **2004**, *64*, 1386–1398. [CrossRef]
26. Metcalfe, N.B.; Steele, G.I. Changing nutritional status causes a shift in the balance of nocturnal to diurnal activity in European minnows. *Funct. Ecol.* **2001**, *15*, 304–309. [CrossRef]
27. David, B.O.; Closs, G.P. Seasonal variation in diel activity and microhabitat use of an endemic New Zealand stream-dwelling galaxiid fish. *Freshw. Biol.* **2003**, *48*, 1765–1781. [CrossRef]
28. Heggenes, J.; Borgstrøm, R. Effect of mink, *Mustela vison* Schreber, predation on cohorts of juvenile Atlantic salmon, *Salmo salar* L., and brown trout, *S. trutta* L., in three small streams. *J. Fish Biol.* **1988**, *33*, 885–894. [CrossRef]
29. Prowse, T.D. River-ice ecology. II: Biological aspects. *J. Cold Reg. Eng.* **2001**, *15*, 17–33. [CrossRef]
30. Finstad, A.G.; Forseth, T.; Næsje, T.F.; Ugedal, O. The importance of ice cover for energy turnover in juvenile Atlantic salmon. *J. Anim. Ecol.* **2004**, *73*, 959–966. [CrossRef]
31. Linnansaari, T.; Alfredsen, K.; Stickler, M.; Arnekleiv, J.O.; Harby, A.; Cunjak, R.A. Does ice matter? Site fidelity and movements by Atlantic salmon (*Salmo salar* L.) parr during winter in a substrate enhanced river reach. *River Res. Appl.* **2008**, *25*, 773–787. [CrossRef]
32. Brown, R.S.; Hubert, W.A.; Daly, S.F. A primer on winter, ice, and fish: What fisheries biologists should know about winter ice processes and stream-dwelling fish. *Fisheries* **2011**, *36*, 8–26. [CrossRef]
33. Hedger, R.D.; Næsje, T.F.; Fiske, P.; Ugedal, O.; Finstad, A.G.; Thorstad, E.B. Ice-dependent winter survival of juvenile Atlantic salmon. *Ecol. Evol.* **2013**, *3*, 523–535. [CrossRef]
34. Watz, J.; Bergman, E.; Piccolo, J.J.; Greenberg, L. Ice cover affects the growth of a stream-dwelling fish. *Oecologia* **2016**, *181*, 299–311. [CrossRef] [PubMed]
35. Härkönen, L.; Louhi, P.; Huusko, R.; Huusko, A. Wintertime growth in Atlantic salmon under changing climate: The importance of ice cover for individual growth dynamics. *Can. J. Fish. Aquat. Sci.* **2021**, *78*, 1479–1485. [CrossRef]
36. Watz, J.; Bergman, E.; Calles, O.; Enefalk, Å.; Gustafsson, S.; Hagelin, A.; Nilsson, A.; Norrgård, J.; Nyqvist, D.; Österling, E.M.; et al. Ice cover alters the behavior and stress level of brown trout *Salmo trutta*. *Behav. Ecol.* **2015**, *26*, 820–827. [CrossRef]
37. Watz, J.; Bergman, E.; Piccolo, J.J.; Greenberg, L. 2013. Effects of ice cover on the diel behaviour and ventilation rate of juvenile brown trout. *Freshw. Biol.* **2013**, *58*, 2325–2332. [CrossRef]
38. Jakober, M.J.; McMahon, T.E.; Thurow, R.F.; Clancy, C.G. Role of stream ice on fall and winter movements and habitat use by bull trout and cutthroat trout in Montana headwater streams. *Trans. Am. Fish. Soc.* **1998**, *127*, 223–235. [CrossRef]
39. Helland, I.P.; Finstad, A.G.; Forseth, T.; Hesthagen, T.; Ugedal, O. 2011. Ice-cover effects on competitive interactions between two fish species. *J. Anim. Ecol.* **2011**, *80*, 539–547. [CrossRef]
40. Watz, J. Stress responses of juvenile brown trout under winter conditions in a laboratory stream. *Hydrobiologia* **2017**, *802*, 131–140. [CrossRef]
41. Lima, S.L.; Dill, L.M. Behavioral decisions made under the risk of predation: A review and prospectus. *Can. J. Zool.* **1990**, *68*, 619–640. [CrossRef]
42. Preisser, E.L.; Bolnick, D.I.; Benard, M.F. Scared to death? The effects of intimidation and consumption in predator-prey interactions. *Ecology* **2005**, *86*, 501–509. [CrossRef]
43. Tolonen, A.; Kjellman, J.; Lappalainen, J. Diet overlap between burbot (*Lota lota* (L.)) and whitefish (*Coregonus lavaretus* (L.)) in a subarctic lake. *Ann. Zool. Fenn.* **1999**, *36*, 205–214.
44. Kahilainen, K.; Lehtonen, H. Piscivory and prey selection of four predator species in a whitefish dominated subarctic lake. *J. Fish. Biol.* **2003**, *63*, 659–672. [CrossRef]
45. Hyvärinen, P.; Vehanen, T. Effect of brown trout body size on post-stocking survival and pike predation. *Ecol. Freshw. Fish.* **2004**, *13*, 77–84. [CrossRef]
46. Vehanen, T.; Hamari, S. Predation threat affects behaviour and habitat use by hatchery brown trout (*Salmo trutta* L.) juveniles. *Hydrobiologia* **2004**, *525*, 229–237. [CrossRef]

47. Hawkins, L.A.; Armstrong, J.D.; Magurran, A.E. Predator-induced hyperventilation in wild and hatchery Atlantic salmon fry. *J. Fish. Biol.* **2004**, *65*, 88–100. [CrossRef]
48. Hackney, P.A. Ecology of the Burbot (*Lota lota*) with Special Reference to Its Role in the Lake Opeongo Fish Community. Ph.D. Thesis, University of Toronto, Toronto, ON, Canada, 1973.
49. Öhlund, G.; Hedström, P.; Norman, S.; Hein, C.L.; Englund, G. Temperature dependence of predation depends on the relative performance of predators and prey. *Proc. R. Soc. B Biol. Sci.* **2015**, *282*, 20142254. [CrossRef]
50. Zar, J.H. *Biostatistical Analysis*, 5th ed.; Pearson Hall: Upper Sadle River, NJ, USA, 2010.
51. Littell, R.C.; Pendergast, J.; Natarajan, R. Modelling covariance structure in the analysis of repeated measures data. *Stat. Med.* **2000**, *19*, 1793–1819. [CrossRef]
52. Filipsson, K.; Bergman, E.; Greenberg, L.; Österling, M.; Watz, J.; Erlandsson, A. Temperature and predator-mediated regulation of plasma cortisol and brain gene expression in juvenile brown trout (*Salmo trutta*). *Front. Zool.* **2020**, *17*, 25. [CrossRef]
53. Pulliainen, E.; Korhonen, K. Seasonal changes in condition indices in adult mature and non-maturing burbot, *Lota lota* (L.), in the north-eastern Bothnian bay, northern Finland. *J. Fish Biol.* **1990**, *36*, 251–259. [CrossRef]
54. Edsall, T.A.; Kennedy, G.W.; Horns, W.H. Distribution, abundance, and resting microhabitat of burbot on Julian's Reef, southwestern Lake Michigan. *Trans. Am. Fish. Soc.* **1993**, *122*, 560–574. [CrossRef]
55. Carl, L.M. Sonic tracking of burbot in Lake Opeongo, Ontario. *Trans. Am. Fish. Soc.* **1995**, *124*, 77–83. [CrossRef]
56. Binner, M.; Kloas, W.; Hardewig, I. Energy allocation in juvenile roach and burbot under different temperature and feeding regimes. *Fish Physiol. Biochem.* **2008**, *34*, 103–116. [CrossRef] [PubMed]
57. Watz, J.; Piccolo, J.J.; Bergman, E.; Greenberg, L. Day and night drift-feeding by juvenile salmonids at low water temperatures. *Environ. Biol. Fishes* **2014**, *97*, 505–513. [CrossRef]
58. Vilhunen, S.; Hirvonen, H. Innate antipredator responses of Arctic charr (*Salvelinus alpinus*) depend on predator species and their diet. *Behav. Ecol. Sociobiol.* **2003**, *55*, 1–10. [CrossRef]
59. Rosell, F.; Holtan, L.B.; Thorsen, J.G.; Heggenes, J. Predator-naïve brown trout (*Salmo trutta*) show antipredator behaviours to scent from an introduced piscivorous mammalian predator fed conspecifics. *Ethology* **2013**, *119*, 303–308. [CrossRef]
60. Mikheev, V.N.; Wanzenböck, J.; Pasternak, A.F. Effects of predator-induced visual and olfactory cues on 0+ perch (*Perca fluviatilis* L.) foraging behaviour. *Ecol. Freshw. Fish* **2006**, *15*, 111–117. [CrossRef]
61. Prowse, T.D. River-ice ecology. I: Hydrologic, geomorphic, and water-quality aspects. *J. Cold Reg. Eng.* **2001**, *127*, 201–208. [CrossRef]
62. Smol, J.P.; Wolfe, A.P.; Birks, H.J.B.; Douglas, M.S.V.; Jones, V.J.; Korhola, A.; Pienitz, R.; Ruhland, K.; Sorvari, S.; Antoniades, D.; et al. Climate-driven regime shifts in the biological communities of Arctic lakes. *Proc. Natl. Acad. Sci. USA* **2005**, *102*, 4397–4402. [CrossRef] [PubMed]
63. Kausrud, K.L.; Mysterud, A.; Steen, H.; Vik, J.O.; Østbye, E.; Cazelles, B.; Framstad, E.; Eikeset, A.M.; Mysterud, I.; Solhøy, T.; et al. Linking climate change to lemming cycles. *Nature* **2008**, *456*, 93–97. [CrossRef] [PubMed]

Review

A Comprehensive Review of the Impacts of Climate Change on Salmon: Strengths and Weaknesses of the Literature by Life Stage

Lisa G. Crozier [1,*] and Jared E. Siegel [2,†]

[1] Northwest Fisheries Science Center, NOAA Fisheries, 2725 Montlake Blvd East, Seattle, WA 98112, USA
[2] Ocean Associates, Inc., Contracted to Northwest Fisheries Science Center, NOAA Fisheries, 2725 Montlake Blvd East, Seattle, WA 98112, USA; jareds@paceengrs.com
[*] Correspondence: lisa.crozier@noaa.gov
[†] Current address: PACE Engineers, Inc., 3501 Colby Ave Suite 101, Everett, WA 98201, USA.

Abstract: As we confront novel environmental challenges, a full understanding of the physical and biological processes that govern species responses to climate change will help maintain biodiversity and support conservation measures that are more robust to irreducible uncertainty. However, climate impacts are so complex, and the literature on salmon and trout is so vast that researchers and decision makers scramble to make sense of it all. Therefore, we conducted a systematic literature review of climate impacts on salmon and anadromous trout as a resource for stakeholders, managers, and researchers. We reviewed studies published from 2010 to 2021 that address climate impacts on these fish and organized them in a database of 1169 physical and 1853 biological papers. Papers are labeled with keywords across eight categories related to subject matter and study methods. We compared the literature by biological process and life stage and used these comparisons to assess strengths and weaknesses. We then summarized expected phenotypic and genetic responses and management actions by life stage. Overall, we found the largest research gaps related to species interactions, behavioral responses, and effects that carry over across life stages. With this collection of the literature, we can better apply scarce conservation resources, fill knowledge gaps, and make informed decisions that do not ignore uncertainty.

Keywords: anadromous fish; ecological responses to climate change; future projections; global change; ocean; freshwater

Key Contribution: This paper summarizes an enormous literature across sub-disciplines revealing the intricacy and complexity of climate impacts on anadromous salmon and trout. Highlighting strengths and weaknesses of each life stage, we provide useful information for decision makers on what to expect with climate change, and what tools are at their disposal for saving these iconic species.

Citation: Crozier, L.G.; Siegel, J.E. A Comprehensive Review of the Impacts of Climate Change on Salmon: Strengths and Weaknesses of the Literature by Life Stage. *Fishes* **2023**, *8*, 319. https://doi.org/10.3390/fishes8060319

Academic Editor: Bror Jonsson

Received: 5 May 2023
Revised: 14 June 2023
Accepted: 14 June 2023
Published: 16 June 2023

1. Introduction

Across the globe, climate change is forcing species to respond to unprecedented conditions. Pressures from climate change are overlaid on numerous other stressors, such as habitat degradation, resource exploitation, and depleted genetic diversity, which have already reduced the natural resilience of many native species [1–3]. Mass mortality events are increasingly attributed to warming temperatures [4,5]. Climate-driven extirpations, range shifts, and changes in phenology and productivity are reported at increasing rates every year [6–8].

The ability to predict the biological consequences of climate allows for proactive planning, which greatly reduces the risk of not meeting societal goals such as ecological, economic or social stability [9]. Quantitative projections of the biological consequences of climate change often rely on broad-scale statistical models that include relatively few predictor variables (e.g., [10,11]). This modelling approach assumes stationarity in the underlying correlation structure of relevant factors, creating a risk of model failure should the ecosystem shift. Predicting responses under conditions that are far outside our historical reference period is likely to be more accurate when it is grounded in more mechanistic understanding of how different biological processes interact [12–14].

Anticipating future change and reducing the extinction risk depends on untangling a web of physiological, ecological, and evolutionary responses [15]. Although our understanding of these complex and interacting processes will never be complete, it is worth considering how to assess our distance from this goal by quantitatively examining the areas of research in well-studied taxonomic groups. By synthesizing what is understood about climate impacts on anadromous salmon and trout (*Oncorhynchus* and *Salmo* spp.) with supporting information from other trout and charr *(Salvelinus* spp.), we show how research can be organized by biological process to evaluate data gaps in order to focus future research and management actions most effectively.

Warming temperatures have been strongly associated with mass mortality events in both cold-water and warm-water inland fish populations [4]. In Atlantic salmon (*Salmo salar*), declines have been attributed to climate change [16–18], with fisheries also to blame [19]. In Pacific salmon (*Oncorhynchus* spp.), large mortality events of critically endangered populations [20–22] and population declines [23–25] have been attributed, in part, to climatic conditions that are becoming more frequent with climate change [26–33]. Preserving these ecologically, economically and culturally significant species will require a rapid change in our current trajectory.

Anadromous salmon and trout face especially complex threats because their life history exposes them to a diverse set of interacting stressors in terrestrial/freshwater systems as well as the marine environment (Figure 1, and e.g., [33,34]). Furthermore, these fishes need to match migration timing with distinct seasonal patterns in biotic and abiotic conditions through diverse habitats. Populations will evolve genetically in both adaptive and maladaptive directions [35], but historical and ongoing declines in genetic diversity [3] and habitat quality and accessibility have greatly reduced the natural adaptive capacity of these fishes. Thus, a primary question facing managers is how to prioritize conservation actions to improve their adaptive capacity.

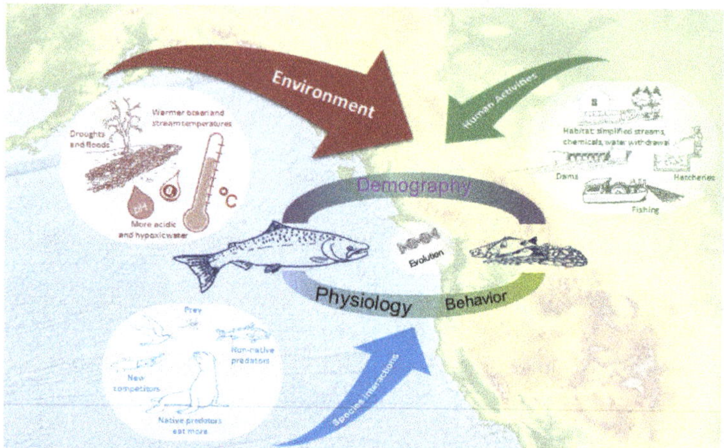

Figure 1. Multiple biological processes shape the impact of climate change on salmon. The primary factors that were represented as drivers are highlighted with arrows and icons. Additional processes are emphasized within the salmon life cycle. Process font size roughly represents the relative frequency of different processes among either drivers or responses in our review. The map shows the western U.S. and Canada and part of the marine migration range of PNW populations of *Onchorynchus*.

A vast and growing body of literature exists on climate impacts to salmon and trout, but the breadth and scope of this collection makes it difficult to evaluate and use. We reviewed and categorized a selection of relevant studies to provide a practical resource for both researchers and conservation decision makers. We included research from around the world, although we focused most on papers that were relevant to threatened and endangered salmon populations in the Columbia River Basin, located in the U.S. Pacific Northwest (PNW).

As a product of this review we provide (1) a categorized, searchable database of the literature as a resource for scientists and managers; (2) a synthesis of the main areas of research on climate impacts on salmon and trout by process and life stage, with relative data strengths and weaknesses identified and with summaries of expected phenotypic or genetic changes and areas of management evaluated for each stage; (3) a summary of future projections that identify phenomena established in the general literature but not included in the projections available for management decisions; and (4) recommendations that address management priorities, research gaps, and dealing with uncertainty.

2. Methods

2.1. Study Region

The Columbia Basin covers 668,000 km^2 and drains more water into the Pacific Ocean than any other river in North or South America. It straddles British Columbia, Canada and seven U.S. states, although much of the upper Columbia and Snake River Basins are inaccessible to anadromous fish due to impassable mainstem dams in Washington and Idaho. Human development has led to extirpation in an estimated 179 of 290 historical populations (62%) of salmon and steelhead trout in the Columbia Basin [36]. Of the remaining populations, five Pacific salmon species, as well as steelhead, include one or more stocks listed as threatened or endangered under the U.S. Endangered Species Act [37]. Some of these listed stocks complete arduous freshwater migrations, migrating distances of up to 1400 km and scaling extraordinary vertical ascents over 2000 m [38].

2.2. Literature Collection

For the overall dataset, we synthesized and augmented results from annual literature reviews conducted since 2010. These reviews were intended to identify new scientific findings relevant to the prediction and mitigation of climate change impacts on federally protected salmon and steelhead (*Oncorhynchus* spp.) from the Columbia River Basin [39–48]. For each annual review, our search focused on peer-reviewed scientific journals included in the *Web of Science* Core Collection database.

We conducted four searches of this database using publication year and (Boolean operators used in the search are shown in boldface) salmon, *Oncorhynchus*, or steelhead, and (1) prespawn mortality, (2) ocean acidification, (3) climate(the wildcard (*) was used to search using "climat*" to capture all forms of the word "climate"), temperature, streamflow, flow, snowpack, precipitation, or PDO, (4) marine, sea level, hyporheic, or groundwater and climate. Additional searches involved physical/climatic terms and geographic terms, without the biological requirement. The physical/climatic terms were: climate, temperature, streamflow, flow, snowpack, precipitation, PDO, marine, sea level, hyporheic, groundwater, upwelling, estuary or ocean acidification. Geographic terms were: Pacific Northwest, Pacific, California Current, Columbia River, Puget Sound or Salish Sea.

We supplemented these results with technical reports from state or federal agencies involving Columbia and Snake River populations. Combined, the number of physical and biological study was reduced to 3022 papers (1169 physical, 1853 biological) that are listed in our database (Data S1). For this synthesis, we quantified characteristics using only the 1853 papers that were primarily relevant to biological impacts on salmon, although we briefly summarize the most pertinent physical results.

2.3. Classification Strategy

We assigned labels within a set of eight categories to all papers based on the primary focus of the study. These categories included *species, life stage, region, sub-region, study type, study duration, drivers* and *responses* (see Supplementary File S1: Table S1 Description of the criteria used to assign each label). A *driver* represented a natural or anthropogenic explanatory variable, i.e., a factor represented as potentially forcing change on salmon. The *response* was defined as biological outcomes measured or described in salmon as the outcome of an explanatory variable, such as a change in survival or behavior (loosely based on definitions suggested by [49]). Where appropriate, papers were assigned multiple labels within a category (e.g., multiple species studied).

In addition to these categorical labels, we also identified papers by biological process. To predict and prevent the most damaging aspects of climate change, Urban et al. [15] identified six biological processes which should be accounted for in projection modeling: *environmental conditions, physiology, demography, species interactions, dispersal and evolution*. It is important to consider all processes, because any one of them could alter the net response to a changing climate. As described below, we refined the definitions of these processes somewhat to reflect topics that are most relevant for salmon. To explore how frequently each process is documented in the salmon literature, we grouped *drivers* and *responses* into these process categories as shown in Table 1.

In general, *environmental* process papers were those focused on environmental conditions as drivers and how they interact to affect the other five processes. Climate indices were often used to represent combinations of physical drivers that are difficult to separate empirically. *Physiology* studies focused on condition metrics, growth, maturation, performance, and morphology; *demography* papers focused on population dynamics, life history and phenology; and *species interactions* papers focused on ecosystem indices, salmon prey, competitors, predators, and dynamics associated with disease. Ecosystem indices may not always reflect direct species interactions, but do reflect some cumulative properties of multiple species. Because the region of primary interest is nested within the larger range of salmon species, and salmon are largely identified by watershed, the *dispersal* process referred mostly to changes in habitat use and migration behavior, and less often to dispersal outside the current range, although colonization was also discussed [50].

The *evolution* process label was assigned to any paper that used evolutionary methods of analysis (e.g., genetic assays, common garden or phylogenetically driven comparisons, estimates of selection). A few papers with this label used more general concepts of adaptation through non-genetic mechanisms. We indicated these papers in the database so that they can be excluded from analyses if desired. Finally, we added a *management* category, which consists of papers that directly addressed anthropogenic actions that are actively managed, such as fisheries, restoration actions, and the impact of changes in salmon as a resource for communities.

2.4. Synthesis

While physical climate change papers were not the main focus of this review, we first provided a summary of the observed and expected physical consequences of climate change. Next, we categorized the research highlights by individual life stage and full life cycle (population-level) analyses. We organized papers by life stage because of the distinctiveness of research for each and to emphasize the biological and management needs for each part of the life cycle. For each life stage, we characterized focal areas of research by clarifying the main drivers and responses examined. We also looked at the number or percentage of papers within each life stage that were assigned a given label and compared these values with the respective values for the relevant label across the database as a whole and across all categories to examine the relative frequency of different topics in research. We then highlighted papers that projected future biological responses to climate changes. Note that papers often addressed multiple life stages, and thus were considered in all of the stages that received particular attention or analysis.

Table 1. Biological processes, labels and their definitions that were assigned to papers within the driver and response categories. Each label was considered either a driver or a response, and they are ordered as such in the table. Methods papers were not assigned to a biological process.

Process		Definition
	Driver	
Environment	Acidity	Ocean and freshwater acidification
	Climate indices	Large-scale climate indices (PDO, ENSO)
	Contaminants	Contaminants such as heavy metals and PCBs
	Dissolved oxygen	Dissolved oxygen concentrations, hypoxic waters
	Environment	Other environmental drivers (e.g., salinity, upwelling)
	Flow	Freshwater flow levels
	Habitat	Physical habitat characteristics
	Marine temp	Marine water temperatures
	Freshwater temp	Freshwater water temperatures
Species interactions	Density food	Density, competition, and food availability
	Disease	Impacts of disease
	Ecosystem	Relevant ecosystem interactions not focused on salmon
	Invasives	Invasive species competing or preying on salmon
	Predators	Salmon predators
	Prey	Salmon prey
Evolution	Genetics	Genetically derived traits or genetic diversity
Management	Fisheries	Impacts from fisheries
	Management	Management levers (e.g., flow, hatchery, policy, framework)
	Restoration	Habitat restoration as a driver
	Methods	Methodologies for science or management
	Response	
Physiology	Growth	Salmon growth
	Immune	Immune system responses
	Maturation	Sexual maturation
	Morphology	Physical morphology
	Performance	Fish performance metrics (e.g., swim speed)
	Physiology/condition	Internal physiological responses/fish condition metrics
Demography	Carryover	Carryover impacts from one life stage to another
	Life history	Life history changes, often demographic
	Mortality	Mortality rates
	Phenology	Timing of life history events (e.g., spawning, migration)
	Population resiliency	Resilience of entire populations
	Productivity	Population productivity
Dispersal	Behavior	Changes in behavior
	Habitat distribution	Distribution within available habitat
Species interactions	Diet	Diet composition as a response
Evolution	Genetic adaptation	Change in genotype or adaptive response (phenotype)
Management	Livelihood	Human economic or subsistence

Finally, we provide a synthesis of the strengths and weaknesses of the literature by life stage. Particular areas of research or types of study that were more heavily represented quantitatively within the reviewed literature were defined as candidate strengths, and those less represented were defined as candidate weaknesses. However, numerical representation is not equal to depth of understanding. We therefore refined this list to more specific study areas using our judgment to ensure that identified strengths were relatively well studied and understood, while weaknesses represented gaps in the scientific knowledge that limit our ability to manage for and predict the consequences of climate change's impacts on Pacific salmon. Note that some of these results are based on the frequencies of representation across categories in our database not detailed in the main text, but are reported in Supplementary File S1: Quantitative analysis of categories across the entire database.

While we provide a large number of references in this review, in the interest of space we did not cite all of the supporting literature, but rather offer the database itself for additional references on each topic. Here, we prioritize some of the lesser-known topics to draw attention to them, while avoiding repetition of widely known information.

3. Results

3.1. Observed and Projected Physical Impacts of Climate Change

3.1.1. Global

Historical trends and projected future trajectories of climate change are summarized at the global and regional scales in the Sixth Assessment Report (AR6) of the Intergovernmental Panel on Climate Change [51] and the most recent Synthesis *Report*. This assessment report contains the most recent, authoritative, and comprehensive summary of our global knowledge on climate impacts and represents a monumental achievement. Over 234 authors and 517 contributing authors from 66 nations contributed to the report and its conclusion that "it is unequivocal that human influence has warmed the atmosphere, ocean and land. Widespread and rapid changes in the atmosphere, ocean, cryosphere and biosphere have occurred." (IPCC 2021; Summary for Policymakers. Pages 3–32).

The AR6 documents numerous observed changes; here we list just a few examples.

- Global surface temperature in the last decade (2011–2020) was 1.09 °C higher than in 1850–1900
- Due to human influences, global average land precipitation has increased, changing near-surface salinity
- Glaciers have been retreating, Arctic sea ice in September has decreased about 40% from 1979–1988 to 2010–2019, and spring snow cover has decreased in the northern hemisphere.
- The global upper ocean (0–700 m) has warmed, the ocean surface is more acidic, oxygen levels have dropped in many upper ocean regions since the mid-20th century, and the global mean sea level has increased by 0.2 m between 1901 and 2018, and at a rate twice as fast as the long-term average from 2006 to 2018.
- Human influence is the main driver of more frequent and more intense terrestrial and marine heat waves, and concurrent events among heatwaves, droughts, wildfires and flooding.

All of these trends are expected to continue and intensify over the next century globally and in the Pacific Northwest (PNW). Here, we provide a summary of specific studies addressing the essential physical environmental changes expected in the PNW and California Current.

As climate science has progressed, analytical tools have been developed to quantitatively attribute extreme events to climate change. For example, the American Meteorological Society (AMS) now publishes an annual special report documenting the causes of extreme climate events around the world (e.g., [52]).

Scientists have long predicted that climate change would create meteorological and oceanographic conditions outside the range of our historic records. However, the AMS special report of 2016 was the "first of these reports to find that some extreme events were not possible in a preindustrial climate" [52]. Specific events identified included the North Pacific marine heat wave known as the Blob, which had major impacts on regional marine mammal, bird, and fish populations, including salmon [52].

Anthropogenic climate change has already impacted the northwestern U.S., and observations of change have been consistent with past projections [53,54]. In Washington [55] and Idaho [56], average temperatures have risen by about 1.1 °C since 1900, while in Oregon, warming has been more extreme at, 1.4 °C [57]. In addition, the seasonal duration of the freeze-free season has declined, and potential evapotranspiration has increased, leading to larger water deficits [58].

Temperature increases have been observed throughout the year, while precipitation increases have been seen primarily in spring, while showing mixed trends during other parts of the year. As new temperature records are set, studies in North America have increasingly attributed changes to increases in atmospheric greenhouse gases [58–60].

3.1.2. Freshwater Impacts

As air temperatures have increased, the spring snowpack has declined throughout the western U.S. during the 20th century [61–64]. Heavy snowfall events have also decreased in frequency in the PNW and California in the period of 1930–2007, [65]. With snowpack decline and warmer air temperatures, glaciers have retreated in North America [66] and in the PNW specifically [67–69]. The snowpack has also begun to melt earlier in the year, and peak stream flows during spring have congruently shifted timing [62,70,71], although water management has compensated for this change in some managed river reaches [72,73]. These factors have combined to make summer drought and low flow conditions more common [70,74]. Declines in summer precipitation have led to lower minimum streamflows, and long-term trends in low flow extremes have occurred throughout the West [75–77]. Shifts in the timing of and extremes of the hydrological cycle have implications for fish behavior, growth rates, and risk of mortality from stranding or overcrowding.

These regime changes have led to a higher frequency of forest fires [78], although poor forest management has contributed to this problem [79]. Forest fires are a dynamic part of the natural landscape, and have complex effects on streams. Initially, there is often sediment input from erosion, and fish passage may be blocked. Over time, the sediment is redistributed and can lead to an increase in nutrients in the stream, boosting productivity. Long-term effects depend on overall landscape processes. from Stream temperatures, which are impacted by climate conditions via solar radiation, precipitation, and snowpack accumulation/melt, have also increased [80]. Higher temperatures are associated with lower oxygen concentration, which can lead to hypoxic conditions. These increases are expected to continue alongside increasing air temperature, declining snowpacks, and decreasing canopy cover from land use and forest fires [31,71,81].

3.1.3. Marine Impacts

Oceans have absorbed much of the heat created by human-produced greenhouse gases, leading to a steep and persistent upward trend in global ocean temperatures [82]. Chen et al. [82] reported increases from the sea surface to the 2000 m depth since 1985, with record temperatures in 2020, although increases in ocean temperatures have generally been more extreme at the surface [83]. In the North Pacific, sea surface temperatures also increased during the 20th century [84–86]. However, these increases may be partly a consequence of regional shifts in wind as opposed to direct forcing from increases in air temperature [84,87].

While higher surface temperatures are mitigated by the upwelling of cold, deep sea waters, warming of ~0.7 °C has occurred since 1900 in PNW coastal areas [88]. Acidic (low pH) and hypoxic (low oxygen) waters commonly occur in coastal waters of the California Current as a consequence of upwelling. However, research suggests that increases in ocean acidification from the uptake of atmospheric CO_2 have increased the spatial extent of acidic and poorly oxygenated waters [89]. Combined with recent increases in upwelling, acidic and hypoxic waters have impacted nearshore areas in particular [90]. Trends in ocean acidification consistent with increases in atmospheric CO_2 have been documented in Puget Sound and the Strait of Georgia, which are isolated from upwelling dynamics [71,91].

A large body of research has focused on the causes and impacts of a major marine heatwave in the North Pacific Ocean known colloquially as the Blob. The Blob persisted from 2014 to 2016, when sea surface temperature anomalies were observed to exceed 3 standard deviations (~4.5 °C) above normal [92,93]. Its development was attributed to multiple co-occurring natural drivers exacerbated by global warming [94]. Persistence of the Blob may have been maintained for multiple years through teleconnections between the North Pacific and El Niño [95].

Novel environmental conditions during this anomalous marine heatwave led to numerous ecological disturbances in the California Current, including a historically unique community composition, including plankton, pyrosomes, crabs, and fish that are generally observed in more southern oceans [96–98]. In addition, the lower energy content in the plankton community had cascading trophic impacts that resulted in seabird die-offs [5,99], reduced pelagic fish conditions [100–102], and shifts in fish species distributions and spawn timings [99,103].

Climate model simulations indicate that extreme conditions such as those experienced during the 2014–2016 marine heatwave are likely to occur more frequently as climate change progresses [95]. Consistent with this prediction, anomalously warm ocean conditions returned to the North Pacific in 2019–2020 [83].

3.2. Life-Stage Specific Research

Salmon and trout spawn in freshwater streams. Generally, spawning occurs in fall or spring, eggs incubate over the following weeks to months, and then emerge from the gravel. In anadromous populations, the freshwater rearing stage lasts until juveniles migrate to the ocean, which is the smolt stage. They then rear in the ocean until they return to spawn. The duration of these stages is extremely variable, across both species and populations in species with longer life spans. For example, pink salmon (*O. gorbuscha*) has a relatively fixed life span of 2 years, of which only a few weeks are spent in freshwater. Coho most frequently lives for about 3 years, divided approximately in half between freshwater and ocean stages. Chum (*O. keta*) lives longer, but spends less time in freshwater.

The remaining salmon species can spend many years in freshwater. *O mykiss* and *O. nerka* have fully freshwater resident forms (rainbow trout and kokanee, respectively). These forms retain the ability to resume anadromy under appropriate conditions. In fact, it is often difficult to distinguish resident and anadromous contributions to reproduction in some populations, which raises challenges for population dynamics modeling.

Salmon life stages are associated with distinct physical and biological needs, and will likely respond differently to future climate forcing. Furthermore, life stages may occur in completely different spatial regions, require different methods of study, and are tractable for answering different questions. We compared the relative frequency of papers studying different life stages. The juvenile rearing stages in freshwater (33%) and saltwater (28%) were the stages that were addressed most commonly. Migration stages were highlighted in 10% and 14% of the papers for the downstream and upstream directions, respectively. Egg, spawning and population level analyses constituted 6%, 7%, and 9% of the database, respectively. In all stages, environmental factors were the most prevalent driver, but the relative proportion of species interactions (more common in marine studies) and drivers affected by human activities (more common in upstream migration and population-level analyses) depended on the life stage (Figure 2). Among responses, demographic processes were generally most prevalent, but the relative proportion of physiological and behavioral processes varied by life stage (Figure 2).

Study types also varied across life stages (Figure 3). The egg stage was most often studied experimentally. Field studies were most prevalent in research on the two migration stages and the spawning stage. The fewest models, reviews, and projections were performed on the egg and migratory stages. The marine stage had the highest percentage of reviews and meta-analyses, whereas theoretical work and projections of future salmon responses to climate change were most often completed at the population level.

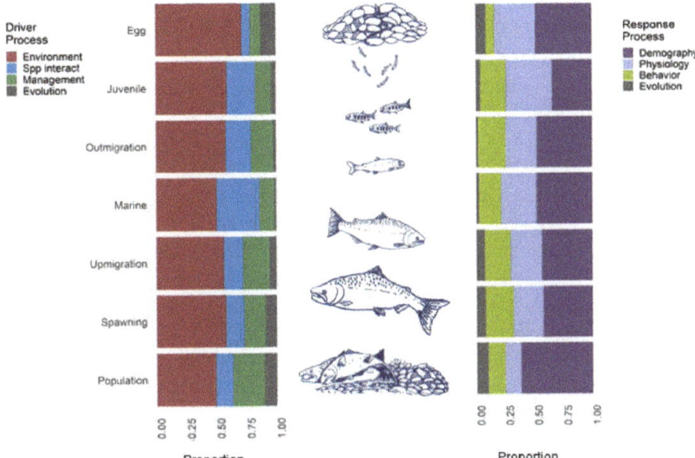

Figure 2. Barcharts show the relative frequency of papers addressing each process among drivers (**left**) and responses (**right**) for each life stage.

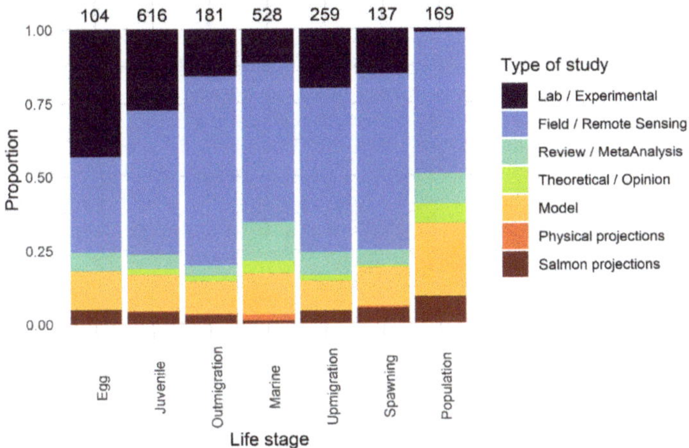

Figure 3. Relative proportion of papers on each life stage that involved different types of studies. The absolute number of papers represented in each column is at the top. The juvenile stage refers to the freshwater rearing portion of the life cycle.

3.2.1. Egg Incubation

Salmon lay eggs in gravel nests, called redds. Incubation periods are as short as a few weeks for low-elevation species, especially pink, chum, fall Chinook, and coho. Incubation may continue for as long as six months in the higher elevation redds of spring Chinook, sockeye, and steelhead. Redds may be laid in very shallow water by beach spawning pink and kokanee, or submerged up to 10 m by summer and fall Chinook [104]. Eggs are also buried within an individual redd at variable depths, which are usually deeper for larger spawners.

Adult female behavior and competition dynamics determines where, when, and how redds are buried, which largely determining the environmental experience of the eggs. After spawning, females will often protect redds from superimposition until they die. However, variability in the river environment over the incubation period plays a large role in determining which spawning habitats are most productive and the rate of development, which determines when alevins emerge from the gravel.

Drivers:

The components of climate change that are most likely to affect egg incubation directly are changes in stream and groundwater temperatures and flow (see Supplementary File S1: Table S2 for stage-specific values for all labels). Temperature affects survival when lethal limits are exceeded, but also governs egg development times, egg viability, and sublethal effects that carry over into subsequent life stages. A higher percentage of papers included temperature as a driver in this stage than in any other (61%, Supplementary File S1: Table S2). Changes in flow, which were examined in 29% of the studies, can affect the risk of dewatering, the deposition rate of fine sediment that can suffocate eggs, and the risk of redds being dislodged as a result of scouring during high-flow events.

Responses:

Deposition of fine sediment on redds reduces the availability of oxygen to embryos. During different stages of embryo development, salmon demonstrate variation in sensitivity to the combined impacts of temperature and hypoxia on mortality and growth [105]. Beyond high flows, increased sedimentation can also occur as a consequence of habitat disturbances from natural events, such as wildfires [106], and from anthropogenic activities, such as forestry or land development [107].

Eggs require more oxygen at higher temperatures, and in some cases oxygen limitation may reduce survival at temperatures that would otherwise not be lethal [21]. Tolerance of hypoxia and high temperature are also genetically linked, and hence display correlations both phylogenetically and within organisms with some overlapping physiological mechanisms [108]. Although relatively rare compared with other life stages, declines in population productivity have been linked to climate effects specifically in the egg stage [23,109,110].

Laboratory experiments on early life stages have been common in the hatchery and aquaculture industries, and these industries also facilitate easier access to fish for more general ecology studies. Thus, extensive laboratory work over decades has informed our understanding of the functional response of embryos to various temperatures and levels of dissolved oxygen (53% of all papers in the egg stage). Many of these studies were transgenerational, tracking eggs from spawners to the juvenile stage. Intergenerational studies improved our understanding of the heritability of traits and evolutionary mechanisms in the egg stage (16% of all papers covering egg stage included genetics as a driver, compared with 8% across all life stages). In particular, there is evidence that egg size, thermal tolerance, and development rates have a heritable component, suggesting the possibility for genetic adaptation of these traits [111–118].

Projections:

Projections specific to the egg stage focused on increasing flood intensity, which will likely increase mortality, and rising temperatures, which will hasten emergence timing or increase mortality [119–124]. These estimates are difficult to field test, and the egg stage is often nested within assessments of spawner-to-juvenile productivity. These projections accounted for direct mortality based on observed correlations, and some accounted for a change in incubation timing that would carry over into the next life stage [124]. Neither carryover effects nor processes other than flow and temperature were included in these projections.

3.2.2. Juvenile Freshwater Rearing and Migration

Juvenile salmon freshwater rearing may continue for days to years, varying at the species, population, and individual level. In the Pacific Northwest, Chinook and steelhead exhibit the widest range of freshwater behavior. At the southern edge of the Chinook range, for example, some adults spawn in late fall or winter, and their offspring migrate the following spring. At the northern edge of the range in Alaska, adult Chinook usually spawns in summer, but smolts may not migrate until spring up to two years later. Steelhead and Atlantic salmon can also spend over two years in freshwater before migrating to the ocean, with substantial variation in life history patterns across and within populations [125–127]. Differences in juvenile age at migration, the seasonal timing of migration, and the rate of travel from their natal habitat to the ocean are largely driven by growth rates and habitat conditions interacting with evolved thresholds for behavioral and physiological changes. In the Pacific Northwest, many freshwater habitats are heavily impounded and degraded by human activities. Habitat loss and degradation (which include both biotic and abiotic components) and management actions to mitigate them are therefore a core area of research in the juvenile life stage.

Drivers:

Salmon behavior, growth, and development rates are extremely plastic and respond quickly to habitat conditions. Habitat suitability played a role in a higher fraction of studies in the juvenile rearing (27%) and migration (32%) stages than in any other life stage. Habitat quality depends on many factors, including all components of water quality (especially temperature, flow, and contaminants), as well as biotic conditions.

Physical conditions affect general biotic conditions, including salmon prey, which then affects salmon. For example, the abundance, biomass, mean body size, and carrying capacity of macroinvertebrate and planktonic prey species have each been shown to respond to changes in river temperature and flow [128–130]. The physical changes therefore affect salmon through both direct and indirect pathways.

Variation in flow impacts habitat suitability by altering current speeds and water availability through the inundation of floodplains. Wildfires also can increase summer stream temperatures through the removal of shading from riparian vegetation and can increase productivity through bursts in nutrient inputs from erosion. However, increased prey fueled by extra nutrients may not be enough to compensate for higher salmon metabolic needs at higher temperatures. For steelhead in a California stream, fire resulted in a net decline in salmon biomass [131].

Species interactions as a driver have received substantial attention in juvenile freshwater (35%) and marine stages (52%) because of the intense competition for food and risk of predation in these stages (Figure 4). These studies examined disease (94 papers), predators (96 papers) and competition, mostly with invasive species (32 papers) and among trout [132]. In the Pacific Northwest, the invasive species studied included zooplankton in the lower Columbia River and multiple reservoirs [133–135].

Although the full implications of these invasions are not known, some feeding experiments show that Chinook salmon and other predators ate the Asian calanoid copepod *Pseudodiaptomus forbesi*. However, in other experiments, these fish preferred native prey [135]. Peak abundance of the invasive zooplankton occurred during periods of peak water temperature, especially in late summer and early fall [133,134].

Invasive American shad (*Alosa sapidissima*) interact with salmon as prey and competitors [136–138]. More generally, a depleted prey base was correlated with the abundance of nonnative fish across seven streams [139]. Additional studies addressed competition among trout species [132].

Changes in growth rate may alter the window of time during which juvenile salmon are vulnerable to size-selective predators such as bass [140]. Smallmouth bass also interacts with salmon at multiple trophic levels, mostly as predators, and are expected to increase their range and abundance in a warmer climate [141–143]. Kuehne, et al. [144] suggested that salmon expend extra energy on predator avoidance at warmer temperatures, further reducing growth.

Responses:

The juvenile life stage was the only stage during in which a higher proportion of papers described a physiological response process (49%) than a demographic response process (42%). This ordering largely reflected the importance of growth rates (26%) as a focus of research in this stage. Behavioral responses have also been closely monitored, especially habitat use (20% of juvenile papers) and migration timing (29% of juvenile migration papers), due to the relative ease of studying freshwater rearing streams. In the rearing and migration stage, adequate numbers of study fish can be sampled using smolt traps and fish can be individually tagged and followed by remote electronic detection.

Behavioral studies have focused on food and growth constraints, movement out of natal areas, interactions with other species, and behavior around dams. Juvenile salmon behaviorally thermoregulate to best take advantage of the amount of available food. For example, in a controlled study, Boltana and Sanhueza [145] found that juvenile Atlantic salmon reared with access to a larger range of temperatures within which they could self-regulate had higher growth, survival, and muscle growth compared to fish raised in a more restricted range of temperatures. Like adults, juvenile salmon use cold-water thermal refuges when temperatures are above optimum. However, food resources may become scarce in refuge habitat as densities increase [146].

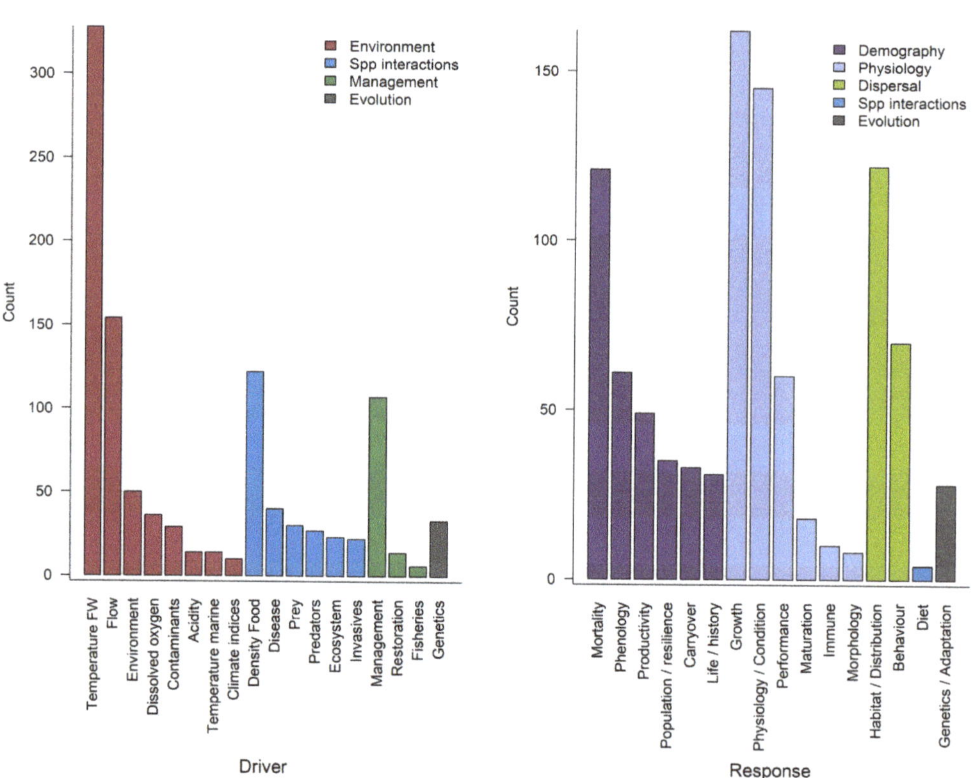

Figure 4. The number of papers covering the juvenile life stage that compared each type of driver (**left**) and each type of response (**right**). Colors reflect the process associated with each label.

Projections:

More work on projecting climate change effects has been completed for the juvenile stage than any other, and many of these juvenile-specific projections are included in full life cycle models. In the PNW, projections using future scenarios for temperature and flow were estimated for the juvenile production of coastal coho [147], dam passage of coho in the Cowlitz River [148], and trout production in northwestern Oregon [149]. In the Columbia Basin, models have estimated steelhead growth in the Yakima Basin [150], smolt survival for Chinook and steelhead in the mainstem Columbia River [151] and Lemhi River [152].

Smallmouth bass are expected to further invade the John Day River [153], and disease risks will likely increase for hatcheries in the upper Columbia [154]. The projection model for early salmon life history with the greatest geographic scope was developed by Beer and Anderson [155] for the whole western U.S. This model predicted higher rates of growth for fish in the coolest streams, but lower rates for fish in warmer streams in the Columbia Basin and California Central Valley leading to smaller sizes at marine entry and thus lower survival. Furthermore, steelhead were predicted to spawn earlier due to their spring spawning behavior, while fall-spawning Chinook are predicted to spawn later at many sites.

3.2.3. Marine Stage

The marine life stage is where explosive growth differentiates anadromous from resident life history types. Salmon smolts enter the marine stage after leaving natal freshwater streams and remain in this stage until they return as adults to spawn in freshwater. Due to the difficulty of directly studying salmon in the ocean, a large proportion of studies on this stage rely on data from freshwater adult returns.

Most of the directly observed information about this life stage comes from oceanic research cruises and fisheries catches. Marine residency may last less than one year, particularly in hatchery fish, but typically lasts 1–3 years for Chinook and coho salmon. Historically, six- and seven-year old Chinook spawners were not uncommon, but now they are extremely rare. Males often return at a younger age than females and make up a higher fraction of the resident component in species that express both anadromous and resident life histories. This difference between males and females is thought to be a consequence of the higher energetic requirements to produce eggs versus sperm; thus, larger size is more directly related to reproductive success in females than in males. Mortality in the marine stage is largely driven by predation and directed salmon fisheries, plus non-targeted catches in other fisheries.

Drivers:

Salmon marine survival is generally correlated with overall ocean productivity. Productivity in the California Current is fueled by cold, nutrient-rich upwelled, deep water, and cooler surface water flowing predominantly from the north, as opposed to warmer water from the south and west. The relative contributions of these different water sources are driven by seasonal and inter-annual wind patterns. Seasonally, the California Current is more productive after the spring transition in wind direction.

At the annual and decadal scale, the Pacific Decadal Oscillation, the North Pacific Gyre Oscillation, and the El Niño Southern Oscillation all affect the strength of currents from north, south or west, feeding the California Current. The complexity of these phenomena and how they interact with the food web explain the high representation (Figure 5) of large-scale climate (14%) and ecosystem (27%) indices over an exclusive reliance on temperature (32% included temperature but not a climate index).

Correlations between salmon productivity and specific ocean indices are typically strong, although their strength varies between species and life history types [156–163]. How climate change will affect drivers of natural ocean variability, on the other hand, is much less certain compared with the data available on climate drivers of freshwater conditions [51,84,164–166]. Moreover, statistical correlations with climate indices tend to break down over time [167–170], which necessitates a more mechanistic understanding to support robust projections with climate change.

Other environmental drivers, such as pH, oxygen, and contaminants, are generally modeled as indirect effects on salmon through ecosystem models [171–173]. There are a few studies on the direct effects of pH on salmon physiology, specifically of pH effects on olfactory systems and the ability to respond to predators [174–176]. Conversely, other work demonstrates that salmon have a relatively higher resilience to changes in pH than other fish [177].

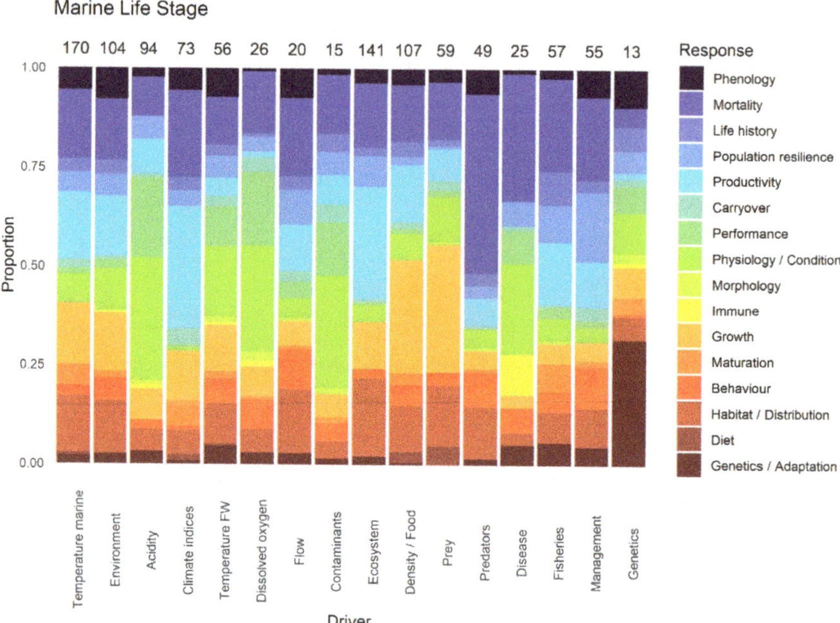

Figure 5. Proportion of papers within the marine life stage that compared each type of driver (x-axis) with each type of response (colors). The absolute number of papers represented in each column is provided at the top.

Forage and groundfish are important as salmon prey when they are small, then competitors and predators as they grow, depending on their spatial and temporal overlap with salmon. However, in California, the excess consumption of anchovy due to lack of other prey in recent years has led to a thiamine deficiency, causing reproductive failure in critically endangered winter-run Chinook salmon [178]. Large schools of forage fish also attract predators, which might then increase or decrease predation on salmon [179,180]. Forage, hake, and groundfish are economically valuable, and thus are the subject of research, monitoring, and climate projections in their own right, which is important for understanding future intersections with salmon [181–183]. Some reviews cover the state of our knowledge on forage fish and their predators in the California Current [184] and specifically in the Columbia River plume [185,186].

End-to-end models attempt to link atmospheric and physical drivers through ocean ecosystem processes all the way to the top predators and fisheries. End-to-end models for the California Current have explored many different aspects of productivity for the food web and fisheries, but relatively few of these models have explicitly focused on salmon. Some examples of Pacific salmon-relevant results include an analysis of the sensitivity of local fish productivity to upwelling intensity [187] and a study on the detrimental effects on salmon from increasing competition from jellyfish [188], which is relevant because jellyfish are generally more tolerant of warmer, more acidic and nutrient-poor conditions associated with climate change and human impacts [189–192]. Another mechanistic approach to modeling the ocean stage is to embed individual based models or models of intermediate complexity into a regional ocean model. This approach has been applied primarily to the southern California Current to understand environmental drivers in salmon prey (krill) [193], salmon growth [194], and salmon predators (sea lions) [195].

Responses:

Largely for the purposes of distributing ocean catch among countries, efforts have evolved to clarify the ocean distribution of salmon. Migration pathways have been revealed by the retrieval of coded wire, acoustic, and passive integrated transponder tags, supplemented by the genetic stock identification of untagged fish. Ocean migration routes of Chinook salmon reflect patterns of both life history (ocean-type subyearling vs stream-type yearling) and population [159,196–198]. Yearling migrants from the interior Columbia and Fraser Basins generally migrate in May and June and move quickly northward along an offshore route. However, subyearling and yearling migrants from warmer natal environments in lower rivers and coastal areas remain in nearshore areas and in the California Current for much longer periods.

Carryover effects from freshwater have received particular attention. A long-standing question of interest is to what extent freshwater conditions affect marine survival, and how management can improve the deleterious consequences of carryover effects. Ocean arrival timing, body size, flow rate and passage route through dams, and wild vs. hatchery origin have often been tested as predictors of marine survival [199,200]. Contaminant exposure in freshwater [201,202], especially aluminum [203], also can reduce marine survival. However, freshwater contaminants have rarely been included in carryover-effect projection studies of Pacific salmon.

Many different toxins are present during harmful algal blooms, and these blooms are increasing worldwide [204,205]. Several fish kills of farmed salmon have been attributed to harmful algal blooms, or red tides [206–208]. Nonetheless, most wild salmon appear to be relatively resilient to domoic acid, the most notorious product of red tides in the California Current [209]. Domoic acid produces amnesic shellfish poison, which is retained in shellfish that feed on contaminated algae, and which then has fatal effects on marine mammals and seabirds that prey on poisoned shellfish [210,211]. The neurotoxin saxitoxin, which causes paralytic shellfish poisoning, is also expected to increase with climate change [212,213]. Saxitoxin negatively impacts a variety of fish, including Atlantic salmon and rainbow trout [214], and it has been isolated from naturally migrating Japanese chum [215].

Understanding species interactions has been especially challenging in the marine life stage. Individual tracking studies can clarify when and where ocean mortality occurs (e.g., before or after good growth), but these studies are rare compared with studies of freshwater movement. Nonetheless, they have produced valuable information on where ocean mortality occurs [216,217] and in some cases, information about the specific predator from depth- or temperature-recording tags. Characteristic body temperatures in different species can be strong identifiers, as can as behaviors such as diving to preferred depths or haul-outs onto land, when tag data can be recovered after predation occurs [218].

Correlations between prey abundance, salmon growth, and salmon survival have been inconsistent across populations and over time. Several papers studying California fall Chinook have described a correspondence between prey abundance and salmon survival [219–221]. Similarly, Columbia River studies have found strong correlations between climate variables, salmon diet or size, and adult returns [222–227].

However, patterns in early marine size and survival do not necessarily match. Inconsistent patterns in growth and early ocean survival (e.g., [228]) have been attributed to higher size-selective mortality in years with lower survival [222,229,230], or alternatively, higher energetic demands in a warmer ocean that ultimately lead to mortality despite better growth conditions [231,232].

Some of the most important trends over time for the marine stage occurred in age at maturation and size at age. Comparisons between Paleolithic and present-day Atlantic salmon showed that spending two and three years at sea was much more common historically than it is now [233], and this trend has continued recently [125,234]. Age at maturity has a strong genetic basis in Atlantic salmon, and is associated with a single locus with a large effect [235]. The strong fishery-induced selection for this trait has been attributed to both indirect effects stemming from harvest of salmon prey and direct effects relating to the harvest of larger salmon [236,237].

In Pacific salmon, declining age at maturation and size of older salmon have also been widespread [238,239]. While fishing is likely to be responsible for some of this trend [240], additional factors are also involved. Warming environmental conditions are also associated with younger age at maturity across many diverse taxa independently from fisheries [241]. Larger smolts at marine entry and faster growing fish in the marine stage generally mature after fewer years in the ocean, explaining another component of the observed trends [125,242].

An additional factor may be changes in predation pressure. Long-term trends of increasing density in marine mammal populations have been blamed for lowering salmon survival in the California Current [243], in Puget Sound [158–160,244,245] and in Europe [246]. Killer whales in particular favor larger Chinook salmon [238,247]. Salmon sharks have also been identified as late-ocean stage predators that could be driving down the average age at return [218,248].

Projections:

Our database included 29 papers that provided projections for salmon or ecosystem components on which they depend. The salmon-focused papers used a number of different approaches to capture ecological mechanisms. For example, temperature indices were used as a covariate in regression models predicting survival [27,249], as a driver of bioenergetic consumption rates that predict growth rates [250], or as a correlate of observed spatial distributions [10,251–254].

Many food web models include salmon as a functional group and predict changes in biomass as they respond to changes in bottom-up and top-down interactions [171,173,255–258]. A few models focused on particular predators that might be directly relevant for salmon [259,260] and their prey, often with a focus on the impacts of ocean acidification [261–265]. There is also a large body of work on lower-trophic-level responses to climate change.

3.2.4. Adult Migration and Spawning

Salmon are considered upstream migrants from the time they enter freshwater until they reach potential spawning grounds. The relative importance of the migration stages (upstream and downstream) differs greatly between coastal populations in undammed rivers and interior populations. Pacific Northwest interior populations migrate past up to nine hydroelectric projects. Thus migration can take just a few days for coastal populations or an entire year for some interior steelhead. Fish may interrupt upstream movement and "hold" or linger downstream of or at spawning grounds for hours, days or months.

The environmental characteristics of the system determine the run type and species that may be present in a given watershed. Salmon typically consume little to no food during their upstream migration and spawning stages, so their bioenergetic reserves are limited by their condition upon entering freshwater. The physiological demands of migration therefore range from relatively unimportant to prohibitive. Anthropogenic barriers to migration have been devastating for some populations and are a major focus of research and management.

Drivers:

Temperature and flow are the main environmental determinants of the timing and success of upstream migration and spawning. Physiological responses to high temperature and flow, fisheries handling, and other stressors have been extensively examined in cardiac tissue because of the importance of this organ during strenuous exertion.

Genetic variation has also been examined in traits such as aerobic scope and optimal swim temperature and speed [266], with such examinations contributing to the relatively large proportion of studies that include evolutionary processes (11%). Once salmon reach holding areas, they may face density-dependent mortality related to multiple stressors, such as temperature, oxygen levels, parasite loads, contaminants, and disease transmission rates [28,267–269]. Bioenergetic constraints driven largely by temperature often co-occur with low flows, a combination that stresses fish physiologically through low dissolved oxygen and crowding. Crowding also tends to amplify the spread and virulence of diseases [270]. Thus, fish exposure to both pathogens and parasites is extended at a time when their vulnerability is already increased because of other stressors. A higher proportion of research focused on the upstream migration stage assessed disease agents as a driver (16%) compared to any other life stage.

Upstream migration studies also constituted the majority of papers in the species interactions process. In the spawning stage, the density of conspecifics (11%) was the most-studied driver within this process. A small number of studies investigated other species that track migrating salmon (predator/prey) [271–274], injuries associated with escape from a predator or fishing net [275], and potential for migration interference by other species such as invasive American shad [137,138,276].

Responses:

Relative to other life stages, migration and spawning had the highest proportion of papers studying behavior (25%) and particularly effects on mortality and evolutionary processes (e.g., selection on run timing or migration rate). Temperature and flow change systematically by season and location and have large impacts on behaviors such as the timing of freshwater entry and migration speed, including the tendency to stop migrating or stray into a temporary habitat. These response processes have a profound impact on the cumulative bioenergetic cost of adult migration, and greatly affect fecundity and carryover effects on life stage in the subsequent generation (maternal effects on eggs).

A large body of work tracking individual fish in the Columbia River has provided detailed information on behavior and survival, largely in relation to dams and use of thermal refuges. Holding in thermal refuges can increase risks from disease [277] and capture in fisheries [278]. Population differences in the use of thermal refuges during migration are thought to reflect the environmental conditions encountered, namely the relative availability of cool habitat, as well as the amount of time available before spawning, or the urgency of reaching upriver sites [279–286]. For example, sumer-run steelhead make extensive use of cool habitats that are beyond the migration route to spawning areas ("overshoot") because they do not need to spawn until the following spring [282,287].

Studies of carryover effects from previous life stages on spawning success include examinations of origin (hatchery vs. wild) and smolt age [288], juvenile experience during migration and river environment [289–291], and the upstream migration experience through dams [268,292]. For Atlantic salmon ocean conditions have been included in carry-over effect studies [17,293], including from the marine to the subsequent generation egg stage [117,293]. A few studies described how the placement and density of redds varied with environmental conditions, having carryover effects for egg survival and development time [155,294].

Projections:

The projected impacts of climate change on the adult stage frequently found the most severe effects were from stressful temperatures that either lowered aerobic scope during migration or increased mortality during migration or holding [27–31,295]. Changes in flow are expected to result from shifts in precipitation during the adult migration period via changes in the frequency of storms, which in part determine the adult migration window [296,297]. Anthropogenic actions could either exacerbate or mitigate these impacts in flow-limited and flooded systems [123,296,298,299].

Populations differ in their exposure and their projected responses to climate change depending on spatial and temporal factors. While increases in water temperature are generally thought to increase migration stress in most PNW populations, some may experience positive effects, such as improved spawning habitat due to changing fire frequency [294] or the increased use of habitats that were previously below the optimum [300,301].

3.2.5. Population-Level Analyses

Papers focusing on population responses (as opposed to those focusing on individual life-stages) were largely represented by three types of studies: time series analyses of spawner abundance; productivity in terms of smolt or adult recruits per spawner; or the cumulative effects of stage-specific characteristics, such as survival, growth, and migration timing. Therefore, most papers given the *population* label had demographic response variables (85 vs. 46% across the database as a whole; Supplementary File S1: Figure S1) and were more likely than average to use a modeling approach, which tracked population abundance or life history characteristics as a function of environmental factors (31 vs. 16% of whole database).

Of the 19 studies that projected future salmon abundance over the entire life cycle, 58% based them on estimates of habitat suitability or carrying capacity, 21% involved population dynamics models with environmental drivers, while 21% involved qualitative assessments of risk across the full life cycle. Another set of papers examined population dynamics from a theoretical perspective or assessed the ability to detect climate-driven changes using simulation studies (Supplementary File S1: Table S2).

Species interactions were represented in 20% of the papers with the population label. Freshwater species interactions focused largely on non-native trout species affecting native trout [32,302,303] or showed declines in prey availability associated with the presence of non-native fish [304,305]. Marine species interactions included indices of ecosystem condition, generally assumed to drive prey availability [306–308], predators [109], and competition among salmon species [309,310].

Drivers:

The drivers that emerged as most important to salmon populations depended on the modeling approach used. Papers seemed to fall into one of two general categories. The first school of thought argues that the spatial and temporal scale of the most important drivers should match the scale of variation in salmon productivity. These studies focused on characterizing the temporal and spatial correlation structure across many populations. They often found that populations with different freshwater conditions, life histories, and individual management impacts showed similar interannual variation or long-term trends, in either adult recruits per spawner or population abundance. They concluded that the primary driver regulating productivity is operates at an oceanographic regional scale, as typically captured by broad-scale climate indices [309,311,312].

The second modeling approach that encompassed many of the papers in our review focused on freshwater drivers. These papers were usually associated with individual populations for which habitat condition, smolt counts, or other stage-survival estimates were available. These populations were often of concern in a freshwater management decision that needed quantitative support. Life cycle models have been built using a range of approaches, from habitat capacity [313,314] to integrated population models [315]. Some of these approaches incorporated drivers that influence freshwater stages only or that use freshwater indices in addition to marine indices.

Overall, freshwater temperature and flow were the most common individual factors driving population responses (included in 33% and 30% of studies, respectively, followed by habitat (25%). Ocean temperatures and climate indices were included in 14 and 12% of the papers, respectively. Other environmental factors usually included other marine indices (e.g., sea ice melt date, upwelling indices, and salinity) and accounted for an additional 11% of papers.

Responses:

A number of modeling studies tracked changes in population abundance and the likelihood of extinction, as well as changes in life history characteristics, especially body size and age at migration/maturation [238,239,316]. Studies also identified changes in phenology [317], effective population size [318], and genetic variation for particular traits [319,320]. Theoretical papers primarily focused on extinction probability and consistency in abundance (stability). Using this approach, several demonstrated that larger population aggregates are more stable than individual populations, called the "portfolio effect" [321,322].

Projections:

Four out of the nineteen papers that were considered projections were vulnerability assessments that used a variety of qualitative and quantitative methods [305,323–325]. Six papers focused on inland trout species [32,149,326–329], while nine focused on salmon. Quantitative projection models for anadromous species captured Atlantic salmon [330,331], masou salmon (*O. masou*) [332], coho salmon (*O. kisutch*) [314,333], steelhead trout [314,334], and most frequently, Chinook salmon [27,314,335,336].

Climate drivers in these projections focused mostly on freshwater temperatures and flows and their associated effects on habitat; only occasionally did they include climate change in both the freshwater and marine stages [27,327,330]. The salmon models accounted for changes in various processes affecting growth and survival implicitly, using correlations with total production or carrying capacities. Models that focused on Atlantic salmon addressed more processes explicitly, including competition with invasive species and evolution [327,330,331].

3.3. Strengths, Weakness, Expected Responses and Management Opportunities

Life stages differed in both the types of studies that predominated, and the specific emphasis of research. In this section, we identify and summarize the strengths and weaknesses of information on each life stage that affects our ability to understand climate change impacts. We also summarize management actions addressed in each life stage and any expected phenotypic or genetic changes based on the studied responses to environmental drivers (Table 2).

Table 2. Comparison across life stages of strengths, weaknesses, the predominant expected form of phenotypic change or genetic adaptation to climate change, and the most prevalent management actions addressed for that life stage.

Strength	Weakness	Expected Change	Management Options
Population level			
Population-scale demographic processes; carryover effects in life history	Missing processes, esp. evolution and species interactions; carryover effects (other than life history); disease	Smaller, younger, fewer spawners; higher pathogen loads	Fisheries, flow management, dams, habitat restoration, forest management, stocking, invasive species management
Egg stage			
Mechanistic understanding, evolutionary processes	Carryover effects in projection models; micro-climatic habitat projections	Egg size, emergence timing, alevin condition, epigenetic effects	Flow management, contaminants, adult condition, habitat condition
Juvenile stage			
Widest understanding, adaptive management and hypothesis testing	Species interactions; carryover effects from eggs; data-poor watersheds (validate remote data)	Life history strategy (timing and growth rate, size and age at migration, migration timing)	Habitat; contaminants; disease; invasive species; dams; juvenile transportation; management of forests, fires, and flows
Marine stage			
Many populations; broad spatial and temporal scales of analysis	Experiments and direct observations; behavior; species interactions; projections grounded in mechanistic understanding	Age and size at maturation; health of adult migrants	Fisheries targeting salmon, forage fish, groundfish, and predators; estuary and bottom habitat; contaminants and pathogens; hatcheries and aquaculture
Adult stage			
Individual behavior, physiology	Less- studied systems outside main rivers; carryover effects from the ocean	Timing, condition, disease tolerance, prespawn mortality, spawn behavior	Fisheries; migration barriers; dam operations; influences on temperature; flow management; fish transportation; marine mammal predators; hatcheries; contaminant and disease reduction

3.3.1. Egg Incubation

Strengths. The primary strength of research focused on egg incubation is a mechanistic understanding of the physiological and evolutionary processes of egg development. For example, Pankhurst and Munday [337] reviewed a set of these physiological mechanisms in detail. They explain how temperature impacts egg development in particular, but also discussed how other stressors such as O_2 and pCO_2 during spawning and the early life stages affect eggs and later life stages. If environmental conditions actually experienced by the egg are known, this baseline knowledge can support predictions of development rates, emergence timing and condition, and survival.

Weaknesses. Despite the large number of experimental studies on egg development and survival, field tests of predictions for naturally deposited eggs are relatively difficult to preform. The non-destructive monitoring of egg survival and condition in the field is challenging, as is the attainment of physical measurements and the hydrological modeling of in situ subsurface flows and temperatures. Lack of field studies creates a weakness in the ability to relate laboratory results to population consequences. For example, in one of the few studies that measured subsurface conditions, Tuor and Shrimpton [338] suggest that there are systematic discrepancies between surface and inter-gravel temperatures across latitudinal and longitudinal gradients. These discrepancies presumably reflected differences in groundwater intrusion, a variable typically not accounted for in landscape-level models.

Furthermore, stream bed habitats are highly dynamic and can be altered suddenly by flooding, sediment deposition, or water diversion, causing radically different survival from year to year despite similar precipitation [339]. Mating behavior can also be more influential than physical characteristics [339]. Therefore, population level projections extrapolated from mechanistic knowledge of egg temperature and oxygen requirements, and even egg-box survival, require field-based corrections.

Individual and interannual variation in adult site selection and redd-building behavior can drastically alter population-level consequences for egg survival. In some cases, such behavior has produced less ecological variation in emergence timing and survival than expected from thermal differences between streams alone [122,340,341]. Finally, a number of research papers on the egg stage documented carryover effects from maternal condition/behavior on egg survival and from egg development to later life stages. However, these epigenetic effects are still poorly understood and can be difficult to assess at the population-level. These complications were largely ignored in the few projections studies for this life stage.

Expected change. Climate impacts acting on the egg stage are expected to cause phenotypic changes throughout the life cycle due to carryover effects. The high number of multi-generational egg studies allows for the assessment of carryover effects, which were more commonly studied in the egg stage than any other (16 vs, 2% across all other life stages). Carryover effects during the egg stage were studied both from the previous stage (spawner-to-egg), and to subsequent stages (egg-to-juvenile or later stages).

Non-genetic carryover effects from spawner to egg are known as maternal effects. In early development, maternal effects typically have large impacts compared with the influence of additive and non-additive genetic variation [342–344]. Various mechanisms were proposed for this, especially hormone-mediated and energetic processes, but increasing work is focusing on DNA methylation and epigenetic mechanisms [343,345–347]. Toxicants can also influence methylation state, and therefore epigenetic dynamics [348]. Other forms of adult stress, such as from migration exposure to fisheries and environmental conditions, can reduce the aerobic scope and other indices of performance in offspring through various mechanisms associated with maternal effects [349].

Factors such as organic pollutants, which affect eggs during incubation [350], can continue to affect later life stages. Alevins that hatch earlier due to higher temperatures tend to be less developmentally advanced, have lower energetic stores [344], and exhibit reduced burst swimming performance [351]. They also exhibit reduced growth rates, altered behavior [352] and higher rates of malformations [353]. These factors all likely reduce their probability of survival as juveniles. However, perhaps counterbalancing some of these effects, incubation temperatures have also been related to diverse traits such as future adult egg and gonad size, growth rate, age at maturation, and adult migration timing in Atlantic salmon [354,355]. These responses appear to occur through plastic epigenetic mechanisms, and the patterns revealed in a review by Jonsson and Jonsson [354] seemed to be adaptive, such that they may help salmon respond to climate change.

Management. Management actions that affected the egg stage included improving flow and thermal regimes by altering dam operations [100,356–360]. Habitat restoration options [361,362] included removing barriers to historic habitat, augmenting gravel for spawning, and reintroductions to newly accessible or restored habitats.

3.3.2. Juvenile Freshwater Rearing and Migration

Strengths. Research on the juvenile stage reflects a vast body of field work and in-stream monitoring that has been conducted using a wide range of sampling techniques. The freshwater juvenile life stage was the most heavily represented in our database, with 616 papers (33%, Figure 4). Advances in tagging technology have produced abundant data on individual size, movement, and mortality, spanning demographic, physiological and behavioral processes.

Spatial analyses of patterns in freshwater temperature and flow are also relatively abundant compared to the spatial analysis of marine conditions ([29], e.g., [323,363–365]). Physiological tolerances are relatively well defined due largely to the feasibility of obtaining and holding young salmon, although some important gaps remain [366]. Many restoration activities have been implemented to improve rearing habitat, although monitoring of their effectiveness is relatively infrequent. Benefits from restoration are difficult to quantify in part because location-specific changes interact with processes in other life stages through carryover effects, and benefits at the population-level can be difficult to detect. Nonetheless, intensively monitored watersheds offer standout examples of tests of restoration effectiveness [367,368].

Weaknesses. Given the consideration of growth rate as a primary constraint in the juvenile life stage, a major weakness for this stage is presented by the lack of existing projections for bottom-up food-web processes. Projections of competitors and predators, particularly from invasive or warm-adapted species, were also sparse compared to ecosystem modelling approaches in the marine environment. There were a few studies on abundant invasive species such as smallmouth bass *Micropterus dolomieu* [369], largemouth bass *Micropterus salmoides* [370] and American shad *Alosa sapidissima* [136,138], but the broader extent of the potential impacts from range shifts is largely unknown.

Although we expect that increases in water temperature and declines in summer flow will further restrict habitat suitability in many places that are already capacity-limited (e.g., [371]), most research does not account for additional processes, such as changes in prey quality. As water temperatures increase, salmon metabolic rates rise exponentially. Therefore, if salmon are to maintain current growth rates, then prey quality or availability must increase in tandem with temperature (e.g., [372,373]). Temperature directly impacts the metabolic, developmental and population growth rates of prey with consequences for potential prey abundance and quality (e.g., [374]). Furthermore, invertebrate species distributions are changing in complex ways at the same time as salmonid habitat availability is being altered by changing environmental conditions and other anthropogenic impacts (e.g., [375,376]). However, no projection studies in our database attempted to quantify these responses in prey as drivers of a response in salmon.

Systematic patterns in smolt timing, age, body size, and habitat requirements are relatively well characterized in relation to environmental conditions in the Columbia Basin. Many of these factors are included in a few projection models (e.g., [155,377]), but this was uncommon in the literature. There remains an ongoing need to account for how these patterns will be altered with climate change [378]. Specific gaps include the need to separate density-dependent from density-independent drivers of life history strategy, and how changes in juvenile strategy will affect marine survival. There is also a need to extrapolate this information to less well-studied streams, to validate habitat metrics for carrying capacities, to study changes in prey, and to examine interactions with competitors and predators (e.g., [370]).

Expected change. Expected climate responses in the juvenile stage include changes in the phenology of growth, size at age and stage, timing of fry, parr and smolt movement and specific habitat use (e.g., [33,155]). There are significant anthropogenic constraints on behavioral adaptation for many populations due to habitat simplification, which limits heterogeneity in stream habitats, barriers to migration such as road culverts, as well as a lack of physical habitats that are currently unoccupied to move into. Diverse stream habitats are needed for thermoregulation and optimal swimming strategies (e.g., [145,379,380]). The relative importance of density-dependent (e.g., growth driven) and density-independent (e.g., flow driven) triggers for fry and parr movement are not completely understood, but both are important.

Smolt migration timing has a strong plastic (e.g., [381]) and genetic ([382], e.g., [383]) control. At least one recent study has demonstrated rapid genetic adaptation in response to selection on smolt timing [384]. The genetic basis of migration timing is related to growth potential because the threshold body size that triggers migration differs across streams [385]. Thermal tolerance may also be related to growth potential [386]. Many genes and physiological traits are involved in thermal tolerance, and some of these traits, such as cardiac arrhythmic temperature, have much less variation than others and may constrain the overall rates of evolution [387].

Management. An extensive body of work on projecting habitat conditions takes into account the extent to which management actions can mitigate further habitat deterioration due to climate change [362], (e.g., [388]). Decision trees have been developed to help managers determine which habitat actions are most likely to ameliorate particular types of climate impacts (e.g., [123,389]). Explicitly modeled actions in climate change projections involved flow regulation [151,152,390–393]; dam passage [148]; riparian vegetation [153]; and, less frequently, forest management [149].

3.3.3. Marine Stage

Strengths. Research addressing the marine stage is strengthened by broad spatial scales and multi-population and ecosystem perspectives. An additional strength is the long time series of historical ocean catch data, which clearly demonstrate the enormous implications of past climate change for salmon population abundance. This attribute helps to differentiate robust, large-scale patterns in the data from idiosyncrasies amongst individual populations.

Weaknesses. A primary weakness of the existing literature on the marine life stage is the lack of direct observations and experimental manipulation due to the difficulty of such studies in the marine environment. As a result, our mechanistic understanding of the processes restricting productivity is limited and researchers are forced to conjecture based on indirect evidence. For example, one might conclude through indirect evidence that the dominant mode of climate sensitivity during this stage is unlikely to be exceedance of physiological tolerances or starvation. First, the ranges of potential physiological stressors typically encountered in the ocean, such as temperature and pH, are far narrower than those encountered in freshwater. Second, for the majority of Columbia River salmon populations, most interannual variation in survival is established during their first summer in the ocean [394], when growth rates are high compared to those in freshwater, even in "bad ocean" years [231].

With starvation and exposure being considered unlikely major drivers of marine mortality, the primary mechanism of marine mortality is assumed to be predation. However, it is not clear exactly why predation would be higher in a warmer ocean. One possibility is that it is mediated through behavioral responses of multiple species. For example, seabirds in California alter their foraging locations in response to changes in primary prey, which leads to greater overlap with and consumption of salmon in warmer years [230]. However, an analogous mechanism has not been identified for Columbia River populations, despite several studies exploring the role of freshwater plume size and interactions with forage fish and predators (e.g., [179]]. Therefore, we conclude that the most significant weakness in the body of knowledge regarding the marine stage is in our understanding of species interactions.

The few direct observations that exist of the marine stage come primarily from ocean fisheries, which are supplemented by a handful of dedicated surveys that collect information on size and diet of multiple species. These data generally account for a small proportion of individuals and are limited in their spatial and temporal distribution across the ocean, limiting conclusions that can be drawn from analyses. Although some effort have been devoted to tracking individually tagged fish through parts of their marine migration [395], the majority of the ocean stage is not studied directly. Therefore, most climate-focused studies on this stage rely on estimates of survival from juvenile to adult stages based on counts in freshwater.

Large-scale climate indices often account for greater variation than more mechanistic indicators of specific interactions [394]. This is likely because many species respond in their own way to these large-scale shifts in ocean conditions, and various combinations of these behavior sets affect salmon survival. Consequently, predictive models of these processes are often correlative instead of mechanistic, making them more likely to break down over time [167,169,396–398].

Individual-based models, which have been developed for Chinook salmon from the Columbia River [399,400] and forage fish in the California Current [195], hold promise for exploring hypotheses relating to more complex species interactions, and assessing their support in the data. However, the open habitat and scale of the marine ecosystem creates complex interactions that are difficult to model, especially combined with intersecting migrations of forage fish, larger predatory fish, seabirds and marine mammals. This inherent complexity creates major challenges for modeling in a non-stationary climate [167,398].

Expected change. Ocean migration routes presumably affect marine survival and are highly heritable, based on different survival rates for different populations, and present relatively little interannual variation [159,196,401]. However, we have no information either on how optimal migration routes might change with climate change nor on the genetic basis of this trait. Therefore, we cannot yet model how quickly this trait might respond to selection due to climate change. Other potentially heritable traits that could respond to selection, but are also highly plastic, are age at return and growth rates. Anticipating how sex-linked age at return responds to selection has received attention in the context of fishery-induced selection [402]. Responses to climate change are thus largely expected to consist of more widespread reductions in survival, in population abundance, and in age and size at return. These are the responses observed during anomalously warm ocean years, which are expected to develop more frequently with climate change [403].

Management. The area of management most frequently quantified in the marine environment was fisheries, and 11% of papers on the marine stage addressed impacts of fisheries in some way. Additional management actions addressed included hatchery production, farmed salmon, aquaculture, modification of shorelines, and nearshore and ocean-bottom habitat restoration, as well as planning and characterizing the efficacy of marine protected areas.

3.3.4. Adult Migration and Spawning

Strengths. The strength of the body adult migration literature is that individual migration histories are extremely well documented in certain systems. These studies are often in larger migration corridors where future projections of temperature and flow are relatively robust (e.g., [26,404]). Systems such as the Columbia River have large fisheries and dams at which adults can be enumerated, and these tend to have the most adult monitoring. However, smaller streams and un-dammed rivers have much less individual data.

Population estimates in such systems, if made at all, often come from spawning ground or carcass surveys, which are less precise, particularly in terms of timing. Although our search criteria focused on the Columbia River and thus did not capture all papers on smaller streams, 65% of papers on upstream migration were based in three major river systems: the Columbia, the Fraser, and the Sacramento. In comparison, the percentage based on these systems was only 29% of the dataset as a whole. The spawning stage, similarly, had a large quantity of redd survey data, adult counts, and descriptions of habitat needs. Management of the adult life stage benefits from the ability to restore migration corridors and to use management levers through fisheries and dams. Field studies have excellent spatial and temporal data coverage.

Weaknesses. Weaknesses in projections for the adult stage stem largely from a poor ability to account for carryover effects. Specific gaps include carryover effects from the ocean stage; behavioral responses to novel conditions in both migration and spawning, especially in smaller streams; and evolutionary responses to increased disease pressures. Uncertainty in predicting responses at the spawning stage relate to the behavioral responses of adults in site selection, stress tolerance, and the consequences of sublethal stressors, such as the need to increase endurance when confronted with higher bioenergetic costs. Our knowledge of behavioral responses is based almost entirely on correlations in field data rather than experimental manipulations. Therefore, it is difficult to extrapolate beyond observed conditions and separate effects from different environmental drivers that tend to co-vary (e.g., temperature and flow).

Expected change. The primary means of adaptation in the adult stage that was reported was a change in run and spawn timing through both plastic and genetic change [405,406]. Genetic mechanisms governing phenology are especially well studied in this stage [407–410]. Additional studies have characterized historical adaptation to physiological challenges, such as evolution in aerobic scope and cardiac performance [411–418]. However, no future projections were available for physiological adaptations in this stage.

Phenological traits are both highly plastic and strongly heritable and thus extremely likely to change [383,406]. Temperature, flow, and photoperiod (as an index of date) are the primary cues thought to drive plastic responses in phenology, with flow being the strongest cue for migration timing [27,296], and temperature being the primary factor driving spawn timing [155]. Shifts in habitat use, where possible, are also likely through a plastic response [294,419].

Management. A relatively large proportion of studies on adult migration involved management (31%). The majority of management options involved hydrosystem and flow management, fisheries, and influences on temperature. Many studies demonstrated carry-over effects (9%) from hatchery practices or juvenile transportation on adult survival, homing, and overall fitness, demonstrating potential ways to improve management (e.g.,[289]). Adult transportation to spawning grounds is also employed for endangered Snake River sockeye, threatened Willamette River Chinook and other populations that spawn upstream of impassable barriers; the success of these programs is highly variable [268,420,421].

3.3.5. Population-Level Analyses

Strengths. Ultimately, the primary concern of most salmon researchers is the risk of population decline or extirpation with climate change. The strength of population-level analyses is that they often directly address our primary concern, which is extinction risk. They also have the potential to include any stage-specific issues that have been studied within a more holistic perspective. A few studies included multiple processes of concern in population dynamic models, such as the combination of evolutionary, behavioral, and demographic processes [330,331]; but such an interdisciplinary approach was rare overall. Population models most often accounted for change in multiple processes affecting growth and survival implicitly, using correlations with total production or carrying capacities. Modeling studies that focused on Atlantic salmon more frequently explicitly addressed multiple processes, including competition with invasive species and evolution [327,330,331].

Weaknesses. A weakness in population-level studies for salmon (but not trout), is species interactions, as well as the absence of information on evolution in disease tolerance and carryover effects other than life history characteristics. Population dynamics models are potentially an appropriate tool for integrating cumulative effects across life stages and interactions among processes, in their respective modes of impact on population viability. However, modeling approaches are still very heterogeneous and face technical challenges. For example, the data and processing requirements can be high for integrated population models. On the other hand, combining results from different studies can lose track of correlated effects and cumulative uncertainty across life stages. While these difficulties can be overcome, they are current hindrances to date.

Carryover effects [422], in particular, involve influences on one life stage that may not directly induce mortality in the observed life stage but do induce mortality or loss of fecundity in later stages. For example, fish may survive poor juvenile growth in freshwater but experience higher predation upon marine entry as a result. Adults returning in poor condition may have low disease resistance and depleted energetic reserves with which to migrate, compete, and produce eggs; thus, despite surviving the marine environment, their reproductive success is limited. These effects were not addressed in any of the projections in our database. Delayed effects of passage through dams, on the other hand, were widely discussed, with radically divergent conclusions.

Expected change. Many population models documented historical changes in abundance and likelihood of extinction, as well as changes in life history characteristics, with climate variability. For some of the reasons described above, body size and age at migration/maturation [238,239,316] may continue to decline, while phenology [317], effective population size [318], and genetic variation for particular traits [319,320] continue to reflect climate trends.

We found 19 projection papers at the population level. Climate drivers in these projections focused mostly on freshwater temperatures and flows and associated effects on freshwater habitat; only occasionally did they include climate change quantitatively in both freshwater and marine stages [27,330,331], but qualitative summaries generally addressed both environments [33,34]. Considering threats in both environments increased the overall estimated vulnerability of anadromous populations [325,423–425].

Management. Human impacts that were modeled as drivers included fisheries as the largest single factor (35%), as well as flow regulation [330,426–428]; dam removal [332]; habitat restoration [336]; forest management, including harvest and wildfire control [149,429]; stocking [430]; and the control of invasive species [328]. Across the broader category of 448 papers assigned to management drivers in general, rather than just those that we associated with the population level, fisheries, habitat, and dams/flow management each accounted for similar proportions (~25%), with hatcheries/fish farms/stocking activities, policy/framework analysis and 'other' each accounting for about 9%. This consistency in frequency indicates a fairly even representation across sectors as major impacts on salmon and trout.

4. Discussion

In cataloging 1853 studies and reviewing several hundred additional studies published recently, we have comprehensively explored the impacts of climate change on anadromous salmon and trout. Their most generally expected responses to climate change are shown in Figure 6. This synthesis showcases salmon as a case study where an extensive body of knowledge is ready for integration into better-informed projections of population and species responses to climate. Skillful incorporation of this information should ensure a robust characterization of likely biological responses and facilitate management that is more effective and better able to anticipate changing conditions.

Even so, the complexity of the salmon life cycle causes particular challenges in predicting climate impacts. Our review identified extensive laboratory and experimental work, which has elucidated many of the physiological processes that shape functional relationships and provide mechanistic understanding. Unfortunately, these relationships and mechanisms can rarely be scaled up to population-level responses directly because of the complex processes and heterogeneous environmental conditions involved across the salmon life cycle. Furthermore, within an individual study, they are usually considered in either freshwater or marine environments, but not both. We found only one modelling team in our review that explicitly accounted for evolutionary, demographic, and physiological processes across multiple life stages: the individual-based demo-genetic model of Atlantic salmon growth and maturation [330,331]. Consequently, managers and scientists have generally depended on population-specific correlations to account for missing parameters, processes, and life stages, and have frequently ignored life stages that lie outside their management domain.

Within the literature, we observed a highly heterogeneous representation of processes across the life cycle, reflecting the different environments and priorities of research groups. Although the number of papers counted in each category did not perfectly reflect the understanding of each topic, they provided an index of the relative attention devoted to different topic areas. The most mechanistic and physiologically grounded studies, where genetic understanding was best developed, focused on early life stages. As the spatial and temporal scales of salmon life stages expanded, mechanistic relationships became less common, while correlations with synoptic indices of environmental and ecological conditions became more common.

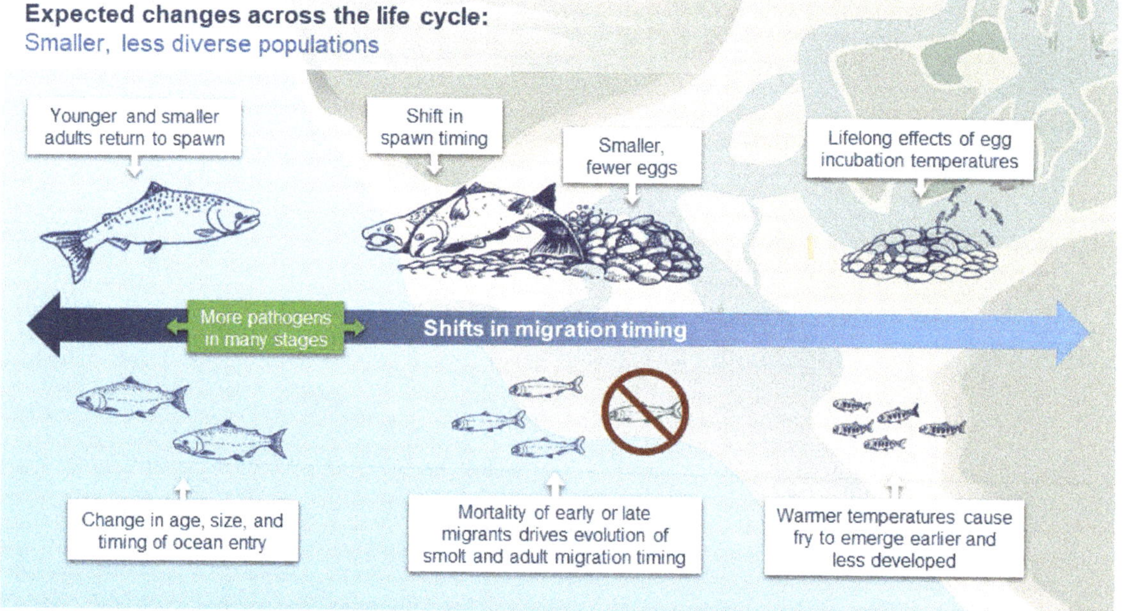

Figure 6. Some of the observed and expected changes in salmon. General shifts in growth and developmental rates lead to behavioral changes. Shifts in migration and spawn timing, as well as in the age and size at maturity, result from both plastic and evolutionary dynamics. Increased prevalence of disease, epigenetic and other carryover effects are less certain.

By synthesizing the rich body of research on different life stages and accounting for process representation explicitly, more of the existing literature can be marshalled toward effective action. The aim is to inform a more holistic approach that considers the whole salmon life cycle. Through this lens research can be refocused on applying new technologies and methodologies to previously intractable questions, and projection modeling can more efficiently address missing processes and other uncertainties. In addition, management can use available resources to the greatest effect by addressing population-limiting factors more directly. Considering this vast body of research, we propose the following path forward for salmon conservation.

(1) Capitalizing on the strengths of research in each life stage: we should use these studies to prioritize actions that reduce human impacts on freshwater and coastal ecosystems with the goal of maximizing the number, diversity, and health of wild smolts and spawners.

Using the breadth of available science, we can target resources toward addressing the factors that most constrain population growth as well as behavioral and genetic diversity. This recommendation is driven by the dominant role of human impacts in freshwater, our corresponding ability to reduce these impacts (dependent on political will), and the strong indication that many freshwater impacts carry over to affect marine survival [431]. For example, improved water quality and reduced burdens from pathogens and contaminants can greatly improve survival in later life stages. Increasing life history diversity through expanding freshwater habitat diversity can also dampen population volatility, which reduces the risk of extinction. In a future climate, productivity may be reduced as a consequence of a warmer and less productive marine environment [432]. In this case, a restored, functional freshwater environment may be the difference between depressed productivity and extinction [433].

Local human activities as well as climate change have significantly reduced ecosystem services and functions around the world ([166], Report sections 4.3.2, 4.3.3, 5.3, 5.4.1, 6.4.2), and the northern California Current is heavily impacted by coastal land and marine activities [434]. The impacts analyzed by Halpern et al. [434] included fishery removals, habitat destruction from fishing, aquaculture, ship traffic and ballast water releases (related to invasive species transport), ocean- and land-based pollution, nutrient and sediment inputs. A similar analysis found that numerous human pressures have increased steadily since the 1980s. Increased pressures included finfish and shellfish aquaculture, seafood demand, fisheries removals, coastal engineering, dredging, freshwater and sediment retention, while those that decreased included oil and gas activities, bottom structures, and light pollution (Figure S27 in [435]). Widespread loss and degradation of estuary habitats [436,437], destructive demersal fishing [438], high nutrient inputs, fishery removals, and aquaculture [435] have had widespread cumulative impacts on salmon and ecosystem capacity.

Nonetheless, freshwater restoration actions can be better targeted and might have more immediate effects because their outcomes have already been researched so extensively. The effects of freshwater habitat conditions on growth, survival, and movement are relatively well understood and sufficiently influential that many recommendations for habitat restoration and flow management can be considered as minimum requirements for persistent salmon populations. These are areas of strength in the literature (Supplementary File S1: Table S2). Because some information is available on nearly all processes, resource decisions can at least qualitatively incorporate potential risk from the less well-studied processes.

Habitat restoration is a tool immediately available to managers with guidance specifically designed to increase resiliency and adaptive capacity in the face of climate change [123,388,439–441]. In many watersheds, degraded, blocked, and disconnected habitats have lowered salmon carrying capacities and homogenized remaining habitat. These losses substantially increase population vulnerability to fluctuations from annual environmental extremes, such as a flood or drought [442]. Restored habitat complexity can augment food webs and is generally beneficial for maximizing growth, as fish use different habitats at different times for different needs such as feeding, digesting, and avoiding predators [380,443]. Thus, well-designed habitat restoration can support a wider diversity of salmon life histories, which could slow the process of reinforce diminishing genetic heterogeneity [444,445]. Population and full life cycle models or methods of similar scope are necessary to verify that population-level benefits are likely [446]. An example of a qualitative framework with specific management recommendations by life stage was completed for Fraser River sockeye salmon [33]. These recommendations can be modified for other species in other locations, ensuring that the full life cycle is considered.

To effectively mitigate long-term climate impacts, habitat must be restored systematically on a larger scale than has been implemented in the past. To support anadromous salmon metapopulations, basin-wide actions are necessary. However, such wide-scale actions face challenges. Region-wide quantification of threats and benefits are needed to guide restoration in the future, but such quantification is often limited to particular components of the environment (e.g., [323,447,448]) because of the difficulty in acquiring data appropriate for larger spatial scales. Perhaps more importantly, putting water back in streams requires whole-basin planning and multi-sector cooperation (e.g., [449]). In the western U.S. and many other places globally, declines in dry season flows are already creating intense competition among users, such as agriculture, urban, recreational and energy sectors. Restored riparian zones, wetlands, and floodplains will not only help salmon populations, but retain runoff, which will increase summer water supplies, providing benefits and security to both humans and wildlife. To understand and ensure these benefits, participation by stakeholders in developing these plans is essential for success [441].

(2) Research should continue to focus on identified gaps in knowledge and model parameters, which limit our ability to predict and counteract climate impacts. Targeting these weaknesses will help managers avoid surprises.

Several knowledge gaps that limit the effectiveness of conservation, management and modeling efforts were identified in our review. The marine life stage was identified as having the largest uncertainty in the mechanistic controls which drive productivity. In particular, new research is needed to improve our understanding of the factors responsible for lower survival in the marine stage during warmer years. The primary research goal should be to identify additional marine management actions, with the secondary goal of improving forecasts of short- and long-term prospects for decision makers. Paleontological, as well as recent studies, have shown that major fluctuations in salmon populations track changes in the marine climate [450]. Therefore, an exclusive focus on freshwater options is unlikely to prevent devastating climate impacts on salmon.

Overall, very few studies addressed more than one or two processes as drivers over the entire life cycle. This general deficit made it difficult to assess the potential influence of less-studied factors and interactions. The magnitude of such influence may be large, judging from analyses that explicitly specify unexplained variation. Random effects that reflect un-modelled processes are typically very large in state space models that estimate process and observation errors explicitly [315,451]. This suggests that there are major factors driving variation in population productivity that we are currently not able to account for.

Our understanding of carryover effects, species interactions, functional relationships, and genetic constraints and opportunities emerged as weaknesses from our analysis. How productivity is impacted by the interacting factors of density, prey, and predators, and how these relationships are modulated by environmental variability remain areas of uncertainty across multiple life stages. As models identify particular parameters or functional relationships that are especially impactful for the focal species, more data are needed to clarify these relationships. These models will require more targeted data collection to fill research gaps, such as those laid out by Wells et al. [452] for the marine stage. These are areas where new technologies and methodological approaches can reduce earlier constraints on research to help overcome the challenges of studying the ocean environment.

New technologies are greatly improving our ability to detect species interactions and habitat use in both freshwater and marine environments. For example, environmental DNA (eDNA) and rapid genetic screening techniques can provide highly targeted information more efficiently and with less handling stress and mortality than traditional survey methods. These and other types of data can be collected by autonomous vehicles, filling holes between (but not replacing) ship-based surveys. These methods provide data in key areas, including population abundance [453], the presence of invasive species [454–456], a history of thermal stress [457], habitat issues such as barriers to passage [458], and community composition [459]. While these techniques have limitations [460], we predict that new genetic approaches will change our entire perspective on freshwater and marine communities and the spatial and temporal overlap among species.

Similarly, new methods of examining satellite data and crowd sourcing drone imagery are available. These techniques can map and track changes in habitat and ecological characteristics, including functional groups at lower trophic levels [461], and could dramatically improve the spatial and temporal resolution of field surveys. New animal tracking technology and video analysis tools are also rapidly improving our ability to record species interaction events and movements. Many new tools and approaches are being coordinated by Global Ocean Observing System (http://goosocean.org, accessed on 1 May 2023) to combine information from different research networks and platforms for a truly global perspective.

Systems-oriented approaches to analyses of community resilience can help to pinpoint communities wherein species interactions are especially important. Naman et al. [462] used a food-web perspective to help address critical riverine management questions. In addition to capacity and productivity questions, Naman et al. [462] pointed out that the introduction of invasive species, as well as the extirpation of native species, can have cascading effects on focal populations and ecosystem processes.

Because the primary tool for anticipating future responses to climate change is the projection model, we discuss factors limiting the utility of projection modeling here. A goal of predictive modeling should be to better identify factors that could cause populations to decline so that management efforts on the ground can be better targeted. To this end a rigorous sensitivity analyses of plausible alternative functional relationships and the full range of reasonable parameter values within a given functional form, followed by real-world testing of model predictions needs to be instituted [463]. Sensitivity analyses should explicitly characterize both the relative uncertainty in different parameters or functional relationships and the biological impact across this range of uncertainty [463]. Our goal should be to more accurately reflect uncertainty in these areas, and then (e.g., [464]) target research on areas with the largest potential effects, given the expected trends in climate. Qualitative network models [465–467] are another way to represent relationships that might not be resolvable empirically or that change frequently. This approach explores a very wide range of parameter combinations more efficiently than fully quantitative models, and in some cases may more honestly reflect our level of uncertainty. Decision makers could then more effectively manage risk.

The ideal approach for salmon would be similar to recommendations for marine science in general in the context of climate change. As detailed by Tommasi et al. [468], we advocate an iterative process of model development and data collection to advance a variety of approaches, including single-species, multi-species, and ecosystem-based models and research in each of the empirical areas mentioned above [469]. Salmon are highly sensitive to lower trophic level processes, so better understanding of climate impacts on prey is essential and requires more study. Still, predators account for most salmon mortality in both freshwater and marine environments. Thus, we need to go beyond the single-species perspective to multi-species experiments and models to probe deeper into how climate drivers affect species interactions and community dynamics.

(3) **Characterizing uncertainty in climate impact projections requires implementing an ensemble modeling approach. A number of models with different assumptions should be used to simulate future conditions, and mechanistic and statistical models should be updated frequently to incorporate different biological processes and new environmental correlations.**

As climate change progresses, and we encounter unprecedented conditions in the atmosphere and ocean, statistical correlations will break down, and model uncertainty will grow. Statistical methods can help extend our knowledge from intensively studied populations to locations with sparse data. However, we must recognize the risk in depending on retrospective correlations for climate change projections: correlations are likely to break down in a non-stationary climate as ecological relationships and animal behaviors shift in response to novel conditions. Novel conditions are already occurring, as demonstrated by the unprecedented warm waters of the Blob that lasted in the North Pacific from 2014 to 2016 [470]. Warm events have occurred repeatedly since the Blob [471], and the frequency of these events is expected to increase [403] and to co-occur with terrestrial droughts in "compound extremes" [472]. Species distribution models trained on data from previous years did not always predict responses to the heatwave well [181]. In future projections, novel conditions constitute from 50% (2060s) to 100% (2090s) of the California Current in an average year under high carbon emission scenarios [473], presenting profound challenges for our modeling tools [474].

Mechanistic models are expected to have longer forecast horizons and perform better at intermediate to long time-scales, assuming the mechanistic drivers are modeled correctly [13]. Exploration of forecast skill in population dynamic modeling is rarely done because short-term uncertainty is expected to be extremely high. However, such analyses would help to clarify the strengths and weaknesses of different modeling approaches. We also need to think carefully about processes that are often ignored: species interactions, changes in behavior, and evolutionary dynamics. These areas need systematic scoping to identify populations and locations where they could alter future outcomes for focal species.

Nonetheless, for short-term forecasting, statistical models will likely continue to have the greatest forecast skill, despite the risk of correlations breaking down over time. Frequent updating of statistical correlations, combined with careful and informed interpretation and forecast skill assessment, can mitigate this risk by recognizing when relationships have broken down and replacing outdated variables with new covariates that better represent the challenges salmon are facing in new environmental conditions. This process in and of itself can lead to greater mechanistic understanding.

In addition, any projection of future conditions should be compared systematically and quantitatively with other projections. Model ensemble averages have been shown to produce better predictions than individual models when rigorously tested. Ensembles also allow for explicit accounting of the uncertainty attributed to model type. Ensemble modeling is necessary to account for the high level of uncertainty in biological as well as physical processes. Such uncertainty follows from weak mechanistic understanding in many research areas combined with limited representation of the full range of biological processes known to affect species responses to climate change. The ensemble approach has been formalized in climate and weather forecasting communities, where a wide range of models are continually under development and regularly compared [51].

In the stock assessment and ecosystem modeling communities, global efforts have been initiated to systematically compare outputs from different models, given a shared set of scenarios within the current climate (e.g., Fisheries and Marine Ecosystem Model Intercomparison Project, FISHMIP [475] and Center for the Advancement of Population Assessment Methodology, CAPAM). Future projections have used a multi-model approach for groundfish [476] and forage fish [477]. However, this practice is not widely established in most ecological applications, despite an urgent need for robust marine planning [9].

In summary, the vast resources that have been devoted to salmon research can be marshalled more effectively by using the strengths of the existing body of knowledge to guide effective management and conservation, address areas of scientific weakness, acknowledge our current level of uncertainty, and more accurately represent the array of potential future conditions. We encourage the community to actively plan for a future of surprises and invoke informed proactive actions to adapt to climate change.

5. Conclusions

The rich literature studying salmon and trout provides a valuable case study demonstrating the complexity of climate impacts on biological systems. Information across biological processes can be marshalled to provide robust projections of responses to climate change. We are at a nexus in which an exquisitely valuable and sensitive taxonomic group is on the brink of catastrophic change. We have an impressive array of knowledge, but widescale cooperation is essential to overcome the enormous challenge of filling knowledge gaps and forestalling the worst outcomes for this resource and the communities who depend on them. Only by combining our collective wisdom and energy can we prevent the loss of this flexible and adaptable set of species while maintaining and augmenting the freshwater and marine ecosystems on which they depend.

Supplementary Materials: The following supporting information can be downloaded at: https://www.mdpi.com/article/10.3390/fishes8060319/s1, Supplementary File S1: Quantitative analysis of categories across the entire database; Supplementary File S1: Table S1. Description of the criteria used to assign each label; Supplementary File S1: Table S2. The number of papers assigned to each label across the database as a whole and within each life stage; Supplementary File S1: Figure S1. Panel plots for each life stage and label, showing the number and percentage of papers in each label; Supplementary File S2: Complete list of studies used in the quantification of categories and the labels assigned to each citation; Code File S1: R code example for reading data file; Data S1: csv file with data for use with R code.

Author Contributions: L.G.C. and J.E.S. designed the study. The literature was searched and compiled by L.G.C. from 2010 to 2016 and by both authors from 2017 to 2021. Both authors designed the categories. J.E.S. assigned the labels and categories. Both authors contributed text and edited the manuscript. All authors have read and agreed to the published version of the manuscript.

Funding: This research received no external funding.

Data Availability Statement: All citations analyzed are provided in Supplementary File S2.

Acknowledgments: We would like to thank NOAA Fisheries West Coast Region Adaptive Management Implementation Plan team for supporting and reviewing the annual literature reviews since 2010, and providing much useful feedback over the years. Blane Bellerud provided salmon images and icon, and Su Kim helped with design in Figures 1, 2 and 6. We also thank Aimee Fullerton, Jeff Jorgensen, Peter Kiffney, Mark Urban, and Beth Sanderson for reading various drafts and providing helpful direction and reviews.

Conflicts of Interest: The authors declare no conflict of interest.

References

1. Pelletier, F.; Coltman, D.W. Will human influences on evolutionary dynamics in the wild pervade the Anthropocene? *BMC Biol.* **2018**, *16*, 7. [CrossRef]
2. Reid, A.J.; Carlson, A.K.; Creed, I.F.; Eliason, E.J.; Gell, P.A.; Johnson, P.T.J.; Kidd, K.A.; MacCormack, T.J.; Olden, J.D.; Ormerod, S.J.; et al. Emerging threats and persistent conservation challenges for freshwater biodiversity. *Biol. Rev.* **2019**, *94*, 849–873. [CrossRef] [PubMed]
3. Exposito-Alonso, M.; Booker, T.R.; Czech, L.; Gillespie, L.; Hateley, S.; Kyriazis, C.C.; Lang, P.L.M.; Leventhal, L.; Nogues-Bravo, D.; Pagowski, V.; et al. Genetic diversity loss in the Anthropocene. *Science* **2022**, *377*, 1431–1435. [CrossRef] [PubMed]
4. Tye, S.P.; Siepielski, A.M.; Bray, A.; Rypel, A.L.; Phelps, N.B.D.; Fey, S.B. Climate warming amplifies the frequency of fish mass mortality events across north temperate lakes. *Limnol. Oceanogr. Lett.* **2022**, *7*, 510–519. [CrossRef]
5. Jones, T.; Parrish, J.K.; Peterson, W.T.; Bjorkstedt, E.P.; Bond, N.A.; Ballance, L.T.; Bowes, V.; Hipfner, J.M.; Burgess, H.K.; Dolliver, J.E.; et al. Massive Mortality of a Planktivorous Seabird in Response to a Marine Heatwave. *Geophys. Res. Lett.* **2018**, *45*, 3193–3202. [CrossRef]
6. Parmesan, C.; Yohe, G. A globally coherent fingerprint of climate change impacts across natural systems. *Nature* **2003**, *421*, 37–42. [CrossRef]
7. Weiskopf, S.R.; Rubenstein, M.A.; Crozier, L.G.; Gaichas, S.; Griffis, R.; Halofsky, J.E.; Hyde, K.J.W.; Morelli, T.L.; Morisette, J.T.; Muñoz, R.C.; et al. Climate change effects on biodiversity, ecosystems, ecosystem services, and natural resource management in the United States. *Sci. Total Environ.* **2020**, *733*, 137782. [CrossRef]
8. Wiens, J.J. Climate-related local extinctions are already widespread among plant and animal species. *PLoS Biol.* **2016**, *14*, e2001104. [CrossRef]

9. Pinsky, M.L.; Rogers, L.A.; Morley, J.W.; Frölicher, T.L. Ocean planning for species on the move provides substantial benefits and requires few trade-offs. *Sci. Adv.* **2020**, *6*, eabb8428. [CrossRef]
10. Morley, J.W.; Selden, R.L.; Latour, R.J.; Frölicher, T.L.; Seagraves, R.J.; Pinsky, M.L. Projecting shifts in thermal habitat for 686 species on the North American continental shelf. *PLoS ONE* **2018**, *13*, e0196127. [CrossRef]
11. Cheung, W.W.L.; Brodeur, R.D.; Okey, T.A.; Pauly, D. Projecting future changes in distributions of pelagic fish species of Northeast Pacific shelf seas. *Prog. Oceanogr.* **2015**, *130*, 19–31. [CrossRef]
12. Buckley, L.B.; Urban, M.C.; Angilletta, M.J.; Crozier, L.G.; Rissler, L.J.; Sears, M.W. Can mechanism inform species' distribution models? *Ecol. Lett.* **2010**, *13*, 1041–1054. [CrossRef] [PubMed]
13. Petchey, O.L.; Pontarp, M.; Massie, T.M.; Kéfi, S.; Ozgul, A.; Weilenmann, M.; Palamara, G.M.; Altermatt, F.; Matthews, B.; Levine, J.M.; et al. The ecological forecast horizon, and examples of its uses and determinants. *Ecol. Lett.* **2015**, *18*, 597–611. [CrossRef] [PubMed]
14. Evans, T.G.; Diamond, S.E.; Kelly, M.W. Mechanistic species distribution modelling as a link between physiology and conservation. *Conserv. Physiol.* **2015**, *3*, cov056. [CrossRef] [PubMed]
15. Urban, M.C.; Bocedi, G.; Hendry, A.P.; Mihoub, J.-B.; Pe'er, G.; Singer, A.; Bridle, J.R.; Crozier, L.G.; De Meester, L.; Godsoe, W.; et al. Improving the forecast for biodiversity under climate change. *Science* **2016**, *353*, 6304. [CrossRef] [PubMed]
16. Mills, K.E.; Pershing, A.J.; Sheehan, T.F.; Mountain, D. Climate and ecosystem linkages explain widespread declines in North American Atlantic salmon populations. *Glob. Chang. Biol.* **2013**, *19*, 3046–3061. [CrossRef]
17. Chaput, G.; Benoit, H.P. Evidence for bottom-up trophic effects on return rates to a second spawning for Atlantic salmon (*Salmo salar*) from the Miramichi River, Canada. *ICES J. Mar. Sci.* **2012**, *69*, 1656–1667. [CrossRef]
18. Nicola, G.G.; Elvira, B.; Jonsson, B.; Ayllon, D.; Almodovar, A. Local and global climatic drivers of Atlantic salmon decline in southern Europe. *Fish. Res.* **2018**, *198*, 78–85. [CrossRef]
19. Dadswell, M.; Spares, A.; Reader, J.; McLean, M.; McDermott, T.; Samways, K.; Lilly, J. The decline and impending collapse of the Atlantic salmon (*Salmo salar*) population in the North Atlantic Ocean: A review of possible causes. *Rev. Fish. Sci. Aquac.* **2021**, *30*, 215–258. [CrossRef]
20. NMFS—National Marine Fisheries Service. 2015 Adult Sockeye Salmon Passage Report. 2016. Available online: https://www.columbiariverkeeper.org/sites/default/files/2017/08/8.pdf (accessed on 14 June 2023).
21. Martin, B.T.; Dudley, P.N.; Kashef, N.S.; Stafford, D.M.; Reeder, W.J.; Tonina, D.; Del Rio, A.M.; Foott, J.S.; Danner, E.M. The biophysical basis of thermal tolerance in fish eggs. *Proc. R. Soc. B-Biol. Sci.* **2020**, *287*, 20201550. [CrossRef]
22. Akbarzadeh, A.; Selbie, D.T.; Pon, L.B.; Miller, K.M. Endangered Cultus Lake sockeye salmon exhibit genomic evidence of hypoxic and thermal stresses while rearing in degrading freshwater lacustrine critical habitat. *Conserv. Physiol.* **2021**, *9*, coab089. [CrossRef] [PubMed]
23. Ward, E.J.; Anderson, J.H.; Beechie, T.J.; Pess, G.R.; Ford, M.J. Increasing hydrologic variability threatens depleted anadromous fish populations. *Glob. Chang. Biol.* **2015**, *21*, 2500–2509. [CrossRef] [PubMed]
24. Welch, D.W.; Porter, A.D.; Rechisky, E.L. A synthesis of the coast-wide decline in survival of West Coast Chinook Salmon (*Oncorhynchus tshawytscha*, Salmonidae). *Fish Fish.* **2020**, *22*, 194–211. [CrossRef]
25. Connors, B.; Malick, M.J.; Ruggerone, G.T.; Rand, P.; Adkison, M.; Irvine, J.R.; Campbell, R.; Gorman, K. Climate and competition influence sockeye salmon population dynamics across the Northeast Pacific Ocean. *Can. J. Fish. Aquat. Sci.* **2020**, *77*, 943–949. [CrossRef]
26. Crozier, L.G.; Siegel, J.E.; Wiesebron, L.E.; Trujillo, E.M.; Burke, B.J.; Sandford, B.P.; Widener, D.L. Snake River sockeye and Chinook salmon in a changing climate: Implications for upstream migration survival during recent extreme and future climates. *PLoS ONE* **2020**, *15*, e0238886. [CrossRef]
27. Crozier, L.G.; Burke, B.J.; Chasco, B.E.; Widener, D.L.; Zabel, R.W. Climate change threatens Chinook salmon throughout their life cycle. *Commun. Biol.* **2021**, *4*, 222. [CrossRef]
28. Bowerman, T.E.; Keefer, M.L.; Caudill, C.C. Elevated stream temperature, origin, and individual size influence Chinook salmon prespawn mortality across the Columbia River Basin. *Fish. Res.* **2021**, *237*, 105874. [CrossRef]
29. FitzGerald, A.M.; John, S.N.; Apgar, T.M.; Mantua, N.J.; Martin, B.T. Quantifying thermal exposure for migratory riverine species: Phenology of Chinook salmon populations predicts thermal stress. *Glob. Chang. Biol.* **2021**, *27*, 536–549. [CrossRef]
30. Martins, E.G.; Hinch, S.G.; Patterson, D.A.; Hague, M.J.; Cooke, S.J.; Miller, K.M.; Lapointe, M.F.; English, K.K.; Farrell, A.P. Effects of river temperature and climate warming on stock-specific survival of adult migrating Fraser River sockeye salmon (*Oncorhynchus nerka*). *Glob. Chang. Biol.* **2011**, *17*, 99–114. [CrossRef]
31. Isaak, D.J.; Luce, C.H.; Horan, D.L.; Chandler, G.L.; Wollrab, S.P.; Nagel, D.E. Global warming of salmon and trout rivers in the northwestern U.S.: Road to ruin or path through purgatory? *Trans. Am. Fish. Soc.* **2018**, *147*, 566–587. [CrossRef]
32. Wenger, S.J.; Isaak, D.J.; Luce, C.H.; Neville, H.M.; Fausch, K.D.; Dunham, J.B.; Dauwalter, D.C.; Young, M.K.; Elsner, M.M.; Rieman, B.E.; et al. Flow regime, temperature, and biotic interactions drive differential declines of trout species under climate change. *Proc. Natl. Acad. Sci. USA* **2011**, *108*, 14175–14180. [CrossRef] [PubMed]
33. Healey, M. The cumulative impacts of climate change on Fraser River sockeye salmon (*Oncorhynchus nerka*) and implications for management. *Can. J. Fish. Aquat. Sci.* **2011**, *68*, 718–737. [CrossRef]
34. Wainwright, T.C.; Weitkamp, L.A. Effects of climate change on Oregon coast Coho salmon: Habitat and life-cycle interactions. *Northwest Sci.* **2013**, *87*, 219–242. [CrossRef]

35. McGaughran, A.; Laver, R.; Fraser, C. Evolutionary Responses to Warming. *Trends Ecol. Evol.* **2021**, *36*, 591–600. [CrossRef] [PubMed]

36. Gustafson, R.G.; Waples, R.S.; Myers, J.M.; Weitkamp, L.A.; Bryant, G.J.; Johnson, O.W.; Hard, J.J. Pacific salmon extinctions: Quantifying lost and remaining diversity. *Conserv. Biol.* **2007**, *21*, 1009–1020. [CrossRef]

37. NOAA Fisheries. Pacific Salmon and Steelhead. Available online: https://www.fisheries.noaa.gov/species/pacific-salmon-and-steelhead (accessed on 1 January 2021).

38. Waples, R.S.; Gustafson, R.G.; Weitkamp, L.A.; Myers, J.M.; Johnson, O.W.; Busby, P.J.; Hard, J.J.; Bryant, G.J.; Waknitz, F.W.; Neely, K.; et al. Characterizing diversity in salmon from the Pacific Northwest. *J. Fish Biol.* **2001**, *59*, 1–41.

39. Crozier, L. Impacts of Climate Change on Columbia River Salmon: A review of the scientific literature published in 2010. In *Pages D1–D50 in Endangered Species Act Section 7(a)(2) Supplemental Biological Opinion: Consultation on Remand for Operation of the Federal Columbia River Power System*; U.S. National Marine Fisheries Service: West Coast Regin, 1201 NE Lloyd Blvd, Ste 110, Portland, OR 97232, USA, 2011. [CrossRef]

40. Crozier, L. Impacts of Climate Change on Columbia River Salmon: A review of the scientific literature published in 2011. In *Pages D1–D50 in Endangered Species Act Section 7(a)(2) Supplemental Biological Opinion: Consultation on Remand for Operation of the Federal Columbia River Power System*; U.S. National Marine Fisheries Service: West Coast Regin, 1201 NE Lloyd Blvd, Ste 110, Portland, OR 97232, USA, 2012.

41. Crozier, L. Impacts of Climate Change on Columbia River Salmon: A review of the scientific literature published in 2012. In *Pages D1–D50 in Endangered Species Act Section 7(a)(2) Supplemental Biological Opinion: Consultation on Remand for Operation of the Federal Columbia River Power System*; U.S. National Marine Fisheries Service: West Coast Regin, 1201 NE Lloyd Blvd, Ste 110, Portland, OR 97232, USA, 2013.

42. Crozier, L. Impacts of Climate Change on Columbia River Salmon: A review of the scientific literature published in 2013. In *Pages D1–D50 in Endangered Species Act Section 7(a)(2) Supplemental Biological Opinion: Consultation on Remand for Operation of the Federal Columbia River Power System*; U.S. National Marine Fisheries Service: West Coast Regin, 1201 NE Lloyd Blvd, Ste 110, Portland, OR 97232, USA, 2014. Available online: https://repository.library.noaa.gov/view/noaa/25641/noaa_25641_DS1.pdf (accessed on 14 June 2023).

43. Crozier, L. Impacts of Climate Change on Columbia River Salmon: A review of the scientific literature published in 2014. In *Pages D1–D50 in Endangered Species Act Section 7(a)(2) Supplemental Biological Opinion: Consultation on Remand for Operation of the Federal Columbia River Power System*; U.S. National Marine Fisheries Service: West Coast Regin, 1201 NE Lloyd Blvd, Ste 110, Portland, OR 97232, USA, 2015.

44. Crozier, L. Impacts of Climate Change on Columbia River Salmon: A review of the scientific literature published in 2015. In *Pages D1–D50 in Endangered Species Act Section 7(a)(2) Supplemental Biological Opinion: Consultation on Remand for Operation of the Federal Columbia River Power System*; U.S. National Marine Fisheries Service: West Coast Regin, 1201 NE Lloyd Blvd, Ste 110, Portland, OR 97232, USA, 2016.

45. Crozier, L. Impacts of Climate Change on Columbia River Salmon: A review of the scientific literature published in 2016. In *Pages D1–D50 in Endangered Species Act Section 7(a)(2) Supplemental Biological Opinion: Consultation on Remand for Operation of the Federal Columbia River Power System*; U.S. National Marine Fisheries Service: West Coast Regin, 1201 NE Lloyd Blvd, Ste 110, Portland, OR 97232, USA, 2017.

46. Crozier, L.G.; Siegel, J. Impacts of Climate Change on Columbia River Salmon: A review of the scientific literature published in 2017. In *Pages D1–D50 in Endangered Species Act Section 7(a)(2) Supplemental Biological Opinion: Consultation on Remand for Operation of the Federal Columbia River Power System*; U.S. National Marine Fisheries Service: West Coast Regin, 1201 NE Lloyd Blvd, Ste 110, Portland, OR 97232, USA, 2018. [CrossRef]

47. Siegel, J.; Crozier, L.G. Impacts of climate change on Columbia River salmon: A review of the scientific literature published in 2018. In *Pages D1–D50 in Endangered Species Act Section 7(a)(2) Supplemental Biological Opinion: Consultation on Remand for Operation of the Federal Columbia River Power System*; U.S. National Marine Fisheries Service: West Coast Regin, 1201 NE Lloyd Blvd, Ste 110, Portland, OR 97232, USA, 2019. [CrossRef]

48. Siegel, J.; Crozier, L.G. *Impacts of Climate Change on Columbia River Salmon: A Review of The scientific Literature Published in 2019*; U.S. National Marine Fisheries Service: West Coast Regin, 1201 NE Lloyd Blvd, Ste 110, Portland, OR 97232, USA, 2020. [CrossRef]

49. Oesterwind, D.; Rau, A.; Zaiko, A. Drivers and pressures—Untangling the terms commonly used in marine science and policy. *J. Environ. Manag.* **2016**, *181*, 8–15. [CrossRef]

50. Nielsen, J.L.; Ruggerone, G.T.; Zimmerman, C.E. Adaptive strategies and life history characteristics in a warming climate: Salmon in the Arctic? *Environ. Biol. Fishes* **2013**, *96*, 1187–1226. [CrossRef]

51. IPCC. *Climate Change 2021: The Physical Science Basis. Contribution of Working Group I to the Sixth Assessment Report of the Intergovernmental Panel on Climate Change*; Masson-Delmotte, V., Zhai, P., Pirani, A., Connors, S., Péan, C., Berger, S., Caud, N., Chen, Y., Goldfarb, L., Gomis, M., et al., Eds.; Cambridge University Press: Cambridge, UK, 2021; Available online: https://www.ipcc.ch/report/ar6/wg1/#FullReport (accessed on 1 January 2022).

52. Herring, S.C.; Christidis, N.; Hoell, A.; Kossin, J.P.; Schreck II, C.J.; Stott, P.A. Explaining extreme events of 2016 from a climate perspective. *Bull. Am. Meteorol. Soc.* **2018**, *99*, S1–S157. [CrossRef]

53. Mote, P.; Snover, A.K.; Capalbo, S.; Eigenbrode, S.D.; Glick, P.; Littell, J.; Raymondi, R.; Reeder, S. Northwest. In *Climate Change Impacts in the United States: The Third National Climate Assessment*; Melillo, J.M., Richmond, T.C., Yohe, G.W., Eds.; U.S. Global Change Research Program: Washington, DC, USA, 2014; pp. 487–513. [CrossRef]

54. May, C.; Luce, C.; Casola, J.; Chang, M.; Cuhaciyan, J.; Dalton, M.; Lowe, S.; Morishima, G.; Mote, P.; Petersen, A.; et al. Northwest. In *Climate Change Impacts in the United States: The Fourth National Climate Assessmentt*; Reidmiller, D.R., Avery, C.W., Easterling, D.R., Kunkel, K.E., Lewis, K.L.M., Maycock, T.K., Stewart, B.C., Eds.; U.S. Global Change Research Program: Washington, DC, USA, 2018; Volume II, pp. 1036–1100. [CrossRef]

55. Frankson, R.; Kunkel, K.E.; Champion, S.M.; Easterling, D.R.; Stevens, L.E.; Bumbaco, K.; Bond, N.; Casola, J.; Sweet, W. Washington State Climate Summary 2022. In *NOAA Technical Report NESDIS 150-WA*; NOAA/NESDIS: Silver Spring, MD, USA, 2022; p. 5.

56. Runkle, J.; Kunkel, K.E.; Frankson, R.; Champion, S.M.; Stevens, L.E.; Abatzoglou, J. Idaho State Climate Summary 2022. In *NOAA Technical Report NESDIS 150-ID*; NOAA/NESDIS: Silver Spring, MD, USA, 2022; p. 4.

57. Frankson, R.; Kunkel, K.E.; Champion, S.M.; Stevens, L.E.; Easterling, D.R.; Dello, K.; Dalton, M.; Sharp, D.; O'Neill, L. Oregon State Climate Summary 2022. In *NOAA Technical Report NESDIS 150-OR*; NOAA/NESDIS: Silver Spring, MD, USA, 2022; p. 5.

58. Abatzoglou, J.T.; Rupp, D.E.; Mote, P.W. Seasonal climate variability and change in the Pacific Northwest of the United States. *J. Clim.* **2014**, *27*, 2125–2142. [CrossRef]

59. Mutiibwa, D.; Vavrus, S.J.; McAfee, S.A.; Albright, T.P. Recent spatiotemporal patterns in temperature extremes across conterminous United States. *J. Geophys. Res-Atmos.* **2015**, *120*, 7378–7392. [CrossRef]

60. Mera, R.; Massey, N.; Rupp, D.E.; Mote, P.; Allen, M.; Frumhoff, P.C. Climate change, climate justice and the application of probabilistic event attribution to summer heat extremes in the California Central Valley. *Clim. Chang.* **2015**, *133*, 427–438. [CrossRef]

61. Durre, I.; Squires, M.F.; Vose, R.S.; Yin, X.G.; Arguez, A.; Applequist, S. NOAA's 1981-2010 U.S. climate normals. Monthly precipitation, snowfall, and snow depth. *J. Appl. Meteorol. Climatol.* **2013**, *52*, 2377–2395. [CrossRef]

62. Mao, Y.; Nijssen, B.; Lettenmaier, D.P. Is climate change implicated in the 2013- 2014 California drought? A hydrologic perspective. *Geophys. Res. Lett.* **2015**, *42*, 2805–2813. [CrossRef]

63. Mote, P.W.; Li, S.; Lettenmaier, D.P.; Xiao, M.; Engel, R. Dramatic declines in snowpack in the western US. *NPJ Clim. Atmos. Sci.* **2018**, *1*, 2. [CrossRef]

64. Lute, A.C.; Luce, C.H. Are Model Transferability And Complexity Antithetical? Insights From Validation of a Variable-Complexity Empirical Snow Model in Space and Time. *Water Resour. Res.* **2017**, *53*, 8825–8850. [CrossRef]

65. Kluver, D.; Leathers, D. Regionalization of snowfall frequency and trends over the contiguous United States. *Int. J. Climatol.* **2015**, *35*, 4348–4358. [CrossRef]

66. Malcomb, N.L.; Wiles, G.C. Tree-ring-based reconstructions of North American glacier mass balance through the Little Ice Age—Contemporary warming transition. *Quat. Res.* **2013**, *79*, 123–137. [CrossRef]

67. Riedel, J.L.; Wilson, S.; Baccus, W.; Larrabee, M.; Fudge, T.J.; Fountain, A. Glacier status and contribution to streamflow in the Olympic Mountains, Washington, USA. *J. Glaciol.* **2015**, *61*, 8–16. [CrossRef]

68. Marcinkowski, K.; Peterson, D.L. A 350-year reconstruction of the response of South Cascade Glacier to interannual and interdecadal climatic variability. *Northwest Sci.* **2015**, *89*, 14–33. [CrossRef]

69. Fountain, A.G.; Gray, C.; Glenn, B.; Menounos, B.; Pflug, J.; Riedel, J.L. Glaciers of the Olympic Mountains, Washington—The Past and Future 100 Years. *J. Geophys. Res. Earth Surf.* **2022**, *127*, e2022JF006670. [CrossRef]

70. Dittmer, K. Changing streamflow on Columbia Basin tribal lands-climate change and salmon. *Clim. Chang.* **2013**, *120*, 627–641. [CrossRef]

71. Riche, O.; Johannessen, S.C.; Macdonald, R.W. Why timing matters in a coastal sea: Trends, variability and tipping points in the Strait of Georgia, Canada. *J. Mar. Syst.* **2014**, *131*, 36–53. [CrossRef]

72. Hatcher, K.L.; Jones, J.A. Climate and streamflow trends in the Columbia River Basin: Evidence for ecological and engineering resilience to climate change. *Atmos.-Ocean* **2013**, *51*, 436–455. [CrossRef]

73. Isaak, D.J.; Wollrab, S.; Horan, D.; Chandler, G. Climate change effects on stream and river temperatures across the northwest US from 1980-2009 and implications for salmonid fishes. *Clim. Chang.* **2012**, *113*, 499–524. [CrossRef]

74. Diffenbaugh, N.S.; Swain, D.L.; Touma, D. Anthropogenic warming has increased drought risk in California. *Proc. Natl. Acad. Sci. USA* **2015**, *112*, 3931–3936. [CrossRef]

75. Kormos, P.R.; Charles, H.L.; Seth, J.W.; Wouter, R.B. Trends and Sensitivies of Low Streamflow Extremes to Discharge Timing and Magnitude in Pacific Northwest Mountain Streams. *Water Resour. Res.* **2016**, *52*, 4990–5007. [CrossRef]

76. Leppi, J.C.; Thomas, H.D.; Solomon, W.H.; Steven, W.R. Impacts of Climate Change on August Stream Discharge in the Central-Rocky Mountains. *Clim. Change* **2012**, *112*, 997–1014. [CrossRef]

77. Luce, C.H. and Holden, Z.A. Declining Annual Streamflow Distributions in the Pacific Northwest United States, 1948–2006. *Geophys. Res. Lett.* **2009**, *36*, 1–6. [CrossRef]

78. Reilly, M.J.; Dunn, C.J.; Meigs, G.W.; Spies, T.A.; Kennedy, R.E.; Bailey, J.D.; Briggs, K. Contemporary patterns of fire extent and severity in forests of the Pacific Northwest, USA (1985–2010). *Ecosphere* **2017**, *8*, e01695. [CrossRef]

79. Johnston, J.D.; Dunn, C.J.; Vernon, M.J.; Bailey, J.D.; Morrissette, B.A.; Morici, K.E. Restoring historical forest conditions in a diverse inland Pacific Northwest landscape. *Ecosphere* **2018**, *9*, e02400. [CrossRef]

80. Siegel, J.E.; Volk, C. Accurate spatiotemporal predictions of daily stream temperature from statistical models accounting for interactions between climate and landscape. *PeerJ* **2019**, *7*, e7892. [CrossRef]
81. Islam, S.U.; Hay, R.W.; Dery, S.J.; Booth, B.P. Modelling the impacts of climate change on riverine thermal regimes in western Canada's largest Pacific watershed. *Sci. Rep.* **2019**, *9*, 14. [CrossRef]
82. Cheng, L.; Abraham, J.; Trenberth, K.E.; Fasullo, J.; Boyer, T.; Locarnini, R.; Zhang, B.; Yu, F.; Wan, L.; Chen, X.; et al. Upper Ocean Temperatures Hit Record High in 2020. *Adv. Atmos. Sci.* **2021**, *38*, 523–530. [CrossRef]
83. Johnson, G.C.; Lumpkin, R. 2021: Global Oceans, in "State of the Climate in 2020". *Bull. Am. Meteorol. Soc.* **2021**, *102*, S143–S198. [CrossRef]
84. Johnstone, J.A.; Mantua, N.J. Atmospheric controls on northeast Pacific temperature variability and change, 1900-2012. *Proc. Natl. Acad. Sci. USA* **2014**, *111*, 14360–14365. [CrossRef]
85. Cummins, P.F.; Masson, D. Climatic variability and trends in the surface waters of coastal British Columbia. *Prog. Oceanogr.* **2014**, *120*, 279–290. [CrossRef]
86. Talloni-Alvarez, N.E.; Sumaila, U.R.; Le Billon, P.; Cheung, W.W.L. Climate change impact on Canada's Pacific marine ecosystem: The current state of knowledge. *Mar. Policy* **2019**, *104*, 163–176. [CrossRef]
87. Johnstone, J.A.; Mantua, N.J. Reply to Abatzoglou et al.: Atmospheric controls on northwest United States air temperatures, 1948-2012. *Proc. Natl. Acad. Sci. USA* **2014**, *111*, E5607–E5608. [CrossRef]
88. Jewett, L.; Romanou, A. Ocean acidification and other ocean changes. In *Climate Science Special Report: Fourth National Climate Assessment*; Wuebbles, D.J., Fahey, D.W., Hibbard, K.A., Dokken, D.J., Stewart, B.C., Maycock, T.K., Eds.; U.S. Global Change Research Program: Washington, DC, USA, 2017; Volume I, pp. 364–392. [CrossRef]
89. Feely, R.A.; Sabine, C.L.; Hernandez-Ayon, J.M.; Ianson, D.; Hales, B. Evidence for upwelling of corrosive "acidified" water onto the continental shelf. *Science* **2008**, *320*, 1490–1492. [CrossRef]
90. Turi, G.; Lachkar, Z.; Gruber, N.; Nunnich, M. Climatic modulation of recent trends in ocean acidification in the California Current System. *Environ. Res. Lett.* **2016**, *11*, 17. [CrossRef]
91. Wootton, J.T.; Pfister, C.A. Carbon system measurements and potential climatic drivers at a site of rapidly declining ocean pH. *PLoS ONE* **2012**, *7*, e53396. [CrossRef] [PubMed]
92. Bond, N.A.; Cronin, M.F.; Freeland, H.; Mantua, N. Causes and impacts of the 2014 warm anomaly in the NE Pacific. *Geophys. Res. Lett.* **2015**, *42*, 3414–3420. [CrossRef]
93. Amaya, D.J.; Miller, A.J.; Xie, S.-P.; Kosaka, Y. Physical drivers of the summer 2019 North Pacific marine heatwave. *Nat. Commun.* **2020**, *11*, 1903. [CrossRef]
94. Jacox, M.G.; Alexander, M.A.; Mantua, N.J.; Scott, J.D.; Hervieux, G.; Webb, R.S.; Werner, F.E. Forcing of multiyear extreme ocean temperature that impacted California Current living marine resources in 2016. *Bull. Am. Meteorol. Soc.* **2018**, *99*, S27–S33. [CrossRef]
95. Di Lorenzo, E.; Mantua, N. Multi-year persistence of the 2014/15 North Pacific marine heatwave. *Nat. Clim. Chang.* **2016**, *6*, 1042. [CrossRef]
96. Peterson, W.T.; Fisher, J.L.; Strub, P.T.; Du, X.; Risien, C.; Peterson, J.; Shaw, C.T. The pelagic ecosystem in the Northern California Current off Oregon during the 2014-2016 warm anomalies within the context of the past 20 years. *J. Geophys. Res. Ocean.* **2017**, *122*, 7267–7290. [CrossRef] [PubMed]
97. Sadowski, J.S.; Gonzalez, J.A.; Lonhart, S.I.; Jeppesen, R.; Grimes, T.M.; Grosholz, E.D. Temperature-induced range expansion of a subtropical crab along the California coast. *Mar. Ecol.* **2018**, *39*, e12528. [CrossRef]
98. Brodeur, R.D.; Auth, T.D.; Phillips, A.J. Major Shifts in Pelagic Micronekton and Macrozooplankton Community Structure in an Upwelling Ecosystem Related to an Unprecedented Marine Heatwave. *Front. Mar. Sci.* **2019**, *6*, 15. [CrossRef]
99. Cavole, L.M.; Demko, A.M.; Diner, R.E.; Giddings, A.; Koester, I.; Pagniello, C.; Paulsen, M.L.; Ramirez-Valdez, A.; Schwenck, S.M.; Yen, N.K.; et al. Biological Impacts of the 2013–2015 Warm-Water Anomaly in the Northeast Pacific. *Oceanography* **2016**, *29*, 273–285. [CrossRef]
100. Moreira, M.; Hayes, D.S.; Boavida, I.; Schletterer, M.; Schmutz, S.; Pinheiro, A. Ecologically-based criteria for hydropeaking mitigation: A review. *Sci. Total Environ.* **2018**, *657*, 1508–1522. [CrossRef] [PubMed]
101. Baker, M.R.; Matta, M.E.; Beaulieu, M.; Paris, N.; Huber, S.; Graham, O.J.; Pham, T.; Sisson, N.B.; Heller, C.P.; Witt, A.; et al. Intra-seasonal and inter-annual patterns in the demographics of sand lance and response to environmental drivers in the North Pacific. *Mar. Ecol. Prog. Ser.* **2019**, *617*, 221–244. [CrossRef]
102. Brodeur, R.D.; Hunsicker, M.E.; Hann, A.; Miller, T.W. Effects of warming ocean conditions on feeding ecology of small pelagic fishes in a coastal upwelling ecosystem: A shift to gelatinous food sources. *Mar. Ecol. Prog. Ser.* **2019**, *617*, 149–163. [CrossRef]
103. Auth, T.D.; Daly, E.A.; Brodeur, R.D.; Fisher, J.L. Phenological and distributional shifts in ichthyoplankton associated with recent warming in the northeast Pacific Ocean. *Glob. Chang. Biol.* **2018**, *24*, 259–272. [CrossRef]
104. Swan, G.A. Chinook salmon spawning surveys in deep waters of a large, regulated river. *Regul. Rivers Res. Manag.* **1989**, *4*, 355–370. [CrossRef]
105. Del Rio, A.M.; Mukai, G.N.; Martin, B.T.; Johnson, R.C.; Fangue, N.A.; Israel, J.A.; Todgham, A.E. Differential sensitivity to warming and hypoxia during development and long-term effects of developmental exposure in early life stage Chinook salmon. *Conserv. Physiol.* **2021**, *9*. [CrossRef]

106. Gould, G.K.; Liu, M.L.; Barber, M.E.; Cherkauer, K.A.; Robichaud, P.R.; Adam, J.C. The effects of climate change and extreme wildfire events on runoff erosion over a mountain watershed. *J. Hydrol.* **2016**, *536*, 74–91. [CrossRef]

107. Srivastava, A.; Brooks, E.S.; Dobre, M.; Elliot, W.J.; Wu, J.Q.; Flanagan, D.C.; Gravelle, J.A.; Link, T.E. Modeling forest management effects on water and sediment yield from nested, paired watersheds in the interior Pacific Northwest, USA using WEPP. *Sci. Total Environ.* **2020**, *701*, 134877. [CrossRef]

108. Anttila, K.; Dhillon, R.S.; Boulding, E.G.; Farrell, A.P.; Glebe, B.D.; Elliott, J.A.K.; Wolters, W.R.; Schulte, P.M. Variation in temperature tolerance among families of Atlantic salmon (*Salmo salar*) is associated with hypoxia tolerance, ventricle size and myoglobin level. *J. Exp. Biol.* **2013**, *216*, 1183–1190. [CrossRef]

109. Friedman, W.R.; Martin, B.T.; Wells, B.K.; Warzybok, P.; Michel, C.J.; Danner, E.M.; Lindley, S.T. Modeling composite effects of marine and freshwater processes on migratory species. *Ecosphere* **2019**, *10*, 21. [CrossRef]

110. Weinheimer, J.; Anderson, J.H.; Downen, M.; Zimmerman, M.; Johnson, T. Monitoring climate impacts: Survival and migration timing of summer chum salmon in Salmon Creek, Washington. *Trans. Am. Fish. Soc.* **2017**, *146*, 983–995. [CrossRef]

111. Narum, S.R.; Campbell, N.R.; Kozfkay, C.C.; Meyer, K.A. Adaptation of redband trout in desert and montane environments. *Mol. Ecol.* **2010**, *19*, 4622–4637. [CrossRef] [PubMed]

112. Kavanagh, K.D.; Haugen, T.O.; Gregersen, F.; Jernvall, J.; Vollestad, L.A. Contemporary temperature-driven divergence in a Nordic freshwater fish under conditions commonly thought to hinder adaptation. *BMC Evol. Biol.* **2010**, *10*, 350. [CrossRef] [PubMed]

113. Beer, W.N.; Steel, E.A. Impacts and implications of temperature variability on Chinook salmon egg development and emergence phenology. *Trans. Am. Fish. Soc.* **2018**, *147*, 3–15. [CrossRef]

114. Fuhrman, A.E.; Larsen, D.A.; Steel, E.A.; Young, G.; Beckman, B.R. Chinook salmon emergence phenotypes: Describing the relationships between temperature, emergence timing and condition factor in a reaction norm framework. *Ecol. Freshw. Fish* **2018**, *27*, 350–362. [CrossRef]

115. Lohmus, M.; Sundstrom, L.F.; Bjorklund, M.; Devlin, R.H. Genotype-temperature interaction in the regulation of development, growth, and morphometrics in wild-type, and growth-hormone transgenic coho salmon. *PLoS ONE* **2010**, *5*, e9980. [CrossRef]

116. Whitney, C.K.; Hinch, S.G.; Patterson, D.A. Provenance matters: Thermal reaction norms for embryo survival among sockeye salmon *Oncorhynchus nerka* populations. *J. Fish Biol.* **2013**, *82*, 1159–1176. [CrossRef]

117. Braun, D.C.; Patterson, D.A.; Reynolds, J.D. Maternal and environmental influences on egg size and juvenile life-history traits in Pacific salmon. *Ecol. Evol.* **2013**, *3*, 1727–1740. [CrossRef]

118. Drinan, D.P.; Zale, A.V.; Webb, M.A.H.; Shepard, B.B.; Kalinowski, S.T. Evidence of local adaptation in westslope cutthroat trout. *Trans. Am. Fish. Soc.* **2012**, *141*, 872–880. [CrossRef]

119. Goode, J.R.; Buffington, J.M.; Tonina, D.; Isaak, D.J.; Thurow, R.F.; Wenger, S.; Nagel, D.; Luce, C.; Tetzlaff, D.; Soulsby, C. Potential effects of climate change on streambed scour and risks to salmonid survival in snow-dominated mountain basins. *Hydrol. Process.* **2013**, *27*, 750–765. [CrossRef]

120. MacDonald, R.J.; Boon, S.; Byrne, J.M.; Robinson, M.D.; Rasmussen, J.B. Potential future climate effects on mountain hydrology, stream temperature, and native salmonid life history. *Can. J. Fish. Aquat. Sci.* **2014**, *71*, 189–202. [CrossRef]

121. Meyers, E.M.; Dobrowski, B.; Tague, C.L. Climate change impacts on flood frequency, intensity, and timing may affect trout species in Sagehen Creek, California. *Trans. Am. Fish. Soc.* **2010**, *139*, 1657–1664. [CrossRef]

122. Sparks, M.M.; Falke, J.A.; Quinn, T.P.; Adkison, M.D.; Schindler, D.E.; Bartz, K.; Young, D.; Westley, P.A.H. Influences of spawning timing, water temperature, and climatic warming on early life history phenology in western Alaska sockeye salmon. *Can. J. Fish. Aquat. Sci.* **2019**, *76*, 123–135. [CrossRef]

123. Beechie, T.; Imaki, H.; Greene, J.; Wade, A.; Wu, H.; Pess, G.; Roni, P.; Kimball, J.; Stanford, J.; Kiffney, P.; et al. Restoring salmon habitat for a changing climate. *River Res. Appl.* **2012**, *29*, 939–960. [CrossRef]

124. Santiago, J.M.; Alonso, C.; de Jalon, D.G.; Solana-Gutierrez, J.; Munoz-Mas, R. Effects of climate change on the life stages of stream-dwelling brown trout (*Salmo trutta* Linnaeus, 1758) at the rear edge of their native distribution range. *Ecohydrology* **2020**, *13*, e2241. [CrossRef]

125. Erkinaro, J.; Czorlich, Y.; Orell, P.; Kuusela, J.; Falkegard, M.; Lansman, M.; Pulkkinen, H.; Primmer, C.R.; Niemela, E. Life history variation across four decades in a diverse population complex of Atlantic salmon in a large subarctic river. *Can. J. Fish. Aquat. Sci.* **2019**, *76*, 42–55. [CrossRef]

126. Satterthwaite, W.H.; Beakes, M.P.; Collins, E.M.; Swank, D.R.; Merz, J.E.; Titus, R.G.; Sogard, S.M.; Mangel, M. State-dependent life history models in a changing (and regulated) environment: Steelhead in the California Central Valley. *Evol. Appl.* **2010**, *3*, 221–243. [CrossRef]

127. Kendall, N.W.; McMillan, J.R.; Sloat, M.R.; Buehrens, T.W.; Quinn, T.P.; Pess, G.R.; Kuzishchin, K.V.; McClure, M.M.; Zabel, R.W. Anadromy and residency in steelhead and rainbow trout (*Oncorhynchus mykiss*): A review of the processes and patterns. *Can. J. Fish. Aquat. Sci.* **2015**, *72*, 319–342. [CrossRef]

128. Yvon-Durocher, G.; Montoya, J.M.; Trimmer, M.; Woodward, G. Warming alters the size spectrum and shifts the distribution of biomass in freshwater ecosystems. *Glob. Chang. Biol.* **2010**, *17*, 1681–1694. [CrossRef]

129. Beveridge, O.S.; Petchey, O.L.; Humphries, S. Direct and indirect effects of temperature on the population dynamics and ecosystem functioning of aquatic microbial ecosystems. *J. Anim. Ecol.* **2010**, *79*, 1324–1331. [CrossRef]

130. Jones, I.D.; Page, T.; Alex Elliott, J.; Thackeray, S.J.; Louise Heathwaite, A. Increases in lake phytoplankton biomass caused by future climate-driven changes to seasonal river flow. *Glob. Chang. Biol.* **2010**, *17*, 1809–1820. [CrossRef]

131. Beakes, M.P.; Moore, J.W.; Hayes, S.A.; Sogard, S.M. Wildfire and the effects of shifting stream temperature on salmonids. *Ecosphere* **2014**, *5*, art63. [CrossRef]

132. Howell, P.J. Changes in native bull trout and non-native brook trout distributions in the upper Powder River basin after 20 years, relationships to water temperature and implications of climate change. *Ecol. Freshw. Fish* **2018**, *27*, 710–719. [CrossRef]

133. Dexter, E.; Bollens, S.M.; Rollwagen-Bollens, G.; Emerson, J.; Zimmerman, J. Persistent vs. ephemeral invasions: 8.5 years of zooplankton community dynamics in the Columbia River. *Limnol. Oceanogr.* **2015**, *60*, 527–539. [CrossRef]

134. Emerson, J.E.; Bollens, S.M.; Counihan, T.D. Seasonal dynamics of zooplankton in Columbia-Snake River reservoirs, with special emphasis on the invasive copepod Pseudodiaptomus forbesi. *Aquat. Invasions* **2015**, *10*, 25–40. [CrossRef]

135. Adams, J.B.; Bollens, S.M.; Bishop, J.G. Predation on the Invasive Copepod, Pseudodiaptomus forbesi, and Native Zooplankton in the Lower Columbia River: An Experimental Approach to Quantify Differences in Prey-Specific Feeding Rates. *PLoS ONE* **2015**, *10*, e0144095. [CrossRef]

136. Haskell, C.A.; Beauchamp, D.A.; Bollens, S.M. Linking functional response and bioenergetics to estimate juvenile salmon growth in a reservoir food web. *PLoS ONE* **2017**, *12*, e0185933. [CrossRef]

137. Hasselman, D.J.; Hinrichsen, R.A.; Shields, B.A.; Ebbesmeyer, C.C. American shad of the Pacific Coast: A harmful invasive species or benign introduction? *Fisheries* **2012**, *37*, 115–122. [CrossRef]

138. Hinrichsen, R.A.; Hasselman, D.J.; Ebbesmeyer, C.C.; Shields, B.A. The Role of Impoundments, Temperature, and Discharge on Colonization of the Columbia River Basin, USA, by Nonindigenous American Shad. *Trans. Am. Fish. Soc.* **2013**, *142*, 887–900. [CrossRef]

139. Hughes, R.M.; Herlihy, A.T. Patterns in catch per unit effort of native prey fish and alien piscivorous fish in 7 Pacific Northwest USA rivers. *Fisheries* **2012**, *37*, 201–211. [CrossRef]

140. Christensen, D.R.; Moore, B.C. Largemouth bass consumption demand on hatchery rainbow trout in two Washington lakes. *Lake Reserv. Manag.* **2010**, *26*, 200–211. [CrossRef]

141. Lawrence, D.J.; Olden, J.D.; Torgersen, C.E. Spatiotemporal patterns and habitat associations of smallmouth bass (*Micropterus dolomieu*) invading salmon-rearing habitat. *Freshw. Biol.* **2012**, *57*, 1929–1946. [CrossRef]

142. Tabor, R.A.; Footen, B.A.; Fresh, K.L.; Celedonia, M.T.; Mejia, F.; Low, D.L.; Park, L. Smallmouth bass and largemouth bass predation on juvenile Chinook salmon and other salmonids in the Lake Washington Basin. *N. Am. J. Fish. Manag.* **2007**, *27*, 1174–1188. [CrossRef]

143. Lawrence, D.J.; Beauchamp, D.A.; Olden, J.D. Life-stage-specific physiology defines invasion extent of a riverine fish. *J. Anim. Ecol.* **2015**, *84*, 879–888. [CrossRef]

144. Kuehne, L.M.; Olden, J.D.; Duda, J.J. Costs of living for juvenile Chinook salmon (*Oncorhynchus tshawytscha*) in an increasingly warming and invaded world. *Can. J. Fish. Aquat. Sci.* **2012**, *69*, 1621–1630. [CrossRef]

145. Boltana, S.; Sanhueza, N.; Aguilar, A.; Gallardo-Escarate, C.; Arriagada, G.; Valdes, J.A.; Soto, D.; Quinones, R.A. Influences of thermal environment on fish growth. *Ecol. Evol.* **2017**, *7*, 6814–6825. [CrossRef]

146. Brewitt, K.S.; Danner, E.M.; Moore, J.W. Hot eats and cool creeks: Juvenile Pacific salmonids use mainstem prey while in thermal refuges. *Can. J. Fish. Aquat. Sci.* **2017**, *74*, 1588–1602. [CrossRef]

147. Ohlberger, J.; Buehrens, T.W.; Brenkman, S.J.; Crain, P.; Quinn, T.P.; Hilborn, R. Effects of past and projected river discharge variability on freshwater production in an anadromous fish. *Freshw. Biol.* **2018**, *63*, 331–340. [CrossRef]

148. Kock, T.J.; Liedtke, T.L.; Rondorf, D.W.; Serl, J.D.; Kohn, M.; Bumbaco, K.A. Elevated streamflows increase dam passage by juvenile coho salmon during winter: Implications of climate change in the Pacific Northwest. *N. Am. J. Fish. Manag.* **2012**, *32*, 1070–1079. [CrossRef]

149. Penaluna, B.E.; Dunham, J.B.; Railsback, S.F.; Arismendi, I.; Johnson, S.L.; Bilby, R.E.; Safeeq, M.; Skaugset, A.E. Local Variability Mediates Vulnerability of Trout Populations to Land Use and Climate Change. *PLoS ONE* **2015**, *10*, e0135334. [CrossRef] [PubMed]

150. Hardiman, J.M.; Mesa, M.G. The effects of increased stream temperatures on juvenile steelhead growth in the Yakima River Basin based on projected climate change scenarios. *Clim. Chang.* **2014**, *124*, 413–426. [CrossRef]

151. Zhang, X.; Li, H.Y.; Deng, Z.Q.D.; Leung, L.R.; Skalski, J.R.; Cooke, S.J. On the variable effects of climate change on Pacific salmon. *Ecol. Model.* **2019**, *397*, 95–106. [CrossRef]

152. Walters, A.W.; Bartz, K.K.; McClure, M.M. Interactive effects of water diversion and climate change for juvenile Chinook salmon in the Lemhi River Basin (U.S.A). *Conserv. Biol.* **2013**, *27*, 1179–1189. [CrossRef]

153. Lawrence, D.J.; Stewart-Koster, B.; Olden, J.D.; Ruesch, A.S.; Torgersen, C.E.; Lawler, J.J.; Butcher, D.P.; Crown, J.K. The interactive effects of climate change, riparian management, and a nonnative predator on stream-rearing salmon. *Ecol. Appl.* **2014**, *24*, 895–912. [CrossRef] [PubMed]

154. Hanson, K.C.; Peterson, D.P. Modeling the potential impacts of climate change on Pacific salmon culture programs: An example at Winthrop National Fish Hatchery. *Environ. Manag.* **2014**, *54*, 433–448. [CrossRef] [PubMed]

155. Beer, W.N.; Anderson, J.J. Sensitivity of salmonid freshwater life history in western US streams to future climate conditions. *Glob. Chang. Biol.* **2013**, *19*, 2547–2556. [CrossRef] [PubMed]

156. Kilduff, D.P.; Di Lorenzo, E.; Botsford, L.W.; Teo, S.L.H. Changing central Pacific El Niños reduce stability of North American salmon survival rates. *Proc. Natl. Acad. Sci. USA* **2015**, *112*, 10962–10966. [CrossRef] [PubMed]

157. Mantua, N.J. Shifting patterns in Pacific climate, West Coast salmon survival rates, and increased volatility in ecosystem services. *Proc. Natl. Acad. Sci. USA* **2015**, *112*, 10823–10824. [CrossRef] [PubMed]
158. Sharma, R.; Liermann, M. Using hierarchical models to estimate effects of ocean anomalies on north-west Pacific Chinook salmon *Oncorhynchus tshawytscha* recruitment. *J. Fish Biol.* **2010**, *77*, 1948–1963. [CrossRef]
159. Sharma, R.; Quinn, T.P. Linkages between life history type and migration pathways in freshwater and marine environments for Chinook salmon, *Oncorhynchus tshawytscha*. *Acta Oecol. -Int. J. Ecol.* **2012**, *41*, 1–13. [CrossRef]
160. Sharma, R.; Velez-Espino, L.A.; Wertheimer, A.C.; Mantua, N.; Francis, R.C. Relating spatial and temporal scales of climate and ocean variability to survival of Pacific Northwest Chinook salmon (*Oncorhynchus tshawytscha*). *Fish. Oceanogr.* **2013**, *22*, 14–31. [CrossRef]
161. Stachura, M.M.; Mantua, N.J.; Scheuerell, M.D. Oceanographic influences on patterns in North Pacific salmon abundance. *Can. J. Fish. Aquat. Sci.* **2014**, *71*, 226–235. [CrossRef]
162. Tucker, S.; Thiess, M.E.; Morris, J.F.T.; Mackas, D.; Peterson, W.T.; Candy, J.R.; Beacham, T.D.; Iwamoto, E.M.; Teel, D.J.; Peterson, M.; et al. Coastal Distribution and Consequent Factors Influencing Production of Endangered Snake River Sockeye Salmon. *Trans. Am. Fish. Soc.* **2015**, *144*, 107–123. [CrossRef]
163. Malick, M.J.; Cox, S.P.; Mueter, F.J.; Peterman, R.M. Linking phytoplankton phenology to salmon productivity along a north-south gradient in the Northeast Pacific Ocean. *Can. J. Fish. Aquat. Sci.* **2015**, *72*, 697–708. [CrossRef]
164. Li, S.; Wu, L.; Yang, Y.; Geng, T.; Cai, W.; Gan, B.; Chen, Z.; Jing, Z.; Wang, G.; Ma, X. The Pacific Decadal Oscillation less predictable under greenhouse warming. *Nat. Clim. Chang.* **2020**, *10*, 30–34. [CrossRef]
165. Osborne, E.B.; Thunell, R.C.; Gruber, N.; Feely, R.A.; Benitez-Nelson, C.R. Decadal variability in twentieth-century ocean acidification in the California Current Ecosystem. *Nat. Geosci.* **2020**, *13*, 43–49. [CrossRef]
166. IPCC. *IPCC Special Report on the Ocean and Cryosphere in a Changing Climate*; Pörtner, H.-O., Roberts, D.C., Masson-Delmotte, V., Zhai, P., Tignor, M., Poloczanska, E., Mintenbeck, K., Alegría, A., Nicolai, M., Okem, A., et al., Eds.; Cambridge University Press: Cambridge, UK; New York, NY, USA, 2019; p. 755. [CrossRef]
167. Litzow, M.A.; Ciannelli, L.; Cunningham, C.J.; Johnson, B.; Puerta, P. Nonstationary effects of ocean temperature on Pacific salmon productivity. *Can. J. Fish. Aquat. Sci.* **2019**, *76*, 1923–1928. [CrossRef]
168. Kilduff, D.P.; Botsford, L.W.; Teo, S.L.H. Spatial and temporal covariability in early ocean survival of Chinook salmon (*Oncorhynchus tshawytscha*) along the west coast of North America. *ICES J. Mar. Sci.* **2014**, *71*, 1671–1682. [CrossRef]
169. Gosselin, J.L.; Crozier, L.G.; Burke, B.J. Shifting signals: Correlations among freshwater, marine and climatic indices often investigated in Pacific salmon studies. *Ecol. Indic.* **2021**, *121*, 107167. [CrossRef]
170. Wainwright, T.C. Ephemeral relationships in salmon forecasting: A cautionary tale. *Prog. Oceanogr.* **2021**, *193*, 102522. [CrossRef]
171. Alava, J.J.; Cisneros-Montemayor, A.M.; Sumaila, U.R.; Cheung, W.W.L. Projected amplification of food web bioaccumulation of MeHg and PCBs under climate change in the Northeastern Pacific. *Sci. Rep.* **2018**, *8*, 13460. [CrossRef]
172. Busch, D.S.; McElhany, P. Estimates of the direct effect of seawater pH on the survival rate of species groups in the California Current Ecosystem. *PLoS ONE* **2016**, *11*, 28. [CrossRef]
173. Marshall, K.N.; Kaplan, I.C.; Hodgson, E.E.; Hermann, A.; Busch, D.S.; Mcelhany, P.; Essington, T.E.; Harvey, C.J.; Fulton, E.A. Risks of ocean acidification in the California Current food web and fisheries: Ecosystem model projections. *Glob. Chang. Biol.* **2017**, *23*, 1525–1539. [CrossRef]
174. Williams, C.R.; Dittman, A.H.; McElhany, P.; Busch, D.S.; Maher, M.T.; Bammler, T.K.; MacDonald, J.W.; Gallagher, E.P. Elevated CO2 impairs olfactory-mediated neural and behavioral responses and gene expression in ocean-phase coho salmon (*Oncorhynchus kisutch*). *Glob. Chang. Biol.* **2019**, *25*, 963–977. [CrossRef]
175. McCormick, S.D.; Regish, A.M. Effects of ocean acidification on salinity tolerance and seawater growth of Atlantic salmon *Salmo salar* smolts. *J. Fish Biol.* **2018**, *93*, 560–566. [CrossRef] [PubMed]
176. Ou, M.; Hamilton, T.J.; Eom, J.; Lyall, E.M.; Gallup, J.; Jiang, A.; Lee, J.; Close, D.A.; Yun, S.S.; Brauner, C.J. Responses of pink salmon to CO2-induced aquatic acidification. *Nat. Clim. Chang.* **2015**, *5*, 950–955. [CrossRef]
177. Frommel, A.Y.; Carless, J.; Hunt, B.P.V.; Brauner, C.J. Physiological resilience of pink salmon to naturally occurring ocean acidification. *Conserv. Physiol.* **2020**, *8*, coaa059. [CrossRef]
178. Keinänen, M.; Nikonen, S.; Käkelä, R.; Ritvanen, T.; Rokka, M.; Myllylä, T.; Pönni, J.; Vuorinen, P.J. High Lipid Content of Prey Fish and n−3 PUFA Peroxidation Impair the Thiamine Status of Feeding-Migrating Atlantic Salmon (*Salmo salar*) and Is Reflected in Hepatic Biochemical Indices. *Biomolecules* **2022**, *12*, 526. [CrossRef]
179. Phillips, E.M.; Horne, J.K.; Zamon, J.E. Characterizing juvenile salmon predation risk during early marine residence. *PLoS ONE* **2021**, *16*, e0247241. [CrossRef] [PubMed]
180. Moore, M.E.; Berejikian, B.A.; Greene, C.M.; Munsch, S. Environmental fluctuation and shifting predation pressure contribute to substantial variation in early marine survival of steelhead. *Mar. Ecol. Prog. Ser.* **2021**, *662*, 139–156. [CrossRef]
181. Muhling, B.A.; Brodie, S.; Smith, J.A.; Tommasi, D.; Gaitan, C.F.; Hazen, E.L.; Jacox, M.G.; Auth, T.D.; Brodeur, R.D. Predictability of species distributions deteriorates under novel environmental conditions in the California Current System. *Front. Mar. Sci.* **2020**, *7*, 589. [CrossRef]
182. Tolimieri, N.; Wallace, J.; Haltuch, M. Spatio-temporal patterns in juvenile habitat for 13 groundfishes in the California Current Ecosystem. *PLoS ONE* **2020**, *15*, e0237996. [CrossRef] [PubMed]

183. Malick, M.J.; Hunsicker, M.E.; Haltuch, M.A.; Parker-Stetter, S.L.; Berger, A.M.; Marshall, K.N. Relationships between temperature and Pacific hake distribution vary across latitude and life-history stage. *Mar. Ecol. Prog. Ser.* **2020**, *639*, 185–197. [CrossRef]

184. Brodeur, R.D.; Buchanan, J.C.; Emmett, R.L. Pelagic and demersal fish predators on juvenile and adult forage fish predators in the northern California Current: Spatial and temporal variations. *Calif. Coop. Ocean. Fish. Investig. Rep.* **2014**, *55*, 96–116.

185. Litz, M.N.C.; Emmett, R.L.; Bentley, P.J.; Claiborne, A.M.; Barcelo, C. Biotic and abiotic factors influencing forage fish and pelagic nekton community in the Columbia River plume (USA) throughout the upwelling season 1999–2009. *ICES J. Mar. Sci.* **2014**, *71*, 5–18. [CrossRef]

186. Kaltenberg, A.M.; Emmett, R.L.; Benoit-Bird, K.J. Timing of forage fish seasonal appearance in the Columbia River plume and link to ocean conditions. *Mar. Ecol. Prog. Ser.* **2010**, *419*, 171–184. [CrossRef]

187. Ruzicka, J.J.; Brink, K.H.; Gifford, D.J.; Bahr, F. A physically coupled end-to-end model platform for coastal ecosystems: Simulating the effects of climate change and changing upwelling characteristics on the Northern California Current ecosystem. *Ecol. Model.* **2016**, *331*, 86–99. [CrossRef]

188. Ruzicka, J.J.; Daly, E.A.; Brodeur, R.D. Evidence that summer jellyfish blooms impact Pacific Northwest salmon production. *Ecosphere* **2016**, *7*, 22. [CrossRef]

189. Goldstein, J.; Steiner, U.K. Ecological drivers of jellyfish blooms—The complex life history of a 'well-known' medusa (*Aurelia aurita*). *J. Anim. Ecol.* **2020**, *89*, 910–920. [CrossRef]

190. Siddique, A.; Purushothaman, J.; Madhusoodhanan, R.; Raghunathan, C. The rising swarms of jellyfish in Indian waters: The environmental drivers, ecological, and socio-economic impacts. *J. Water Clim. Chang.* **2022**, *13*, 3747–3759. [CrossRef]

191. Heneghan, R.F.; Everett, J.D.; Blanchard, J.L.; Sykes, P.; Richardson, A.J. Climate-driven zooplankton shifts cause large-scale declines in food quality for fish. *Nat. Clim. Chang.* **2023**, *13*, 470–477. [CrossRef]

192. Lee, S.H.; Tseng, L.; Yoon, Y.H.; Ramirez-Romero, E.; Hwang, J.S.; Molinero, J.C. The global spread of jellyfish hazards mirrors the pace of human imprint in the marine environment. *Environ. Int.* **2023**, *171*, 107699. [CrossRef] [PubMed]

193. Fiechter, J.; Santora, J.A.; Chavez, F.; Northcott, D.; Messie, M. Krill Hotspot Formation and Phenology in the California Current Ecosystem. *Geophys. Res. Lett.* **2020**, *47*, e2020GL088039. [CrossRef] [PubMed]

194. Fiechter, J.; Huff, D.D.; Martin, B.T.; Jackson, D.W.; Edwards, C.A.; Rose, K.A.; Curchitser, E.N.; Hedstrom, K.S.; Lindley, S.T.; Wells, B.K. Environmental conditions impacting juvenile Chinook salmon growth off central California: An ecosystem model analysis. *Geophys. Res. Lett.* **2015**, *42*, 2910–2917. [CrossRef]

195. Fiechter, J.; Huckstadt, L.A.; Rose, K.A.; Costa, D.P. A fully coupled ecosystem model to predict the foraging ecology of apex predators in the California Current. *Mar. Ecol. Prog. Ser.* **2016**, *556*, 273–285. [CrossRef]

196. Shelton, A.O.; Satterthwaite, W.H.; Ward, E.J.; Feist, B.E.; Burke, B. Using hierarchical models to estimate stock-specific and seasonal variation in ocean distribution, survivorship, and aggregate abundance of fall run Chinook salmon. *Can. J. Fish. Aquat. Sci.* **2019**, *76*, 95–108. [CrossRef]

197. Weitkamp, L.A.; Teel, D.J.; Liermann, M.; Hinton, S.A.; Van Doornik, D.M.; Bentley, P.J. Stock-Specific Size and Timing at Ocean Entry of Columbia River Juvenile Chinook Salmon and Steelhead: Implications for Early Ocean Growth. *Mar. Coast. Fish.* **2015**, *7*, 370–392. [CrossRef]

198. Teel, D.J.; Burke, B.J.; Kuligowski, D.R.; Morgan, C.A.; Van Doornik, D.M. Genetic Identification of Chinook Salmon: Stock-Specific Distributions of Juveniles along the Washington and Oregon Coasts. *Mar. Coast. Fish.* **2015**, *7*, 274–300. [CrossRef]

199. Gosselin, J.L.; Buhle, E.R.; Van Holmes, C.; Beer, W.N.; Iltis, S.; Anderson, J.J. Role of carryover effects in conservation of wild Pacific salmon migrating regulated rivers. *Ecosphere* **2021**, *12*, e03618. [CrossRef]

200. Haeseker, S.L.; McCann, J.A.; Tuomikoski, J.; Chockley, B. Assessing Freshwater and Marine Environmental Influences on Life-Stage-Specific Survival Rates of Snake River Spring-Summer Chinook Salmon and Steelhead. *Trans. Am. Fish. Soc.* **2012**, *141*, 121–138. [CrossRef]

201. Russell, I.C.; Aprahamian, M.W.; Barry, J.; Davidson, I.C.; Fiske, P.; Ibbotson, A.T.; Kennedy, R.J.; Maclean, J.C.; Moore, A.; Otero, J.; et al. The influence of the freshwater environment and the biological characteristics of Atlantic salmon smolts on their subsequent marine survival. *ICES J. Mar. Sci.* **2012**, *69*, 1563–1573. [CrossRef]

202. Lundin, J.I.; Spromberg, J.A.; Jorgensen, J.C.; Myers, J.M.; Chittaro, P.M.; Zabel, R.W.; Johnson, L.L.; Neely, R.M.; Scholz, N.L. Legacy habitat contamination as a limiting factor for Chinook salmon recovery in the Willamette Basin, Oregon, USA. *PLoS ONE* **2019**, *14*, e0214399. [CrossRef] [PubMed]

203. Thorstad, E.B.; Uglem, I.; Finstad, B.; Kroglund, F.; Einarsdottir, I.E.; Kristensen, T.; Diserud, O.; Arechavala-Lopez, P.; Mayer, I.; Moore, A.; et al. Reduced marine survival of hatchery-reared Atlantic salmon post-smolts exposed to aluminium and moderate acidification in freshwater. *Estuar. Coast. Shelf Sci.* **2013**, *124*, 34–43. [CrossRef]

204. Wells, M.L.; Trainer, V.L.; Smayda, T.J.; Karlson, B.S.; Trick, C.G.; Kudela, R.M.; Ishikawa, A.; Bernard, S.; Wulff, A.; Anderson, D.M.; et al. Harmful algal blooms and climate change: Learning from the past and present to forecast the future. *Harmful Algae* **2015**, *49*, 68–93. [CrossRef] [PubMed]

205. Huisman, J.; Codd, G.A.; Paerl, H.W.; Ibelings, B.W.; Verspagen, J.M.H.; Visser, P.M. Cyanobacterial blooms. *Nat. Rev. Microbiol.* **2018**, *16*, 471–483. [CrossRef] [PubMed]

206. Esenkulova, S.; Suchy, K.D.; Pawlowicz, R.; Costa, M.; Pearsall, I.A. Harmful Algae and Oceanographic Conditions in the Strait of Georgia, Canada Based on Citizen Science Monitoring. *Front. Mar. Sci.* **2021**, *8*, 725092. [CrossRef]

207. Rensel, J.E.J.; Haigh, N.; Tynan, T.J. Fraser river sockeye salmon marine survival decline and harmful blooms of *Heterosigrna akashiwo. Harmful Algae* **2010**, *10*, 98–115. [CrossRef]
208. Burridge, L.E.; Martin, J.L.; Lyons, M.C.; LeGresley, M.M. Lethality of microalgae to farmed Atlantic salmon (*Salmo salar*). *Aquaculture* **2010**, *308*, 101–105. [CrossRef]
209. Lefebvre, K.A.; Frame, E.R.; Kendrick, P.S. Domoic acid and fish behavior: A review. *Harmful Algae* **2012**, *13*, 126–130. [CrossRef]
210. McCabe, R.M.; Hickey, B.M.; Kudela, R.M.; Lefebvre, K.A.; Adams, N.G.; Bill, B.D.; Gulland, F.M.; Thomson, R.E.; Cochlan, W.P.; Trainer, V.L. An unprecedented coastwide toxic algal bloom linked to anomalous ocean conditions. *Geophys. Res. Lett.* **2016**, *43*, 10366–10376. [CrossRef]
211. Akmajian, A.M.; Scordino, J.J.; Acevedo-Gutierrez, A. Year-round algal toxin exposure in free-ranging sea lions. *Mar. Ecol. Prog. Ser.* **2017**, *583*, 243–258. [CrossRef]
212. Moore, S.K.; Johnstone, J.A.; Banas, N.S.; Salathe, E.P. Present-day and future climate pathways affecting Alexandrium blooms in Puget Sound, WA, USA. *Harmful Algae* **2015**, *48*, 1–11. [CrossRef] [PubMed]
213. Ralston, D.K.; Moore, S.K. Modeling harmful algal blooms in a changing climate. *Harmful Algae* **2020**, *91*, 101729. [CrossRef] [PubMed]
214. Costa, P.R. Impact and effects of paralytic shellfish poisoning toxins derived from harmful algal blooms to marine fish. *Fish Fish.* **2016**, *17*, 226–248. [CrossRef]
215. Sato, S.; Ogata, T.; Kodama, M. Trace amounts of saxitoxins in the viscera of chum salmon Oncorhynchus keta. *Mar. Ecol. Prog. Ser.* **1998**, *175*, 295–298. [CrossRef]
216. Brosnan, I.G.; Welch, D.W.; Rechisky, E.L.; Porter, A.D. Evaluating the influence of environmental factors on yearling Chinook salmon survival in the Columbia River plume (USA). *Mar. Ecol. Prog. Ser.* **2014**, *496*, 181–196. [CrossRef]
217. Rechisky, E.L.; Porter, A.D.; Clark, T.D.; Furey, N.B.; Gale, M.K., Hinch, S.G.; Welch, D.W. Quantifying survival of age-2 Chilko Lake sockeye salmon during the first 50 days of migration. *Can. J. Fish. Aquat. Sci.* **2019**, *76*, 136–152. [CrossRef]
218. Seitz, A.C.; Courtney, M.B.; Evans, M.D.; Manishin, K. Pop-up satellite archival tags reveal evidence of intense predation on large immature Chinook salmon (*Oncorhynchus tshawytscha*) in the North Pacific Ocean. *Can. J. Fish. Aquat. Sci.* **2019**, *76*, 1608–1615. [CrossRef]
219. Friedman, W.R.; Santora, J.A.; Schroeder, I.D.; Huff, D.D.; Brodeur, R.D.; Field, J.C.; Wells, B.K. Environmental and geographic relationships among salmon forage assemblages along the continental shelf of the California Current. *Mar. Ecol. Prog. Ser.* **2018**, *596*, 181–198. [CrossRef]
220. Henderson, M.; Fiechter, J.; Huff, D.D.; Wells, B.K. Spatial variability in ocean-mediated growth potential is linked to Chinook salmon survival. *Fish. Oceanogr.* **2018**, *28*, 334–344. [CrossRef]
221. Wells, B.K.; Santora, J.A.; Schroeder, I.D.; Mantua, N.; Sydeman, W.J.; Huff, D.D.; Field, J.C. Marine ecosystem perspectives on Chinook salmon recruitment: A synthesis of empirical and modeling studies from a California upwelling system. *Mar. Ecol. Prog. Ser.* **2016**, *552*, 271–284. [CrossRef]
222. Dale, K.E.; Daly, E.A.; Brodeur, R.D. Interannual variability in the feeding and condition of subyearling Chinook salmon off Oregon and Washington in relation to fluctuating ocean conditions. *Fish. Oceanogr.* **2017**, *26*, 1–16. [CrossRef]
223. Daly, E.A.; Auth, T.D.; Brodeur, R.D.; Peterson, W.T. Winter ichthyoplankton biomass as a predictor of early summer prey fields and survival of juvenile salmon in the northern California Current. *Mar. Ecol. Prog. Ser.* **2013**, *484*, 203–217. [CrossRef]
224. Beaugrand, G.; Reid, P.C. Relationships between North Atlantic salmon, plankton, and hydroclimatic change in the Northeast Atlantic. *ICES J. Mar. Sci.* **2012**, *69*, 1549–1562. [CrossRef]
225. Thompson, S.A.; Sydeman, W.J.; Santora, J.A.; Black, B.A.; Suryan, R.M.; Calambokidis, J.; Peterson, W.T.; Bograd, S.J. Linking predators to seasonality of upwelling: Using food web indicators and path analysis to infer trophic connections. *Prog. Oceanogr.* **2012**, *101*, 106–120. [CrossRef]
226. Tomaro, L.M.; Teel, D.J.; Peterson, W.T.; Miller, J.A. When is bigger better? Early marine residence of middle and upper Columbia River spring Chinook salmon. *Mar. Ecol. Prog. Ser.* **2012**, *452*, 237–252. [CrossRef]
227. Trueman, C.N.; MacKenzie, K.M.; Palmer, M.R. Stable isotopes reveal linkages between ocean climate, plankton community dynamics, and survival of two populations of Atlantic salmon (*Salmo salar*). *ICES J. Mar. Sci.* **2012**, *69*, 784–794. [CrossRef]
228. Sabal, M.C.; Huff, D.D.; Henderson, M.J.; Fiechter, J.; Harding, J.A.; Hayes, S.A. Contrasting patterns in growth and survival of Central Valley fall run Chinook salmon related to hatchery and ocean conditions. *Environ. Biol. Fishes* **2016**, *99*, 949–967. [CrossRef]
229. Woodson, L.E.; Wells, B.K.; Weber, P.K.; MacFarlane, R.B.; Whitman, G.E.; Johnson, R.C. Size, growth, and origin-dependent mortality of juvenile Chinook salmon Oncorhynchus tshawytscha during early ocean residence. *Mar. Ecol. Prog. Ser.* **2013**, *487*, 163–175. [CrossRef]
230. Wells, B.K.; Santora, J.A.; Henderson, M.J.; Warzybok, P.; Jahncke, J.; Bradley, R.W.; Huff, D.D.; Schroeder, I.D.; Nelson, P.; Field, J.C.; et al. Environmental conditions and prey-switching by a seabird predator impact juvenile salmon survival. *J. Mar. Syst.* **2017**, *174*, 54–63. [CrossRef]
231. Daly, E.A.; Brodeur, R.D. Warming Ocean Conditions Relate to Increased Trophic Requirements of Threatened and Endangered Salmon. *PLoS ONE* **2015**, *10*, e0144066. [CrossRef] [PubMed]
232. Farley, E.V.; Starovoytov, A.; Naydenko, S.; Heintz, R.; Trudel, M.; Guthrie, C.; Eisner, L.; Guyon, J.R. Implications of a warming eastern Bering Sea for Bristol Bay sockeye salmon. *ICES J. Mar. Sci.* **2011**, *68*, 1138–1146. [CrossRef]

233. Turrero, P.; Horreo, J.L.; Garcia-Vazquez, E. Same old *Salmo*? Changes in life history and demographic trends of North Iberian salmonids since the Upper Palaeolithic as revealed by archaeological remains and beast analyses. *Mol. Ecol.* **2012**, *21*, 2318–2329. [CrossRef] [PubMed]
234. Chaput, G. Overview of the status of Atlantic salmon (*Salmo salar*) in the North Atlantic and trends in marine mortality. *ICES J. Mar. Sci.* **2012**, *69*, 1538–1548. [CrossRef]
235. Sinclair-Waters, M.; Nome, T.; Wang, J.; Lien, S.; Kent, M.P.; Saerov, H.; Floro-Larsen, B.; Bolstad, G.H.; Primmer, C.R.; Barson, N.J. Dissecting the loci underlying maturation timing in Atlantic salmon using haplotype and multi-SNP based association methods. *Heredity* **2022**, *129*, 356–365. [CrossRef] [PubMed]
236. Czorlich, Y.; Aykanat, T.; Erkinaro, J.; Orell, P.; Primmer, C.R. Rapid evolution in salmon life history induced by direct and indirect effects of fishing. *Science* **2022**, *376*, 420–423. [CrossRef]
237. Czorlich, Y.; Aykanat, T.; Erkinaro, J.; Orell, P.; Primmer, C.R. Rapid sex-specific evolution of age at maturity is shaped by genetic architecture in Atlantic salmon. *Nat. Ecol. Evol.* **2018**, *2*, 1800–1807. [CrossRef]
238. Ohlberger, J.; Ward, E.J.; Schindler, D.E.; Lewis, B. Demographic changes in Chinook salmon across the Northeast Pacific Ocean. *Fish Fish.* **2018**, *19*, 533–546. [CrossRef]
239. Oke, K.B.; Cunningham, C.J.; Westley, P.A.H.; Baskett, M.L.; Carlson, S.M.; Clark, J.; Hendry, A.P.; Karatayev, V.A.; Kendall, N.W.; Kibele, J.; et al. Recent declines in salmon body size impact ecosystems and fisheries. *Nat. Commun.* **2020**, *11*, 4155. [CrossRef]
240. Kendall, N.W.; Quinn, T.P. Quantifying and comparing size selectivity among Alaskan sockeye salmon fisheries. *Ecol. Appl.* **2012**, *22*, 804–816. [CrossRef]
241. Waples, R.S.; Audzijonyte, A. Fishery-induced evolution provides insights into adaptive responses of marine species to climate change. *Front. Ecol. Environ.* **2016**, *14*, 217–224. [CrossRef]
242. Otero, J.; Jensen, A.J.; L'Abee-Lund, J.H.; Stenseth, N.C.; Storvik, G.O.; Vollestad, L.A. Contemporary ocean warming and freshwater conditions are related to later sea age at maturity in Atlantic salmon spawning in Norwegian rivers. *Ecol. Evol.* **2012**, *2*, 2192–2203. [CrossRef] [PubMed]
243. Chasco, B.E.; Kaplan, I.C.; Thomas, A.C.; Acevedo-Gutiérrez, A.; Noren, D.P.; Ford, M.J.; Hanson, M.B.; Scordino, J.J.; Jeffries, S.J.; Marshall, K.N.; et al. Competing tradeoffs between increasing marine mammal predation and fisheries harvest of Chinook salmon. *Sci. Rep.* **2017**, *7*, 15439. [CrossRef] [PubMed]
244. Moore, M.E.; Berejikian, B.A. Population, habitat, and marine location effects on early marine survival and behavior of Puget Sound steelhead smolts. *Ecosphere* **2017**, *8*, e01834. [CrossRef]
245. Moore, M.E.; Berejikian, B.A.; Goetz, F.A.; Berger, A.G.; Hodgson, S.S.; Connor, E.J.; Quinn, T.P. Multi-population analysis of Puget Sound steelhead survival and migration behavior. *Mar. Ecol. Prog. Ser.* **2015**, *537*, 217–232. [CrossRef]
246. Mantyniemi, S.; Romakkaniemi, A.; Dannewitz, J.; Palm, S.; Pakarinen, T.; Pulkkinen, H.; Grdmark, A.; Karlsson, O. Both predation and feeding opportunities may explain changes in survival of Baltic salmon post-smolts. *ICES J. Mar. Sci.* **2012**, *69*, 1574–1579. [CrossRef]
247. Ford, J.K.B.; Ellis, G.M. Selective foraging by fish-eating killer whales *Orcinus orca* in British Columbia. *Mar. Ecol. Prog. Ser.* **2006**, *316*, 185–199. [CrossRef]
248. Manishin, K.A.; Cunningham, C.J.; Westley, P.A.H.; Seitz, A.C. Can late stage marine mortality explain observed shifts in age structure of Chinook salmon? *PLoS ONE* **2021**, *16*, e0247370. [CrossRef]
249. Shelton, A.O.; Sullaway, G.H.; Ward, E.J.; Feist, B.E.; Somers, K.A.; Tuttle, V.J.; Watson, J.T.; Satterthwaite, W.H. Redistribution of salmon populations in the northeast Pacific ocean in response to climate. *Fish Fish.* **2021**, *22*, 503–517. [CrossRef]
250. Kao, Y.C.; Madenjian, C.P.; Bunnell, D.B.; Lofgren, B.M.; Perroud, M. Temperature effects induced by climate change on the growth and consumption by salmonines in Lakes Michigan and Huron. *Environ. Biol. Fishes* **2015**, *98*, 1089–1104. [CrossRef]
251. Abdul-Aziz, O.I.; Mantua, N.J.; Myers, K.W. Potential climate change impacts on thermal habitats of Pacific salmon (*Oncorhynchus* spp.) in the North Pacific Ocean and adjacent seas. *Can. J. Fish. Aquat. Sci.* **2011**, *68*, 1660–1680. [CrossRef]
252. Cheung, W.W.L.; Dunne, J.; Sarmiento, J.L.; Pauly, D. Integrating ecophysiology and plankton dynamics into projected maximum fisheries catch potential under climate change in the Northeast Atlantic. *ICES J. Mar. Sci. J. Du Cons.* **2011**, *68*, 1008–1018. [CrossRef]
253. Cheung, W.W.L.; Frolicher, T.L. Marine heatwaves exacerbate climate change impacts for fisheries in the northeast Pacific. *Sci. Rep.* **2020**, *10*, 6678. [CrossRef]
254. Cheung, W.W.L.; Lam, V.W.Y.; Sarmiento, J.L.; Kearney, K.; Watson, R.E.G.; Zeller, D.; Pauly, D. Large-scale redistribution of maximum fisheries catch potential in the global ocean under climate change. *Glob. Chang. Biol.* **2010**, *16*, 24–35. [CrossRef]
255. Ainsworth, C.H.; Samhouri, J.F.; Busch, D.S.; Chueng, W.W.L.; Dunne, J.; Okey, T.A. Potential impacts of climate change on Northeast Pacific marine fisheries and food webs. *ICES J. Mar. Sci.* **2011**, *68*, 1217–1229. [CrossRef]
256. Blanchard, J.L.; Jennings, S.; Holmes, R.; Harle, J.; Merino, G.; Allen, J.I.; Holt, J.; Dulvy, N.K.; Barange, M. Potential consequences of climate change for primary production and fish production in large marine ecosystems. *Philos. Trans. R. Soc. B Biol. Sci.* **2012**, *367*, 2979–2989. [CrossRef]
257. Heneghan, R.F.; Galbraith, E.; Blanchard, J.L.; Harrison, C.; Barrier, N.; Bulman, C.; Cheung, W.; Coll, M.; Eddy, T.D.; Erauskin-Extramiana, M.; et al. Disentangling diverse responses to climate change among global marine ecosystem models. *Prog. Oceanogr.* **2021**, *198*, 102659. [CrossRef]

258. Reum, J.C.P.; Blanchard, J.L.; Holsman, K.K.; Aydin, K.; Hollowed, A.B.; Hermann, A.J.; Cheng, W.; Faig, A.; Haynie, A.C.; Punt, A.E. Ensemble Projections of Future Climate Change Impacts on the Eastern Bering Sea Food Web Using a Multispecies Size Spectrum Model. *Front. Mar. Sci.* **2020**, *7*, 124. [CrossRef]

259. Hazen, E.L.; Jorgensen, S.; Rykaczewski, R.R.; Bograd, S.J.; Foley, D.G.; Jonsen, I.D.; Shaffer, S.A.; Dunne, J.P.; Costa, D.P.; Crowder, L.B.; et al. Predicted habitat shifts of Pacific top predators in a changing climate. *Nat. Clim. Chang.* **2012**, *3*, 234–238. [CrossRef]

260. Wolf, S.G.; Snyder, M.A.; Sydeman, W.J.; Doak, D.F.; Croll, D.A. Predicting population consequences of ocean climate change for an ecosystem sentinel, the seabird Cassin's auklet. *Glob. Chang. Biol.* **2010**, *16*, 1923–1935. [CrossRef]

261. Bednarsek, N.; Pelletier, G.; Ahmed, A.; Feely, R.A. Chemical Exposure Due to Anthropogenic Ocean Acidification Increases Risks for Estuarine Calcifiers in the Salish Sea: Biogeochemical Model Scenarios. *Front. Mar. Sci.* **2020**, *7*, 580. [CrossRef]

262. Comeau, S.; Gattuso, J.-P.; Nisumaa, A.-M.; Orr, J. Impact of aragonite saturation state changes on migratory pteropods. *Proc. R. Soc. B Biol. Sci.* **2012**, *279*, 732–738. [CrossRef] [PubMed]

263. Comeau, S.; Jeffree, R.; Teyssie, J.L.; Gattuso, J.P. Response of the arctic pteropod *Limacina helicina* to projected future environmental conditions. *PLoS ONE* **2010**, *5*, e11362. [CrossRef]

264. Flynn, K.J.; Blackford, J.C.; Baird, M.E.; Raven, J.A.; Clark, D.R.; Beardall, J.; Brownlee, C.; Fabian, H.; Wheeler, G.L. Changes in pH at the exterior surface of plankton with ocean acidification. *Nat. Clim. Chang.* **2012**, *2*, 510–513. [CrossRef]

265. Magel, C.L.; Lee, E.M.J.; Strawn, A.M.; Swieca, K.; Jensen, A.D. Connecting Crabs, Currents, and Coastal Communities: Examining the Impacts of Changing Ocean Conditions on the Distribution of US West Coast Dungeness Crab Commercial Catch. *Front. Mar. Sci.* **2020**, *7*, 401. [CrossRef]

266. Anttila, K.; Eliason, E.J.; Kaukinen, K.H.; Miller, K.M.; Farrell, A.P. Facing warm temperatures during migration: Cardiac mRNA responses of two adult *Oncorhynchus nerka* populations to warming and swimming challenges. *J. Fish Biol.* **2014**, *84*, 1439–1456. [CrossRef]

267. Fellman, J.B.; Hood, E.; Nagorski, S.; Hudson, J.; Pyare, S. Interactive physical and biotic factors control dissolved oxygen in salmon spawning streams in coastal Alaska. *Aquat. Sci.* **2019**, *81*, 11. [CrossRef]

268. Keefer, M.L.; Taylor, G.A.; Garletts, D.F.; Gauthier, G.A.; Pierce, T.M.; Caudil, C.C. Prespawn mortality in adult spring Chinook salmon outplanted above barrier dams. *Ecol. Freshw. Fish* **2010**, *19*, 361–372. [CrossRef]

269. Sergeant, C.J.; Bellmore, J.R.; McConnell, C.; Moore, J.W. High salmon density and low discharge create periodic hypoxia in coastal rivers. *Ecosphere* **2017**, *8*, e01846. [CrossRef]

270. Tillotson, M.D.; Quinn, T.P. Climate and conspecific density trigger pre-spawning mortality in sockeye salmon (*Oncorhynchus nerka*). *Fish. Res.* **2017**, *188*, 138–148. [CrossRef]

271. Bell, D.A.; Kovach, R.P.; Vulstek, S.C.; Joyce, J.E.; Tallmon, D.A. Climate-induced trends in predator–prey synchrony differ across life-history stages of an anadromous salmonid. *Can. J. Fish. Aquat. Sci.* **2017**, *74*, 1431–1438. [CrossRef]

272. Rub, A.M.W.; Sandford, B.P. Evidence of a 'dinner bell' effect from acoustic transmitters in adult Chinook salmon. *Mar. Ecol. Prog. Ser.* **2020**, *641*, 1–11. [CrossRef]

273. Rub, A.M.W.; Som, N.A.; Henderson, M.J.; Sandford, B.P.; Van Doornik, D.M.; Teel, D.J.; Tennis, M.J.; Langness, O.P.; van der Leeuw, B.K.; Huff, D.D. Changes in adult Chinook salmon (*Oncorhynchus tshawytscha*) survival within the lower Columbia River amid increasing pinniped abundance. *Can. J. Fish. Aquat. Sci.* **2019**, *76*, 1862–1873. [CrossRef]

274. Sergeant, C.J.; Armstrong, J.B.; Ward, E.J. Predator-prey migration phenologies remain synchronised in a warming catchment. *Freshw. Biol.* **2015**, *60*, 724–732. [CrossRef]

275. Keefer, M.L.; Jepson, M.A.; Naughton, G.P.; Blubaugh, T.J.; Clabough, T.S.; Caudill, C.C. Condition-dependent en route migration mortality of adult Chinook salmon in the Willamette River main stem. *N. Am. J. Fish. Manag.* **2017**, *37*, 370–379. [CrossRef]

276. Keefer, M.L.; Caudill, C.C.; Peery, C.A.; Moser, M.L. Context-dependent diel behavior of upstream-migrating anadromous fishes. *Environ. Biol. Fishes* **2013**, *96*, 691–700. [CrossRef]

277. Benda, S.E.; Naughton, G.P.; Caudill, C.C.; Kent, M.L.; Schreck, C.B. Cool, Pathogen-Free Refuge Lowers Pathogen-Associated Prespawn Mortality of Willamette River Chinook Salmon. *Trans. Am. Fish. Soc.* **2015**, *144*, 1159–1172. [CrossRef]

278. Keefer, M.L.; Caudill, C.C. Estimating thermal exposure of adult summer steelhead and fall Chinook salmon migrating in a warm impounded river. *Ecol. Freshw. Fish* **2016**, *25*, 599–611. [CrossRef]

279. Goetz, F.A.; Quinn, T.P. Behavioral thermoregulation by adult Chinook salmon (*Oncorhynchus tshawytscha*) in estuary and freshwater habitats prior to spawning. *Fish. Bull.* **2019**, *117*, 258–271. [CrossRef]

280. Hasler, C.T.; Cooke, S.J.; Hinch, S.G.; Guimond, E.; Donaldson, M.R.; Mossop, B.; Patterson, D.A. Thermal biology and bioenergetics of different upriver migration strategies in a stock of summer-run Chinook salmon. *J. Therm. Biol.* **2012**, *37*, 265–272. [CrossRef]

281. Keefer, M.L.; Clabough, T.S.; Jepson, M.A.; Bowerman, T.; Caudill, C.C. Temperature and depth profiles of Chinook salmon and the energetic costs of their long-distance homing migrations. *J. Therm. Biol.* **2019**, *79*, 155–165. [CrossRef]

282. Keefer, M.L.; Clabough, T.S.; Jepson, M.A.; Johnson, E.L.; Peery, C.A.; Caudill, C.C. Thermal exposure of adult Chinook salmon and steelhead: Diverse behavioral strategies in a large and warming river system. *PLoS ONE* **2018**, *13*, e0204274. [CrossRef] [PubMed]

283. Lennox, R.J.; Eliason, E.J.; Havn, T.B.; Johansen, M.R.; Thorstad, E.B.; Cooke, S.J.; Diserud, O.H.; Whoriskey, F.G.; Farrell, A.P.; Uglem, I. Bioenergetic consequences of warming rivers to adult Atlantic salmon Salmo salar during their spawning migration. *Freshw. Biol.* **2018**, *63*, 1381–1393. [CrossRef]

284. O'Sullivan, A.M.; Linnansaari, T.; Leavitt, J.; Samways, K.M.; Kurylyk, B.L.; Curry, R.A. The salmon-peloton: Hydraulic habitat shifts of adult Atlantic salmon (*Salmo salar*) due to behavioural thermoregulation. *River Res. Appl.* **2022**, *38*, 107–118. [CrossRef]

285. Siegel, J.E.; Crozier, L.G.; Wiesebron, L.E.; Widener, D.L. Environmentally triggered shifts in steelhead migration behavior and consequences for survival in the mid-Columbia River. *PLoS ONE* **2021**, *16*, e0250831. [CrossRef] [PubMed]

286. Strange, J.S. Upper thermal limits to migration in adult Chinook salmon: Evidence from the Klamath River Basin. *Trans. Am. Fish. Soc.* **2010**, *139*, 1091–1108. [CrossRef]

287. Keefer, M.R.L.; Boggs, C.T.; Peery, C.; Caudill, C.C. Overwintering distribution, behavior, and survival of adult summer steelhead: Variability among Columbia river populations. *N. Am. J. Fish. Manag.* **2008**, *28*, 81–96. [CrossRef]

288. Berejikian, B.A.; Tatara, C.P.; Van Doornik, D.M.; Humling, M.A.; Cooper, M.R.; Pasley, C.R.; Atkins, J.J. Duration in captivity affects competitive ability and breeding success of male but not female steelhead trout (*Oncorhynchus mykiss*). *Can. J. Fish. Aquat. Sci.* **2020**, *77*, 1000–1009. [CrossRef]

289. Bond, M.H.; Westley, P.A.H.; Dittman, A.H.; Holecek, D.; Marsh, T.; Quinn, T.P. Combined effects of barge transportation, river environment, and rearing location on straying and migration of adult Snake River fall-run Chinook salmon. *Trans. Am. Fish. Soc.* **2016**, *146*, 60–73. [CrossRef]

290. Crozier, L.; Burke, B.; Sandford, B.; Axel, G.; Sanderson, B. *Passage and Survival of Adult Snake River Sockeye Salmon within and Upstream from the Federal Columbia River Power System*; Fish Ecology Division, Northwest Fisheries Science Center National Marine Fisheries Service: Seattle, WA, USA, 2014.

291. Pankhurst, N.W.; King, H.R. Temperature and salmonid reproduction: Implications for aquaculture. *J. Fish Biol.* **2010**, *76*, 69–85. [CrossRef]

292. Burnett, N.J.; Hinch, S.G.; Bett, N.N.; Braun, D.C.; Casselman, M.T.; Cooke, S.J.; Gelchu, A.; Lingard, S.; Middleton, C.T.; Minke-Martin, V.; et al. Reducing carryover effects on the migration and spawning success of Sockeye salmon through a management experiment of dam flows. *River Res. Appl.* **2017**, *33*, 3–15. [CrossRef]

293. Todd, C.D.; Friedland, K.D.; MacLean, J.C.; Whyte, B.D.; Russell, I.C.; Lonergan, M.E.; Morrissey, M.B. Phenological and phenotypic changes in Atlantic salmon populations in response to a changing climate. *ICES J. Mar. Sci.* **2012**, *69*, 1686–1698. [CrossRef]

294. Jacobs, G.R.; Thurow, R.F.; Buffington, J.M.; Isaak, D.J.; Wenger, S.J. Climate, fire regime, geomorphology, and conspecifics influence the spatial distribution of Chinook salmon redds. *Trans. Am. Fish. Soc.* **2021**, *150*, 8–23. [CrossRef]

295. Hague, M.J.; Ferrari, M.R.; Miller, J.R.; Patterson, D.A.; Russell, G.L.; Farrell, A.P.; Hinch, S.G. Modelling the future hydroclimatology of the lower Fraser River and its impacts on the spawning migration survival of sockeye salmon. *Glob. Chang. Biol.* **2011**, *17*, 87–98. [CrossRef]

296. Boughton, D.A.; Pike, A.S. Floodplain rehabilitation as a hedge against hydroclimatic uncertainty in a migration corridor of threatened steelhead. *Conserv. Biol.* **2013**, *27*, 1158–1168. [CrossRef] [PubMed]

297. Flitcroft, R.; Lewis, S.; Arismendi, I.; Davis, C.; Giannico, G.; Penaluna, B.; Santelmann, M.; Safeeq, M.; Snyder, J. Using expressed behaviour of coho salmon (*Oncorhynchus kisutch*) to evaluate the vulnerability of upriver migrants under future hydrological regimes: Management implications and conservation planning. *Aquat. Conserv-Mar. Freshw. Ecosyst.* **2019**, *29*, 1083–1094. [CrossRef]

298. Donley, E.E.; Naiman, R.J.; Marineau, M.D. Strategic planning for instream flow restoration: A case study of potential climate change impacts in the central Columbia River basin. *Glob. Chang. Biol.* **2012**, *18*, 3071–3086. [CrossRef]

299. Shanley, C.S.; Albert, D.M. Climate change sensitivity index for Pacific salmon habitat in southeast Alaska. *PLoS ONE* **2014**, *9*, e104799. [CrossRef]

300. Connor, W.P.; Piston, C.E.; Garcia, A.P. Temperature during incubation as one factor affecting the distribution of Snake River Fall Chinook salmon spawning areas. *Trans. Am. Fish. Soc.* **2003**, *132*, 1236–1243. [CrossRef]

301. Pitman, K.J.; Moore, J.W.; Huss, M.; Sloat, M.R.; Whited, D.C.; Beechie, T.J.; Brenner, R.; Hood, E.W.; Milner, A.M.; Pess, G.R.; et al. Glacier retreat creating new Pacific salmon habitat in western North America. *Nat. Commun.* **2021**, *12*, 6816. [CrossRef]

302. Sinnatamby, R.N.; Cantin, A.; Post, J.R. Threats to at-risk salmonids of the Canadian Rocky Mountain region. *Ecol. Freshw. Fish* **2019**, *18*, 477–494. [CrossRef]

303. Al-Chokhachy, R.; Lien, M.K.; Shepard, B.B.; High, B. The interactive effects of stream temperature, stream size, and non-native species on Yellowstone cutthroat trout. *Can. J. Fish. Aquat. Sci.* **2021**, *78*, 1073–1083. [CrossRef]

304. Labonne, J.; Vignon, M.; Prevost, E.; Lecomte, F.; Dodson, J.J.; Kaeuffer, R.; Aymes, J.C.; Jarry, M.; Gaudin, P.; Davaine, P.; et al. Invasion dynamics of a fish-free landscape by brown trout (*Salmo trutta*). *PLoS ONE* **2013**, *8*, e71052. [CrossRef] [PubMed]

305. Moyle, P.; Kiernan, J.; Crain, P.; Quiñones, R. Climate Change Vulnerability of Native and Alien Freshwater Fishes of California: A Systematic Assessment Approach. *PLoS ONE* **2013**, *8*, e63883. [CrossRef] [PubMed]

306. Almodovar, A.; Ayllon, D.; Nicola, G.G.; Jonsson, B.; Elvira, B. Climate-driven biophysical changes in feeding and breeding environments explain the decline of southernmost European Atlantic salmon populations. *Can. J. Fish. Aquat. Sci.* **2019**, *76*, 1581–1595. [CrossRef]

307. Cunningham, C.J.; Westley, P.A.H.; Adkison, M.D. Signals of large scale climate drivers, hatchery enhancement, and marine factors in Yukon River Chinook salmon survival revealed with a Bayesian life history model. *Glob. Chang. Biol.* **2018**, *24*, 4399–4416. [CrossRef] [PubMed]

308. Honkanen, H.M.; Boylan, P.; Dodd, J.A.; Adams, C.E. Life stage-specific, stochastic environmental effects overlay density dependence in an Atlantic salmon population. *Ecol. Freshw. Fish* **2019**, *28*, 156–166. [CrossRef]

309. Irvine, J.R.; Michielsens, C.J.G.; O'Brien, M.; White, B.A.; Folkes, M. Increasing dominance of odd-year returning pink salmon. *Trans. Am. Fish. Soc.* **2014**, *143*, 939–956. [CrossRef]

310. Jeffrey, K.M.; Côté, I.M.; Irvine, J.R.; Reynolds, J.D. Changes in body size of Canadian Pacific salmon over six decades. *Can. J. Fish. Aquat. Sci.* **2017**, *74*, 191–201. [CrossRef]

311. Dorner, B.; Catalano, M.J.; Peterman, R.M. Spatial and temporal patterns of covariation in productivity of Chinook salmon populations of the northeastern Pacific Ocean. *Can. J. Fish. Aquat. Sci.* **2018**, *75*, 1082–1095. [CrossRef]

312. Friedland, K.D.; Shank, B.V.; Todd, C.D.; McGinnity, P.; Nye, J.A. Differential response of continental stock complexes of Atlantic salmon (*Salmo salar*) to the Atlantic Multidecadal Oscillation. *J. Mar. Syst.* **2014**, *133*, 77–87. [CrossRef]

313. Jorgensen, J.C.; Nicol, C.; Fogel, C.; Beechie, T.J. Identifying the potential of anadromous salmonid habitat restoration with life cycle models. *PLoS ONE* **2021**, *16*, e0256792. [CrossRef]

314. Nicol, C.L.; Jorgensen, J.C.; Fogel, C.B.; Timpane-Padgham, B.; Beechie, T.J. Spatially overlapping salmon species have varied population response to early life history mortality from increased peak flows. *Can. J. Fish. Aquat. Sci.* **2022**, *79*, 342–351. [CrossRef]

315. Scheuerell, M.D.; Ruff, C.P.; Anderson, J.H.; Beamer, E.M. An integrated population model for estimating the relative effects of natural and anthropogenic factors on a threatened population of steelhead trout. *J. Appl. Ecol.* **2021**, *58*, 114–124. [CrossRef]

316. Olmos, M.; Massiot-Granier, F.; Prevost, E.; Chaput, G.; Bradbury, I.R.; Nevoux, M.; Rivot, E. Evidence for spatial coherence in time trends of marine life history traits of Atlantic salmon in the North Atlantic. *Fish Fish.* **2019**, *20*, 322–342. [CrossRef]

317. Madsen, R.P.A.; Jacobsen, M.W.; O'Malley, K.G.; Nygaard, R.; Praebel, K.; Jonsson, B.; Pujolar, J.M.; Fraser, D.J.; Bernatchez, L.; Hansen, M.M. Genetic population structure and variation at phenologyrelated loci in anadromous Arctic char (*Salvelinus alpinus*). *Ecol. Freshw. Fish* **2020**, *29*, 170–183. [CrossRef]

318. Lehnert, S.J.; Kess, T.; Bentzen, P.; Kent, M.P.; Lien, S.; Gilbey, J.; Clement, M.; Jeffery, N.W.; Waples, R.S.; Bradbury, I.R. Genomic signatures and correlates of widespread population declines in salmon. *Nat. Commun.* **2019**, *10*, 10. [CrossRef] [PubMed]

319. Kannry, S.H.; O'Rourke, S.M.; Kelson, S.J.; Miller, M.R. On the ecology and distribution of Steelhead (*Oncorhynchus mykiss*) in California's Eel River. *J. Hered.* **2020**, *111*, 548–563. [CrossRef] [PubMed]

320. Kelson, S.J.; Miller, M.R.; Thompson, T.Q.; O'Rourke, S.M.; Carlson, S.M. Temporal dynamics of migration-linked genetic variation are driven by streamflows and riverscape permeability. *Mol. Ecol.* **2020**, *29*, 870–885. [CrossRef]

321. Schindler, D.E.; Armstrong, J.B.; Reed, T.E. The portfolio concept in ecology and evolution. *Front. Ecol. Environ.* **2015**, *13*, 257–263. [CrossRef]

322. Anderson, S.C.; Moore, J.W.; McClure, M.M.; Dulvy, N.K.; Cooper, A.B. Portfolio conservation of metapopulations under climate change. *Ecol. Appl.* **2015**, *25*, 559–572. [CrossRef]

323. Wade, A.A.; Beechie, T.J.; Fleishman, E.; Mantua, N.J.; Wu, H.; Kimball, J.S.; Stoms, D.M.; Stanford, J.A. Steelhead vulnerability to climate change in the Pacific Northwest. *J. Appl. Ecol.* **2013**, *50*, 1093–1104. [CrossRef]

324. Wade, A.A.; Hand, B.K.; Kovach, R.P.; Luikart, G.; Whited, D.C.; Muhlfeld, C.C. Accounting for adaptive capacity and uncertainty in assessments of species' climate-change vulnerability. *Conserv. Biol.* **2017**, *31*, 136–149. [CrossRef] [PubMed]

325. Crozier, L.G.; McClure, M.M.; Beechie, T.; Bograd, S.J.; Boughton, D.A.; Carr, M.; Cooney, T.D.; Dunham, J.B.; Greene, C.M.; Haltuch, M.A.; et al. Climate vulnerability assessment for Pacific salmon and steelhead in the California Current Large Marine Ecosystem. *PLoS ONE* **2019**, *14*, e0217711. [CrossRef] [PubMed]

326. Almodovar, A.; Nicola, G.G.; Ayllon, D.; Elvira, B. Global warming threatens the persistence of Mediterranean brown trout. *Glob. Chang. Biol.* **2012**, *18*, 1549–1560. [CrossRef]

327. Ayllon, D.; Nicola, G.G.; Elvira, B.; Almodovar, A. Climate change will render size-selective harvest of cold-water fish species unsustainable in Mediterranean freshwaters. *J. Appl. Ecol.* **2021**, *58*, 562–575. [CrossRef]

328. Neville, H.M.; Leasure, D.R.; Dauwalter, D.C.; Dunham, J.B.; Bjork, R.; Fesenmyer, K.A.; Chelgren, N.D.; Peacock, M.M.; Luce, C.H.; Isaak, D.J.; et al. Application of multiple-population viability analysis to evaluate species recovery alternatives. *Conserv. Biol.* **2019**, *34*, 482–493. [CrossRef]

329. Roberts, J.J.; Fausch, K.D.; Peterson, D.P.; Hooten, M.B. Fragmentation and thermal risks from climate change interact to affect persistence of native trout in the Colorado River basin. *Glob. Chang. Biol.* **2013**, *19*, 1383–1398. [CrossRef]

330. Piou, C.; Prevost, E. Contrasting effects of climate change in continental vs. oceanic environments on population persistence and microevolution of Atlantic salmon. *Glob. Chang. Biol.* **2013**, *19*, 711–723. [CrossRef]

331. Piou, C.; Taylor, M.H.; Papaix, J.; Prevost, E. Modelling the interactive effects of selective fishing and environmental change on Atlantic salmon demogenetics. *J. Appl. Ecol.* **2015**, *52*, 1629–1637. [CrossRef]

332. Battle, L.; Chang, H.Y.; Tzeng, C.S.; Lin, H.J. Modeling the impact of dam removal on the Formosan landlocked salmon in the context of climate change. *Aquat. Sci.* **2020**, *82*, 3. [CrossRef]

333. Leppi, J.C.; Rinella, D.J.; Wilson, R.R.; Loya, W.M. Linking climate change projections for an Alaskan watershed to future coho salmon production. *Glob. Chang. Biol.* **2014**, *20*, 1808–1820. [CrossRef] [PubMed]

334. McHugh, P.A.; Saunders, W.C.; Bouwes, N.; Wall, C.E.; Bangen, S.; Wheaton, J.M.; Nahorniak, M.; Ruzycki, J.R.; Tattam, I.A.; Jordan, C.E. Linking models across scales to assess the viability and restoration potential of a threatened population of steelhead (*Oncorhynchus mykiss*) in the Middle Fork John Day River, Oregon, USA. *Ecol. Model.* **2017**, *355*, 24–38. [CrossRef]

335. Battin, J.; Wiley, M.W.; Ruckelshaus, M.H.; Palmer, R.N.; Korb, E.; Bartz, K.K.; Imaki, H. Projected impacts of climate change on salmon habitat restoration. *Proc. Natl. Acad. Sci. USA* **2007**, *104*, 6720–6725. [CrossRef]

336. Honea, J.M.; McClure, M.M.; Jorgensen, J.C.; Scheuerell, M.D. Assessing freshwater life-stage vulnerability of an endangered Chinook salmon population to climate change influences on stream habitat. *Clim. Res.* **2016**, *71*, 127–137. [CrossRef]

337. Pankhurst, N.W.; Munday, P.L. Effects of climate change on fish reproduction and early life history stages. *Mar. Freshw. Res.* **2011**, *62*, 1015–1026. [CrossRef]

338. Tuor, K.M.F.; Shrimpton, J.M. Differences in water temperatures experienced by embryo and larval Coho Salmon (*Oncorhynchus kisutch*) across their geographic range in British Columbia. *Environ. Biol. Fishes* **2019**, *102*, 955–967. [CrossRef]

339. Roni, P.; Johnson, C.; De Boer, T.; Pess, G.; Dittman, A.; Sear, D. Interannual variability in the effects of physical habitat and parentage on Chinook salmon egg-to-fry survival. *Can. J. Fish. Aquat. Sci.* **2016**, *73*, 1047–1059. [CrossRef]

340. Campbell, E.Y.; Dunham, J.B.; Reeves, G.H.; Wondzell, S.M. Phenology of hatching, emergence, and end-of-season body size in young-of-year coho salmon in thermally contrasting streams draining the Copper River Delta, Alaska. *Can. J. Fish. Aquat. Sci.* **2019**, *76*, 185–191. [CrossRef]

341. Adelfio, L.A.; Wondzell, S.M.; Mantua, N.J.; Reeves, G.H. Warm winters reduce landscape-scale variability in the duration of egg incubation for coho salmon (*Oncorhynchus kisutch*) on the Copper River Delta, Alaska. *Can. J. Fish. Aquat. Sci.* **2019**, *76*, 1362–1375. [CrossRef]

342. Evans, M.L.; Neff, B.D.; Heath, D.D. Quantitative genetic and translocation experiments reveal genotype-by-environment effects on juvenile life-history traits in two populations of Chinook salmon (*Oncorhynchus tshawytscha*). *J. Evol. Biol.* **2010**, *23*, 687–698. [CrossRef] [PubMed]

343. Venney, C.J.; Love, O.P.; Drown, E.J.; Heath, D.D. DNA Methylation Profiles Suggest Intergenerational Transfer of Maternal Effects. *Mol. Biol. Evol.* **2020**, *37*, 540–548. [CrossRef] [PubMed]

344. Janhunen, M.; Piironen, J.; Peuhkuri, N. Parental effects on embryonic viability and growth in Arctic charr *Salvelinus alpinus* at two incubation temperatures. *J. Fish Biol.* **2010**, *76*, 2558–2570. [CrossRef] [PubMed]

345. Le Luyer, J.; Laporte, M.; Beacham, T.D.; Kaukinen, K.H.; Withler, R.E.; Leong, J.S.; Rondeau, E.B.; Koop, B.F.; Bernatchez, L. Parallel epigenetic modifications induced by hatchery rearing in a Pacific salmon. *Proc. Natl. Acad. Sci. USA* **2017**, *114*, 12964–12969. [CrossRef]

346. Gavery, M.R.; Roberts, S.B. Epigenetic considerations in aquaculture. *PeerJ* **2017**, *5*, e4147. [CrossRef]

347. Eirin-Lopez, J.M.; Putnam, H.M. Marine environmental epigenetics. *Annu. Rev. Mar. Sci.* **2019**, *11*, 335–368. [CrossRef]

348. Bollati, V.; Baccarelli, A. Environmental epigenetics. *Heredity* **2010**, *105*, 105–112. [CrossRef]

349. Banet, A.I.; Healy, S.J.; Eliason, E.J.; Roualdes, E.A.; Patterson, D.A.; Hinch, S.G. Simulated maternal stress reduces offspring aerobic swimming performance in Pacific salmon. *Conserv. Physiol.* **2019**, *7*, coz095. [CrossRef]

350. Daley, J.M.; Leadley, T.A.; Pitcher, T.E.; Drouillard, K.G. Bioamplification and the selective depletion of persistent organic pollutants in Chinook salmon larvae. *Environ. Sci. Technol.* **2012**, *46*, 2420–2426. [CrossRef]

351. Burt, J.M.; Hinch, S.G.; Patterson, D.A. Parental identity influences progeny responses to incubation thermal stress in sockeye salmon *Onchorhynchus nerka*. *J. Fish Biol.* **2012**, *80*, 444–462. [CrossRef]

352. Leblanc, C.A.; Horri, K.; Skulason, S.; Benhaim, D. Subtle temperature increase can interact with individual size and social context in shaping phenotypic traits of a coldwater fish. *PLoS ONE* **2019**, *14*, 21. [CrossRef] [PubMed]

353. Pasquet, A.; Realis-Doyelle, E.; Fontaine, P.; Teletchea, F. What can we learn by studying dead fish fry? *Aquac. Res.* **2019**, *50*, 1824–1833. [CrossRef]

354. Jonsson, B.; Jonsson, N. Phenotypic plasticity and epigenetics of fish: Embryo temperature affects later-developing life-history traits. *Aquat. Biol.* **2019**, *28*, 21–32. [CrossRef]

355. Jonsson, B.; Jonsson, N.; Finstad, A.G. Linking embryonic temperature with adult reproductive investment in Atlantic salmon *Salmo salar*. *Mar. Ecol. Prog. Ser.* **2014**, *515*, 217–226. [CrossRef]

356. Adams, L.E.; Lund, J.R.; Moyle, P.B.; Quiñones, R.M.; Herman, J.D.; O'Rear, T.A. Environmental hedging: A theory and method for reconciling reservoir operations for downstream ecology and water supply. *Water Resour. Res.* **2017**, *53*, 7816–7831. [CrossRef]

357. Hayes, D.S.; Moreira, M.; Boavida, I.; Haslauer, M.; Unfer, G.; Zeiringer, B.; Greimel, F.; Auer, S.; Ferreira, T.; Schmutz, S. Life stage-specific hydropeaking flow rules. *Sustainability* **2019**, *11*, 17. [CrossRef]

358. Langshaw, R.B.; Graf, P.J.; Pearsons, T.N. Hydropower and high productivity in the Hanford Reach: A synthesis of how flow management may benefit fall Chinook Salmon in the Columbia River, USA. *Wiley Interdiscip. Rev. -Water* **2018**, *5*, e1275. [CrossRef]

359. Sapin, J.R.; Saito, L.; Dai, A.; Rajagopalan, B.; Blair Hanna, R.; Kauneckis, D. Demonstration of integrated reservoir operations and extreme hydroclimate modeling of water temperatures for fish sustainability below Shasta Lake. *J. Water Resour. Plan. Manag.* **2017**, *143*, 04017062. [CrossRef]

360. Sellheim, K.; Zeug, S.; Merz, J. Informed water management alternatives for an over-allocated river: Incorporating salmon life stage effects into a decision tree process during drought. *Fish. Manag. Ecol.* **2020**, *27*, 498–516. [CrossRef]

361. Dudley, P.N. Insights from an individual based model of a fish population on a large regulated river. *Environ. Biol. Fishes* **2019**, *102*, 1069–1095. [CrossRef]

362. Roni, P.; Anders, P.J.; Beechie, T.J.; Kaplowe, D.J. Review of tools for identifying, planning, and implementing habitat restoration for Pacific salmon and steelhead. *N. Am. J. Fish. Manag.* **2018**, *38*, 355–376. [CrossRef]

363. Isaak, D.J.; Wenger, S.J.; Peterson, E.E.; Ver Hoef, J.M.; Nagel, D.E.; Luce, C.H.; Hostetler, S.W.; Dunham, J.B.; Roper, B.B.; Wollrab, S.P.; et al. The NorWeST summer stream temperature model and scenarios for the western U.S.: A crowd-sourced database and new geospatial tools foster a user community and predict broad climate warming of rivers and streams. *Water Resour. Res.* **2017**, *53*, 9181–9205. [CrossRef]
364. Fullerton, A.H.; Torgersen, C.E.; Lawler, J.J.; Steel, E.A.; Ebersole, J.L.; Lee, S.Y. Longitudinal thermal heterogeneity in rivers and refugia for coldwater species: Effects of scale and climate change. *Aquat. Sci.* **2018**, *80*, 1–15. [CrossRef]
365. Lee, S.Y.; Fullerton, A.H.; Sun, N.; Torgersen, C.E. Projecting spatiotemporally explicit effects of climate change on stream temperature: A model comparison and implications for coldwater fishes. *J. Hydrol.* **2020**, *588*, 125066. [CrossRef]
366. McCullough, D.A.; Bartholow, J.M.; Jager, H.I.; Beschta, R.L.; Cheslak, E.F.; Deas, M.L.; Ebersole, J.L.; Foott, J.S.; Johnson, S.L.; Marine, K.R.; et al. Research in thermal biology: Burning questions for coldwater stream fishes. *Rev. Fish. Sci.* **2009**, *17*, 90–115. [CrossRef]
367. Bennett, S.; Pess, G.; Bouwes, N.; Roni, P.; Bilby, R.E.; Gallagher, S.; Ruzycki, J.; Buehrens, T.; Krueger, K.; Ehinger, W.; et al. Progress and Challenges of Testing the Effectiveness of Stream Restoration in the Pacific Northwest Using Intensively Monitored Watersheds. *Fisheries* **2016**, *41*, 92–103. [CrossRef]
368. Bilby, R.; Johnson, A.; Foltz, J.R.; Puls, A.L. *Management Implications from Pacific Northwest Intensively Monitored Watersheds*; U.S. Geological Survey: Reston, VA, USA, 2022; 99p, Available online: https://www.pnamp.org/document/15207 (accessed on 1 June 2022).
369. Erhardt, J.M.; Tiffan, K.F.; Connor, W.P. Juvenile Chinook salmon mortality in a Snake River reservoir: Smallmouth bass pedation revisited. *Trans. Am. Fish. Soc.* **2018**, *147*, 316–328. [CrossRef]
370. Hawkins, B.L.; Fullerton, A.H.; Sanderson, B.L.; Steel, E.A. Individual-based simulations suggest mixed impacts of warmer temperatures and a nonnative predator on Chinook salmon. *Ecosphere* **2020**, *11*, e03218. [CrossRef]
371. Tonina, D.; McKean, J.A.; Isaak, D.; Benjankar, R.M.; Tang, C.; Chen, Q. Climate Change Shrinks and Fragments Salmon Habitats in a Snow-Dependent Region. *Geophys. Res. Lett.* **2022**, *49*, e2022GL098552. [CrossRef] [PubMed]
372. Pennock, C.A.; Budy, P.; Atkinson, C.L.; Barrett, N. Effects of increased temperature on arctic slimy sculpin Cottus cognatus is mediated by food availability: Implications for climate change. *Freshw. Biol.* **2021**, *66*, 549–561. [CrossRef]
373. Spanjer, A.R.; Gendaszek, A.S.; Wulfkuhle, E.J.; Black, R.W.; Jaeger, K.L. Assessing climate change impacts on Pacific salmon and trout using bioenergetics and spatiotemporal explicit river temperature predictions under varying riparian conditions. *PLoS ONE* **2022**, *17*, e0266871. [CrossRef]
374. Benjamin, J.R.; Dunham, J.B.; Johnson, S.L.; Ashkenas, L.; Penaluna, B.E.; Bilby, R.E.; Bateman, D.; Leer, D.; Bellmore, J.R. Pathways of productivity and influences on top consumers in forested streams. *For. Ecol. Manag.* **2022**, *508*, 120046. [CrossRef]
375. Platts, P.J.; Mason, S.C.; Palmer, G.; Hill, J.K.; Oliver, T.H.; Powney, G.D.; Fox, R.; Thomas, C.D. Habitat availability explains variation in climate-driven range shifts across multiple taxonomic groups. *Sci. Rep.* **2019**, *9*, 15039. [CrossRef]
376. Fuchs, H.L.; Chant, R.J.; Hunter, E.J.; Curchitser, E.N.; Gerbi, G.P.; Chen, E.Y. Wrong-way migrations of benthic species driven by ocean warming and larval transport. *Nat. Clim. Chang.* **2020**, *10*, 1052–1056. [CrossRef]
377. Beer, W.N.; Anderson, J.J. Sensitivity of juvenile salmonid growth to future climate trends. *River Res. Appl.* **2011**, *27*, 663–669. [CrossRef]
378. Fullerton, A.H.; Burke, B.J.; Lawler, J.J.; Torgersen, C.E.; Ebersole, J.L.; Leibowitz, S.G. Simulated juvenile salmon growth and phenology respond to altered thermal regimes and stream network shape. *Ecosphere* **2017**, *8*, e02052-23. [CrossRef]
379. Huntsman, B.M.; Falke, J.A. Main stem and off-channel habitat use by juvenile Chinook salmon in a sub-Arctic riverscape. *Freshw. Biol.* **2019**, *64*, 433–446. [CrossRef]
380. Armstrong, J.B.; Fullerton, A.H.; Jordan, C.E.; Ebersole, J.L.; Bellmore, J.R.; Arismendi, I.; Penaluna, B.E.; Reeves, G.H. The importance of warm habitat to the growth regime of cold-water fishes. *Nat. Clim. Chang.* **2021**, *11*, 354–361. [CrossRef] [PubMed]
381. Teichert, N.; Benitez, J.P.; Dierckx, A.; Tetard, S.; de Oliveira, E.; Trancart, T.; Feunteun, E.; Ovidio, M. Development of an accurate model to predict the phenology of Atlantic salmon smolt spring migration. *Aquat. Conserv. -Mar. Freshw. Ecosyst.* **2020**, *30*, 1552–1565. [CrossRef]
382. Spence, B.C.; Dick, E.J. Geographic variation in environmental factors regulating outmigration timing of coho salmon (*Oncorhynchus kisutch*) smolts. *Can. J. Fish. Aquat. Sci.* **2014**, *71*, 56–69. [CrossRef]
383. Carlson, S.M.; Seamons, T.R. A review of quantitative genetic components of fitness in salmonids: Implications for adaptation to future change. *Evol. Appl.* **2008**, *1*, 222–238. [CrossRef] [PubMed]
384. Manhard, C.V.; Joyce, J.E.; Gharrett, A.J. Evolution of phenology in a salmonid population: A potential adaptive response to climate change. *Can. J. Fish. Aquat. Sci.* **2017**, *74*, 1519–1527. [CrossRef]
385. Debes, P.V.; Piavchenko, N.; Erkinaro, J.; Primmer, C.R. Genetic growth potential, rather than phenotypic size, predicts migration phenotype in Atlantic salmon. *Proc. R. Soc. B. Biol. Sci.* **2020**, *287*, 20200867. [CrossRef]
386. Debes, P.V.; Solberg, M.F.; Matre, I.H.; Dyrhovden, L.; Glover, K.A. Genetic variation for upper thermal tolerance diminishes within and between populations with increasing acclimation temperature in Atlantic salmon. *Heredity* **2021**, *127*, 455–466. [CrossRef]
387. Muñoz, N.J.; Farrell, A.P.; Heath, J.W.; Neff, B.D. Adaptive potential of a Pacific salmon challenged by climate change. *Nat. Clim. Chang.* **2015**, *5*, 163–166. [CrossRef]

388. Timpane-Padgham, B.L.; Beechie, T.; Klinger, T. A systematic review of ecological attributes that confer resilience to climate change in environmental restoration. *PLoS ONE* **2017**, *12*, e0173812. [CrossRef]

389. Klein, S.; Herron, H.; Butcher, D. *EPA Region 10 Climate Change and TMDL Pilot—South Fork Nooksack River, Washington*; U.S. Environmental Protection Agency: Washington, DC, USA, 2017.

390. Sundt-Hansen, L.E.; Hedger, R.D.; Ugedal, O.; Diserud, O.H.; Finstad, A.G.; Sauterleute, J.F.; Tofte, L.; Alfredsen, K.; Forseth, T. Modelling climate change effects on Atlantic salmon: Implications for mitigation in regulated rivers. *Sci. Total Environ.* **2018**, *631–632*, 1005–1017. [CrossRef]

391. Hallnan, R.; Saito, L.; Busby, D.; Tyler, S. Modeling Shasta Reservoir Water-Temperature Response to the 2015 Drought and Response under Future Climate Change. *J. Water Resour. Plan. Manag.* **2020**, *146*, 04020018. [CrossRef]

392. Hatten, J.R.; Batt, T.R.; Connolly, P.J.; Maule, A.G. Modeling effects of climate change on Yakima River salmonid habitats. *Clim. Chang.* **2014**, *124*, 427–439. [CrossRef]

393. Jager, H.I.; King, A.W.; Gangrade, S.; Haines, A.; DeRolph, C.; Naz, B.S.; Ashfaq, M. Will future climate change increase the risk of violating minimum flow and maximum temperature thresholds below dams in the Pacific Northwest? *Clim. Risk Manag.* **2018**, *21*, 69–84. [CrossRef]

394. Burke, B.J.; Peterson, W.T.; Beckman, B.R.; Morgan, C.; Daly, E.A.; Litz, M. Multivariate models of adult Pacific salmon returns. *PLoS ONE* **2013**, *8*, e54134. [CrossRef] [PubMed]

395. Rechisky, E.L.; Welch, D.W.; Porter, A.D.; Hess, J.E.; Narum, S.R. Testing for delayed mortality effects in the early marine life history of Columbia River Basin yearling Chinook salmon. *Mar. Ecol. Prog. Ser.* **2014**, *496*, 159–180. [CrossRef]

396. Litzow, M.A.; Ciannelli, L.; Puerta, P.; Wettstein, J.J.; Rykaczewski, R.R.; Opiekun, M. Non-stationary climate-salmon relationships in the Gulf of Alaska. *Proc. R. Soc. B-Biol. Sci.* **2018**, *285*, 20181855. [CrossRef] [PubMed]

397. Litzow, M.A.; Hunsicker, M.E.; Bond, N.A.; Burke, B.J.; Cunningham, C.J.; Gosselin, J.L.; Norton, E.L.; Ward, E.J.; Zador, S.G. The changing physical and ecological meanings of North Pacific Ocean climate indices. *Proc. Natl. Acad. Sci. USA* **2020**, *117*, 7665–7671. [CrossRef]

398. Litzow, M.A.; Malick, M.J.; Bond, N.A.; Cunningham, C.J.; Gosselin, J.L.; Ward, E.J. Quantifying a Novel Climate Through Changes in PDO-Climate and PDO-Salmon Relationships. *Geophys. Res. Lett.* **2020**, *47*, e2020GL087972. [CrossRef]

399. Burke, B.J.; Anderson, J.J.; Miller, J.A.; Tomaro, L.; Teel, D.J.; Banas, N.S.; Baptista, A.M. Estimating behavior in a black box: How coastal oceanographic dynamics influence yearling Chinook salmon marine growth and migration behaviors. *Environ. Biol. Fishes* **2016**, *99*, 671–686. [CrossRef]

400. Snyder, M.N.; Schumaker, N.H.; Ebersole, J.L.; Dunham, J.B.; Comeleo, R.L.; Keefer, M.L.; Leinenbach, P.; Brookes, A.; Cope, B.; Wu, J.; et al. Individual based modeling of fish migration in a 2-D river system: Model description and case study. *Landsc. Ecol.* **2019**, *34*, 737–754. [CrossRef]

401. Seeb, L.W.; Seeb, J.E.; Habicht, C.; Farley, E.V.; Utter, F.M. Single-nucleotide polymorphic genotypes reveal patterns of early juvenile migration of sockeye salmon in the eastern Bering Sea. *Trans. Am. Fish. Soc.* **2011**, *140*, 734–748. [CrossRef]

402. Kuparinen, A.; Hutchings, J.A. Genetic architecture of age at maturity can generate divergent and disruptive harvest-induced evolution. *Philos. Trans. R. Soc. London. Ser. B Biol. Sci.* **2017**, *372*, 20160035. [CrossRef]

403. Frölicher, T.L.; Fischer, E.M.; Gruber, N. Marine heatwaves under global warming. *Nature* **2018**, *560*, 360–364. [CrossRef]

404. Keefer, M.L.; Caudill, C.C.; Peery, C.A.; Boggs, C.T. Non-direct homing behaviours by adult Chinook salmon in a large, multi-stock river system. *J. Fish Biol.* **2008**, *72*, 27–44. [CrossRef]

405. Crozier, L.G.; Hutchings, J.A. Plastic and evolutionary responses to climate change in fish. *Evol. Appl.* **2014**, *7*, 68–87. [CrossRef] [PubMed]

406. Reed, T.E.; Schindler, D.E.; Hague, M.J.; Patterson, D.A.; Meir, E.; Waples, R.S.; Hinch, S.G. Time to evolve? Potential evolutionary responses of Fraser River sockeye salmon to climate change and effects on persistence. *PLoS ONE* **2011**, *6*, e20380. [CrossRef] [PubMed]

407. Muñoz, N.J.; Anttila, K.; Chen, Z.; Heath, J.W.; Farrell, A.P.; Neff, B.D. Indirect genetic effects underlie oxygen-limited thermal tolerance within a coastal population of chinook salmon. *Proc. R. Soc. B Biol. Sci.* **2014**, *281*, 20141082. [CrossRef] [PubMed]

408. O'Malley, K.G.; Ford, M.J.; Hard, J.J. Clock polymorphism in Pacific salmon: Evidence for variable selection along a latitudinal gradient. *Proc. R. Soc. B Biol. Sci.* **2010**, *277*, 3703–3714. [CrossRef]

409. Ford, M.D.; Nichols, K.M.; Waples, R.S.; Anderson, E.C.; Kardos, M.; Koch, I.; McKinney, G.; Miller, M.R.; Myers, J.; Naish, K.; et al. Reviewing and Synthesizing the State of the Science Regarding Associations between Adult Run Timing and Specific Genotypes in Chinook Salmon and Steelhead. In Proceedings of the Report of a Workshop; National Marine Fisheries Service, Northwest Fisheries Science Center: Seattle, WA 98112, USA, 27–28 February 2020. [CrossRef]

410. Collins, E.E.; Hargrove, J.S.; Delomas, T.A.; Narum, S.R. Distribution of genetic variation underlying adult migration timing in steelhead of the Columbia River basin. *Ecol. Evol.* **2020**, *10*, 9486–9502. [CrossRef]

411. Donaldson, M.R.; Hinch, S.G.; Patterson, D.A.; Farrell, A.P.; Shrimpton, J.M.; Miller-Saunders, K.M.; Robichaud, D.; Hills, J.; Hruska, K.A.; Hanson, K.C.; et al. Physiological condition differentially affects the behavior and survival of two populations of sockeye salmon during their freshwater spawning migration. *Physiol. Biochem. Zool.* **2010**, *83*, 446–458. [CrossRef]

412. Eliason, E.J.; Clark, T.D.; Hague, M.J.; Hanson, L.M.; Gallagher, Z.S.; Jeffries, K.M.; Gale, M.K.; Patterson, D.A.; Hinch, S.G.; Farrell, A.P. Differences in thermal tolerance among sockeye salmon populations. *Science* **2011**, *332*, 109–112. [CrossRef]

413. Eliason, E.J.; Clark, T.D.; Hinch, S.G.; Farrell, A.P. Cardiorespiratory collapse at high temperature in swimming adult sockeye salmon. *Conserv. Physiol.* **2013**, *1*, cot008. [CrossRef] [PubMed]

414. Eliason, E.J.; Farrell, A.P. Oxygen uptake in Pacific salmon *Oncorhynchus* spp.: When ecology and physiology meet. *J. Fish Biol.* **2016**, *88*, 359–388. [CrossRef] [PubMed]

415. Eliason, E.J.; Wilson, S.M.; Farrell, A.P.; Cooke, S.J.; Hinch, S.G. Low cardiac and aerobic scope in a coastal population of sockeye salmon *Oncorhynchus nerka* with a short upriver migration. *J. Fish Biol.* **2013**, *82*, 2104–2112. [CrossRef]

416. Farrell, A.P. Environment, antecedents and climate change: Lessons from the study of temperature physiology and river migration of salmonids. *J. Exp. Biol.* **2009**, *212*, 3771–3780. [CrossRef] [PubMed]

417. Farrell, A.P. Aerobic scope and its optimum temperature: Clarifying their usefulness and limitations—Correspondence on. *J. Exp. Biol.* 216, 2771–2782. *J. Exp. Biol.* **2013**, *216*, 4493–4494. [CrossRef] [PubMed]

418. Farrell, A.P.; Hinch, S.G.; Cooke, S.J.; Patterson, D.A.; Crossin, G.T.; Lapointe, M.; Mathes, M.T. Pacific salmon in hot water: Applying aerobic scope models and biotelemetry to predict the success of spawning migrations. *Physiol. Biochem. Zool.* **2008**, *81*, 697–708. [CrossRef] [PubMed]

419. Connor, W.P.; Tiffan, K.F.; Chandler, J.A.; Rondorf, D.W.; Arnsberg, B.D.; Anderson, K.C. Upstream Migration and Spawning Success of Chinook Salmon in a Highly Developed, Seasonally Warm River System. *Rev. Fish. Sci. Aquac.* **2019**, *27*, 1–50. [CrossRef]

420. Lusardi, R.A.; Moyle, P.B. Two-way trap and haul as a conservation strategy for anadromous salmonids. *Fisheries* **2017**, *42*, 478–487. [CrossRef]

421. Colvin, M.E.; Peterson, J.T.; Sharpe, C.; Kent, M.L.; Schreck, C.B. Identifying optimal hauling densities for adult Chinook salmon trap and haul operations. *River Res. Appl.* **2018**, *34*, 1158–1167. [CrossRef]

422. O'Connor, C.M.; Norris, D.R.; Crossin, G.T.; Cooke, S.J. Biological carryover effects: Linking common concepts and mechanisms in ecology and evolution. *Ecosphere* **2014**, *5*, art28. [CrossRef]

423. Moyle, P.B.; Lusardi, R.; Samuel, P.; Katz, J. *State of the Salmonids: Status of California's Emblematic Fishes 2017*; Center for Watershed Sciences, University of California, Davis and California Trout: San Francisco, CA, USA, 2017; p. 579. Available online: http://caltrout.org/sos/ (accessed on 1 January 2019).

424. Hare, J.A.; Morrison, W.E.; Nelson, M.W.; Stachura, M.M.; Teeters, E.J.; Griffis, R.B.; Alexander, M.A.; Scott, J.D.; Alade, L.; Bell, R.J.; et al. A vulnerability assessment of fish and invertebrates to climate change on the northeast U.S. continental shelf. *PLoS ONE* **2016**, *11*, e0146756. [CrossRef]

425. McClure, M.M.; Haltuch, M.A.; Willis-Norton, E.; Huff, D.D.; Hazen, E.L.; Crozier, L.G.; Jacox, M.G.; Nelson, M.W.; Andrews, K.S.; Barnett, L.A.K.; et al. Vulnerability to climate change of managed stocks in the California Current large marine ecosystem. *Front. Mar. Sci.* **2023**, *10*, 3767. [CrossRef]

426. Bjornas, K.L.; Railsback, S.F.; Calles, O.; Piccolo, J.J. Modeling Atlantic salmon (*Salmo salar*) and brown trout (*S. trutta*) population responses and interactions under increased minimum flow in a regulated river. *Ecol. Eng.* **2021**, *162*, 106182. [CrossRef]

427. Huang, B.; Langpap, C.; Adams, R.M. Using instream water temperature forecasts for fisheries management: An application in the Pacific Northwest. *J. Am. Water Resour. Assoc.* **2011**, *47*, 861–876. [CrossRef]

428. Zeug, S.C.; Bergman, P.S.; Cavallo, B.J.; Jones, K.S. Application of a life cycle simulation model to evaluate impacts of water management and conservation actions on an endangered population of Chinook salmon. *Environ. Model. Assess.* **2012**, *17*, 455–467. [CrossRef]

429. Falke, J.A.; Flitcroft, R.L.; Dunham, J.B.; McNyset, K.M.; Hessburg, P.F.; Reeves, G.H. Climate change and vulnerability of bull trout (*Salvelinus confluentus*) in a fire-prone landscape. *Can. J. Fish. Aquat. Sci.* **2015**, *72*, 304–318. [CrossRef]

430. Almodovar, A.; Leal, S.; Nicola, G.G.; Horreo, J.L.; Garcia-Vazquez, E.; Elvira, B. Long-term stocking practices threaten the original genetic diversity of the southernmost European populations of Atlantic salmon *Salmo salar*. *Endanger. Species Res.* **2020**, *41*, 303–317. [CrossRef]

431. Thorstad, E.B.; Bliss, D.; Breau, C.; Damon-Randall, K.; Sundt-Hansen, L.E.; Hatfield, E.M.C.; Horsburgh, G.; Hansen, H.; Maoileidigh, N.O.; Sheehan, T.; et al. Atlantic salmon in a rapidly changing environment-Facing the challenges of reduced marine survival and climate change. *Aquat. Conserv-Mar. Freshw. Ecosyst.* **2021**, *31*, 2654–2665. [CrossRef]

432. Ariza, A.; Lengaigne, M.; Menkes, C.; Lebourges-Dhaussy, A.; Receveur, A.; Gorgues, T.; Habasque, J.; Gutiérrez, M.; Maury, O.; Bertrand, A. Global decline of pelagic fauna in a warmer ocean. *Nat. Clim. Chang.* **2022**, *12*, 928–934. [CrossRef]

433. Hause, C.L.; Singer, G.P.; Buchanan, R.A.; Cocherell, D.E.; Fangue, N.A.; Rypel, A.L. Survival of a threatened salmon is linked to spatial variability in river conditions. *Can. J. Fish. Aquat. Sci.* **2022**, *79*, 2056–2071. [CrossRef]

434. Halpern, B.S.; Kappel, C.V.; Selkoe, K.A.; Micheli, F.; Ebert, C.M.; Kontgis, C.; Crain, C.M.; Martone, R.G.; Shearer, C.; Teck, S.J. Mapping cumulative human impacts to California Current marine ecosystems. *Conserv. Lett.* **2009**, *2*, 138–148. [CrossRef]

435. Andrews, K.S.; Williams, G.D.; Samhouri, J.F.; Marshall, K.N.; Gertseva, V.; Levin, P.S. The legacy of a crowded ocean: Indicators, status, and trends of anthropogenic pressures in the California Current ecosystem. *Environ. Conserv.* **2015**, *42*, 139–151. [CrossRef]

436. Toft, J.D.; Munsch, S.H.; Cordell, J.R.; Siitari, K.; Hare, V.C.; Holycross, B.M.; DeBruyckere, L.A.; Greene, C.M.; Hughes, B.B. Impact of multiple stressors on juvenile fish in estuaries of the northeast Pacific. *Glob. Chang. Biol.* **2018**, *24*, 2008–2020. [CrossRef] [PubMed]

437. Greene, C.M.; Blackhart, K.; Nohner, J.; Candelmo, A.; Nelson, D.M. A national assessment of stressors to estuarine fish habitats in the contiguous USA. *Estuaries Coasts* **2015**, *38*, 782–799. [CrossRef]

438. Teck, S.J.; Halpern, B.S.; Kappel, C.V.; Micheli, F.; Selkoe, K.A.; Crain, C.M.; Martone, R.; Shearer, C.; Arvai, J.; Fischhoff, B.; et al. Using expert judgment to estimate marine ecosystem vulnerability in the California Current. *Ecol. Appl.* **2010**, *20*, 1402–1416. [CrossRef] [PubMed]
439. Beechie, T.J.; Fogel, C.; Nicol, C.; Jorgensen, J.; Timpane-Padgham, B.; Kiffney, P. How does habitat restoration influence resilience of salmon populations to climate change? *Ecosphere* **2022**, *14*, e4402. [CrossRef]
440. Justice, C.; White, S.M.; McCullough, D.A.; Graves, D.S.; Blanchard, M.R. Can stream and riparian restoration offset climate change impacts to salmon populations? *J. Environ. Manag.* **2017**, *188*, 212–227. [CrossRef] [PubMed]
441. White, S.M.; Brandy, S.; Justice, C.; Morinaga, K.A.; Naylor, L.; Ruzycki, J.; Sedell, E.R.; Steele, J.; Towne, A.; Webster, J.G.; et al. Progress towards a comprehensive approach for habitat restoration in the Columbia Basin: Case study in the Grande Ronde River. *Fisheries* **2021**, *46*, 229–243. [CrossRef]
442. Jordan, C.E.; Fairfax, E. Beaver: The North American freshwater climate action plan. *WIREs Water* **2022**, *9*, e1592. [CrossRef]
443. Brennan, S.R.; Schindler, D.E.; Cline, T.J.; Walsworth, T.E.; Buck, G.; Fernandez, D.P. Shifting habitat mosaics and fish production across river basins. *Science* **2019**, *364*, 783–786. [CrossRef]
444. Waples, R.S.; Zabel, R.W.; Scheuerell, M.D.; Sanderson, B.L. Evolutionary responses by native species to major anthropogenic changes to their ecosystems: Pacific salmon in the Columbia River hydropower system. *Mol. Ecol.* **2008**, *17*, 84–96. [CrossRef]
445. Kovach, R.P.; Muhlfeld, C.C.; Wade, A.A.; Hand, B.K.; Whited, D.C.; DeHaan, P.W.; Al-Chokhachy, R.; Luikart, G. Genetic diversity is related to climatic variation and vulnerability in threatened bull trout. *Glob. Chang. Biol.* **2015**, *21*, 2510–2524. [CrossRef]
446. Beechie, T.J.; Nicol, C.; Fogel, C.; Jorgensen, J.; Thompson, J.; Seixas, G.; Chamberlin, J.; Hall, J.; Timpane-Padgham, B.; Kiffney, P.; et al. Chehalis LIfe-Cycle Modeling Phase 1 Report. In *NOAA Contract Report NMFS-NWFSC; CR-2021-01*; United States National Marine Fisheries Service: Sacramento, CA, USA; Northwest Fisheries Science Center (U.S.): Seattle, WA, USA; NMFS (National Marine Fisheries Service): Silver Spring, MD, USA; NWFSC (Northwest Fisheries Science Center): Niceville, FL, USA, 2021.
447. Bond, M.H.; Nodine, T.G.; Beechie, T.J.; Zabel, R.W. Estimating the benefits of widespread floodplain reconnection for Columbia River Chinook salmon. *Can. J. Fish. Aquat. Sci.* **2019**, *76*, 1212–1226. [CrossRef]
448. FitzGerald, A.M.; Martin, B.T. Quantification of thermal impacts across freshwater life stages to improve temperature management for anadromous salmonids. *Conserv. Physiol.* **2022**, *10*, coac013. [CrossRef] [PubMed]
449. Yakima River Basin Water Enhancement Project Workgroup. Yakima Basin Integrated Plan. Available online: https://yakimabasinintegratedplan.org/the-integrated-plan/ (accessed on 1 November 2022).
450. McCarthy, M.D.; Rinella, D.J.; Finney, B.P. Sockeye Salmon Population Dynamics over the Past 4000 Years in Upper Russian Lake, South-Central Alaska. *Journal of Paleolimnology* **2018**, *60*, 67–75. [CrossRef]
451. Chasco, B.; Burke, B.; Crozier, L.; Zabel, R. Differential impacts of freshwater and marine covariates on wild and hatchery Chinook salmon marine survival. *PLoS ONE* **2021**, *16*, e0246659. [CrossRef]
452. Wells, B.K.; Huff, D.D.; Burke, B.J.; Brodeur, R.D.; Santora, J.A.; Field, J.C.; Richerson, K.; Mantua, N.J.; Fresh, K.L.; McClure, M.M.; et al. Implementing Ecosystem-Based Management Principles in the Design of a Salmon Ocean Ecology Program. *Front. Mar. Sci.* **2020**, *7*, 342. [CrossRef]
453. Tillotson, M.D.; Kelly, R.P.; Duda, J.J.; Hoy, M.; Kralj, J.; Quinn, T.P. Concentrations of environmental DNA (eDNA) reflect spawning salmon abundance at fine spatial and temporal scales. *Biol. Conserv.* **2018**, *220*, 1–11. [CrossRef]
454. Rubenson, E.S.; Lawrence, D.J.; Olden, J.D. Threats to rearing juvenile Chinook salmon from non-native smallmouth bass inferred from stable isotope and fatty acid biomarkers. *Trans. Am. Fish. Soc.* **2020**, *149*, 350–363. [CrossRef]
455. Franklin, T.W.; Dysthe, J.C.; Rubenson, E.S.; Carim, K.J.; Olden, J.D.; McKelvey, K.S.; Young, M.K.; Schwartz, M.K. A non-invasive sampling method for detecting non-native smallmouth bass (*Micropterus dolomieu*). *Northwest Sci.* **2018**, *92*, 149–157. [CrossRef]
456. Sepulveda, A.J.; Schmidt, C.; Amberg, J.; Hutchins, P.; Stratton, C.; Mebane, C.; Laramie, M.B.; Pilliod, D.S. Adding invasive species biosurveillance to the US Geological Survey streamgage network. *Ecosphere* **2019**, *10*, e02843. [CrossRef]
457. Akbarzadeh, A.; Gunther, O.P.; Houde, A.L.; Li, S.R.; Ming, T.J.; Jeffries, K.M.; Hinch, S.G.; Miller, K.M. Developing specific molecular biomarkers for thermal stress in salmonids. *BMC Genom.* **2018**, *19*, 1–28. [CrossRef]
458. Consuegra, S.; O'Rorke, R.; Rodriguez-Barreto, D.; Fernandez, S.; Jones, J.; de Leaniz, C.G. Impacts of large and small barriers on fish assemblage composition assessed using environmental DNA metabarcoding. *Sci. Total Environ.* **2021**, *790*, 148054. [CrossRef]
459. Gold, Z.; Curd, E.E.; Goodwin, K.D.; Choi, E.S.; Frable, B.W.; Thompson, A.R.; Walker, H.J.; Burton, R.S.; Kacev, D.; Martz, L.D.; et al. Improving metabarcoding taxonomic assignment: A case study of fishes in a large marine ecosystem. *Mol. Ecol. Resour.* **2021**, *21*, 2546–2564. [CrossRef]
460. Jeffery, N.W.; Lehnert, S.J.; Kess, T.; Layton, K.K.S.; Wringe, B.F.; Stanley, R.R.E. Application of omics tools in designing and monitoring marine protected areas for a sustainable blue economy. *Front. Genet.* **2022**, *13*, 886494. [CrossRef]
461. Dauwalter, D.C.; Fesenmyer, K.A.; Bjork, R.; Leasure, D.R.; Wenger, S.J. Satellite and Airborne Remote Sensing Applications for Freshwater Fisheries. *Fisheries* **2017**, *42*, 526–537. [CrossRef]
462. Naman, S.; White, S.; Bellmore, J.; McHugh, P.; Kaylor, M.; Baxter, C.; Danehy, R.; Naiman, R.; Puls, A. Food web perspectives and methods for riverine fish conservation. *Wiley Interdiscip. Rev. Water* **2022**, *9*, e1590. [CrossRef]
463. Koenigstein, S.; Dahlke, F.T.; Stiasny, M.H.; Storch, D.; Clemmesen, C.; Portner, H.O. Forecasting future recruitment success for Atlantic cod in the warming and acidifying Barents Sea. *Glob. Chang. Biol.* **2018**, *24*, 526–535. [CrossRef]

464. Dietze, M.C.; Fox, A.; Beck-Johnson, L.M.; Betancourt, J.L.; Hooten, M.B.; Jarnevich, C.S.; Keitt, T.H.; Kenney, M.A.; Laney, C.M.; Larsen, L.G.; et al. Iterative near-term ecological forecasting: Needs, opportunities, and challenges. *Proc. Natl. Acad. Sci. USA* **2018**, *115*, 1424–1432. [CrossRef]

465. Sobocinski, K.L.; Greene, C.M.; Schmidt, M.W. Using a qualitative model to explore the impacts of ecosystem and anthropogenic drivers upon declining marine survival in Pacific salmon. *Environ. Conserv.* **2018**, *45*, 278–290. [CrossRef]

466. Forget, N.L.; Duplisea, D.E.; Sardenne, F.; McKindsey, C.W. Using qualitative network models to assess the influence of mussel culture on ecosystem dynamics. *Ecol. Model.* **2020**, *430*, 109070. [CrossRef]

467. Reum, J.C.P.; McDonald, P.S.; Long, W.C.; Holsman, K.K.; Divine, L.; Armstrong, D.; Armstrong, J. Rapid assessment of management options for promoting stock rebuilding in data-poor species under climate change. *Conserv. Biol.* **2020**, *34*, 611–621. [CrossRef]

468. Tommasi, D.; Stock, C.A.; Hobday, A.J.; Methot, R.; Kaplan, I.C.; Eveson, J.P.; Holsman, K.; Miller, T.J.; Gaichas, S.; Gehlen, M.; et al. Managing living marine resources in a dynamic environment: The role of seasonal to decadal climate forecasts. *Prog. Oceanogr.* **2017**, *152*, 15–49. [CrossRef]

469. Koenigstein, S.; Mark, F.C.; Gossling-Reisemann, S.; Reuter, H.; Poertner, H.O. Modelling climate change impacts on marine fish populations: Process-based integration of ocean warming, acidification and other environmental drivers. *Fish Fish.* **2016**, *17*, 972–1004. [CrossRef]

470. Jacox, M.G.; Hazen, E.L.; Zaba, K.D.; Rudnick, D.L.; Edwards, C.A.; Moore, A.M.; Bograd, S.J. Impacts of the 2015-2016 El Nino on the California Current System: Early assessment and comparison to past events. *Geophys. Res. Lett.* **2016**, *43*, 7072–7080. [CrossRef]

471. Thompson, A.R.; Bjorkstedt, E.P.; Bograd, S.J.; Fisher, J.L.; Hazen, E.L.; Leising, A.; Santora, J.A.; Satterthwaite, E.V.; Sydeman, W.J.; Alksne, M.; et al. State of the California Current Ecosystem in 2021: Winter is coming? *Front. Mar. Sci.* **2022**, *9*, 958727. [CrossRef]

472. Shi, H.; García-Reyes, M.; Jacox, M.G.; Rykaczewski, R.R.; Black, B.A.; Bograd, S.J.; Sydeman, W.J. Co-occurrence of California Drought and Northeast Pacific Marine Heatwaves Under Climate Change. *Geophys. Res. Lett.* **2021**, *48*, e2021GL092765. [CrossRef]

473. Smith, J.A.; Pozo Buil, M.; Fiechter, J.; Tommasi, D.; Jacox, M.G. Projected novelty in the climate envelope of the California Current at multiple spatial-temporal scales. *PLoS Clim.* **2022**, *1*, e0000022. [CrossRef]

474. Brodie, S.; Abrahms, B.; Bograd, S.J.; Carroll, G.; Hazen, E.L.; Muhling, B.A.; Buil, M.P.; Smith, J.A.; Welch, H.; Jacox, M.G. Exploring timescales of predictability in species distributions. *Ecography* **2021**, *44*, 832–844. [CrossRef]

475. Cheung, W.W.L.; Ota, Y.; Cisneros-Montemayor, A. *Predicting Future Oceans: Sustainability of Ocean and Human Systems Amidst Global Environmental Change*; Elsevier: San Diego, CA, USA, 2019; ISBN 978-0-12-817946-8.

476. Holsman, K.K.; Aydin, K. Comparative methods for evaluating climate change impacts on the foraging ecology of Alaskan groundfish. *Mar. Ecol. Prog. Ser.* **2015**, *521*, 217–235. [CrossRef]

477. Kaplan, I.C.; Francis, T.B.; Punt, A.E.; Koehn, L.E.; Curchitser, E.; Hurtado-Ferro, F.; Johnson, K.F.; Lluch-Cota, S.E.; Sydeman, W.J.; Essington, T.E.; et al. A multi-model approach to understanding the role of Pacific sardine in the California Current food web. *Mar. Ecol. Prog. Ser.* **2019**, *617–618*, 307–321. [CrossRef]

MDPI
St. Alban-Anlage 66
4052 Basel
Switzerland
www.mdpi.com

Fishes Editorial Office
E-mail: fishes@mdpi.com
www.mdpi.com/journal/fishes